Springer Proceedings in Materials

Volume 25

Springer Proceedings in Materials publishes the latest research in Materials Science and Engineering presented at high standard academic conferences and scientific meetings. It provides a platform for researchers, professionals and students to present their scientific findings and stay up-to-date with the development in Materials Science and Engineering. The scope is multidisciplinary and ranges from fundamental to applied research, including, but not limited to:

- Structural Materials
- Metallic Materials
- Magnetic, Optical and Electronic Materials
- Ceramics, Glass, Composites, Natural Materials
- Biomaterials
- Nanotechnology
- Characterization and Evaluation of Materials
- Energy Materials
- Materials Processing

To submit a proposal or request further information, please contact one of our Springer Publishing Editors according to your affiliation:

European countries: **Mayra Castro** (mayra.castro@springer.com)
India, South Asia and Middle East: **Priya Vyas** (priya.vyas@springer.com)
South Korea: **Smith Chae** (smith.chae@springer.com)
Southeast Asia, Australia and New Zealand: **Ramesh Nath Premnath** (ramesh.premnath@springer.com)
The Americas: **Michael Luby** (michael.luby@springer.com)
China and all the other countries or regions: **Mengchu Huang** (mengchu.huang@springer.com)

This book series is indexed in **SCOPUS** database.

Bibhu Prasad Swain
Editor

Recent Advances in Materials

Select Proceedings of ICSTE 2023

Springer

Editor
Bibhu Prasad Swain
Department of Physics
National Institute of Technology Manipur
Imphal, Manipur, India

ISSN 2662-3161 ISSN 2662-317X (electronic)
Springer Proceedings in Materials
ISBN 978-981-99-3843-8 ISBN 978-981-99-3844-5 (eBook)
https://doi.org/10.1007/978-981-99-3844-5

This Springer imprint is published by the registered company Springer Nature Singapore Pte Ltd.
The registered company address is: 152 Beach Road, #21-01/04 Gateway East, Singapore 189721, Singapore

Preface

"Recent Advances in Materials" comprises a collection of research and review articles presented at the 1st International Conference on Sci/Tech and Engineering (ICSTE 23). It was organized by the National Institute of Technology (NIT), Manipur, Langol, Imphal, on February 17–18, 2023.

Eminent speakers like Prof. Manoj Gupta, National University of Singapore, Singapore; Prof. Bharat Dahiya, Thammasat University, Thailand; Prof. Surender Kumar, Ex. Prof. (BIT Mesra), India; Prof. Sawan Suman Sinha, IIT Delhi, India; Prof. Abhishek Dey, Indian Association for the Cultivation of Science, Kolkata, India; Prof. Bimlesh Kumar, IIT Guwahati, India, and Dr. Sanjeet Kumar Dwivedi, Green Hydrogen-based Danish MNC, Denmark, shared their knowledge and experience. The conference was attended and enriched by participants from numerous institutes such as IITs, NITs, NEHU, BIT, VIT, MIT Manipur, Manipur University, IIEST Kolkata, and abroad; several speakers deliberated on their research works. In addition, the paper presentations were accompanied by six keynote addresses from leading academic and industrial researchers around the globe. Out of 268 papers, 138 papers were selected for publication in the ICSTE 23 Proceedings.

The review committee has done an excellent job in reviewing articles and approving high-quality research articles to be published in the conference proceedings. The editors are thankful to all the faculty members and students of various committees for their dedication in making it a very successful conference. Thanks to the staff of Springer for making the publication of this book possible. We sincerely hope that it will inspire researchers/scientists to explore the potential in the diverse upcoming research fields.

Imphal, India

Dr. Bibhu Prasad Swain
bpswain@nitmanipur.ac.in

Contents

About the Editor

Dr. Bibhu Prasad Swain is currently an associate professor at the Department of Physics, National Institute of Technology Manipur. He obtained his B.Sc. (Physics) from Utkal University, Bhubaneswar, M.Sc. (Physics) from the National Institute of Technology Rourkela, M.Tech. (Materials Science) from Barkatullah University and Ph.D. from the Indian Institute of Technology, Bombay. His major areas of research interests include large bandgap Semiconductors, Mechanical Hard-materials, Nanostructured materials, Graphene-based Supercapacitor, Photovoltaic materials, Biocompatible coating, and Modeling of advanced MOSFET. He has published 150 papers in respected international journals. Dr. Swain received the Brain Korea 21 Fellow and Japanese Society of promotion of Science postdoctoral Fellow from Materials Research, Seoul National University and Government of Japan in the years 2007 and 2008, respectively. He has supervised many research projects sponsored by AICTE, Government of India; DST SERB, Government of India; DBT, Government of India. Currently, he is an editorial board member of the Nanoscience and Nanotechnology—Asia and guest editor in *Materials* journals in MDPI publications.

Chemical Compositions of Ground and Unground Rice Husk Ash Produced by Uncontrolled Burning

Rajiv Sharma Leihaothabam and Khwairakpam Sachidananda

1 Introduction

With the rapid increase in infrastructural development, there is also an increase in the production of cement. But, cement production leads to environmental pollution due to the emissions of greenhouse gases [1]. So, many researchers are exploring different substitutes for cement. Rice husk ash (RHA) is the ash that is left over after burning rice husk. This rice husk ash possesses high silica content, making it one of the potential pozzolanic material for cement replacement [2]. Additionally, by incorporating RHA into cement, environmental issues caused by the disposal of rice husk waste will be reduced [3].

Amorphous silica contained in RHA contributes to the pozzolanic reaction in cement. C-S-H gel is created when the reactive silica in RHA combines with the calcium hydroxide created during the cement hydration. It undergoes reaction with calcium hydroxide in cement to form silica gel. RHA's pozzolanic activity has an impact on the characteristics of concrete [4–6].

The majority of the research focused on controlled burning and ground RHA after Mehta's investigation in the early 1970s, which claimed that highly reactive RHA can be produced through controlled combustion and that the pozzolanic activity of the RHA depends on its fineness and the temperature in which it is burned. [7]. Many researchers have investigated the usage of RHA as a pozzolanic material [8–10].

While most researchers are focused on the controlled burning of RHA, few have studied the uncontrolled burnt RHA [4, 11–13]. But, no literature is available in the comparative study of ground and unground RHA produced by uncontrolled burning. Ground RHA is ash that is grounded to fine ashes along with the unburnt carbon particles. Unground RHA is ash that is directly sieved without grinding. This paper

R. S. Leihaothabam (✉) · K. Sachidananda
Department of Civil Engineering, National Institute of Technology Manipur, Imphal, Manipur 795001, India
e-mail: rajivleihaothabam@gmail.com

B. P. Swain (ed.), *Recent Advances in Materials*, Springer Proceedings in Materials 25,
https://doi.org/10.1007/978-981-99-3844-5_1

gives a comparative study on the amorphous nature and the chemical compositions of ground and unground RHA produced by uncontrolled burning.

2 Materials and Methods

2.1 Materials

Rice husks, collected from a local rice mill in Imphal, Manipur, were washed with water to remove unwanted particles and sundried for 48 h. Then, they were burnt in an uncontrolled condition using a local earthen furnace (Fig. 1). Two types of RHA were prepared (Fig. 2):

(a) Ground RHA: The ash was ground for 10 min and sieved through 90-micron sieve.
(b) Unground RHA: The ash was sieved directly through 90 microns sieve without grinding.

(a) (b)

Fig. 1 **a** Burning of rice husk in a local earthen furnace **b** RHA left after burning

(a) (b)

Fig. 2 **a** Ground RHA **b** unground RHA

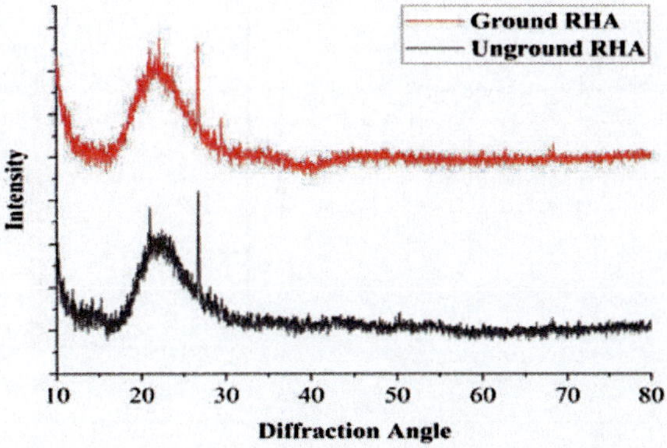

Fig. 3 XRD pattern of ground and unground RHA

2.2 *Methods*

2.2.1 XRD and EDX Analysis

XRD test was done to analyze the crystalline and amorphous nature of RHA. XRD analysis was done with a 2-theta angle limit of 10–60. EDX analysis was done to understand the elemental compositions of the two samples of RHA.

3 Results and Discussion

3.1 *Nature of the Samples*

The XRD plot of the two rice husk ash samples was observed (Fig. 3). The diffraction patterns were almost similar between them. Board peaks spanning a 2-theta angle range of 18–25 indicate that the RHA samples have a high amount of silica which are amorphous in nature.

3.2 *Chemical Compositions*

EDX analysis of the two RHA samples was done. The analysis showed that both of them have high silicon content which agreed with the XRD results indicating the content of high amount of silicon compound (Figs. 4 and 5).

The chemical compositions of ground RHA and unground RHA done by using EDX analysis were given in Table 1.

It was observed that unground RHA have higher amount of silicon than the ground RHA, which might help in the development of strength in concrete. On the other hand,

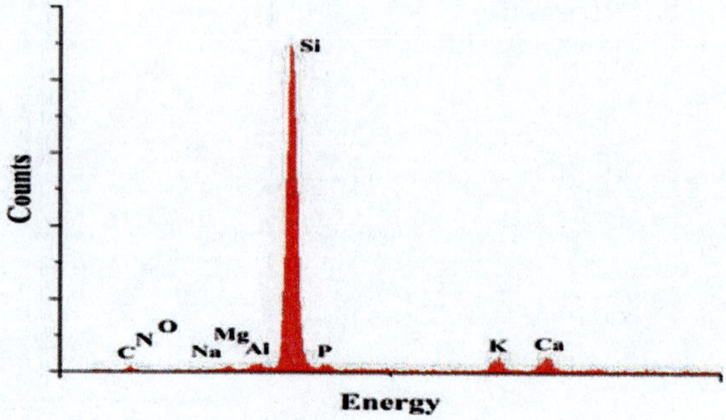

Fig. 4 EDX analysis of ground RHA

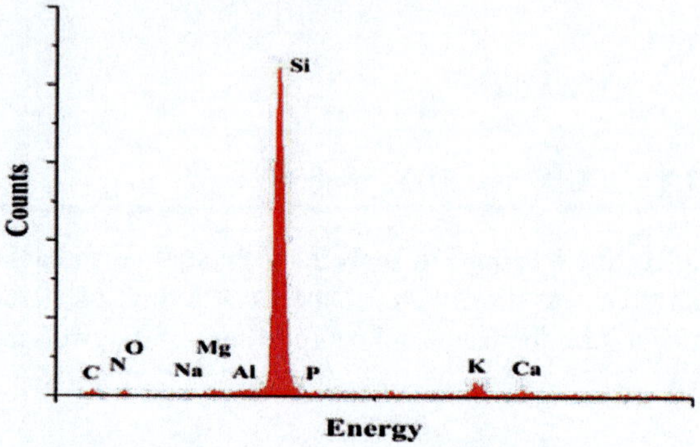

Fig. 5 EDX analysis of unground RHA

Table 1 Chemical compositions of ground RHA and unground RHA

Chemical compositions	Ground RHA (%)	Unground RHA (%)
Si	66.26	73.64
Al	0.79	1.32%
O	4.17	4.80
C	17.16	2.35
K	5.77	6.59
Na	0.13	0.00
N	0.86	0.00
Ca	2.51	7.02
Mg	0.75	0.81
P	1.59	3.47

ground RHA have higher amount of carbon content due to the mixing of unburnt carbon particles while grinding. This might negatively affect the strength of concrete.

4 Conclusion

This comparative study revealed that both of the ground and unground rice husk ashes produced by open burning have high amount of amorphous silica. Additionally, it was found that unground rice husk ash has a larger silicon concentration than ground rice husk ash, which indicates a higher silica percentage. Both the unground and ground rice husk ashes might be suitable for partial cement replacement. Unground RHA might be more favorable compare to ground RHA for partial replacement of cement due to higher silicon compound.

References

1. Zainudeen N, Jeyamathan J (2008) Cement and its effect to the environment: a case study in Sri Lanka. Proc from Int Conf Build Educ Res, 14081416
2. Bui DD (2001) Rice husk ash as a mineral admixture for high performance concrete, p 122
3. Pode R (2016) Potential applications of rice husk ash waste from rice husk biomass powerplant. Renew Sustain Energy Rev 53:1468–1485
4. Hadipramana J, Riza FV, Rahman IA, Loon LY, Adnan SH, Zaidi AMA (2016) Pozzolanic characterization of waste rice husk ash (RHA) from Muar, Malaysia. IOP Conf Ser Mater Sci Eng 160(1)
5. Ramasamy V (2012) Compressive strength and durability properties of rice husk ash concrete. KSCE J Civ Eng 16(1):93–102
6. Habeeb GA, Mahmud HB (2010) Study on properties of rice husk ash and its use as cement replacement material. Mater Res 13(2):185–190
7. Mehta PK (1973) U.S. Patent No. 4,105,459. Washington, DC
8. Rodríguez De Sensale G (2006) Strength development of concrete with rice-husk ash. Cem Concr Compos 28(2):158–160
9. Ramezanianpour AA, Khani MM, Ahmadibeni G (2009) The effect of rice husk ash on mechanical properties and durability of sustainable concretes. Int J Civil Eng 7(2):83–91
10. Nair DG, Fraaij A, Klaassen AAK, Kentgens APM (2008) Structural investigation relating to the pozzolanic activity of rice husk ashes. Cem Concr Res 38:861–869
11. Hadipramana J, et al (2013) Effect of uncontrolled burning rice husk ash in foamed concrete 626:769–775
12. Akinyele JO, Salim RW, Oikelome KO, Olateju OT (2015) The use of rice husk ash as a stabilizing agent in lateritic clay soil 9(11):1418–1422
13. Jonathan KM, Kuria KP, Mwangi GJ, Gichuki NF (2020) Characterization of rice husk ash prepared by open air burning and furnace calcination. J Chem Eng Mat 11(December):24–30

Characteristics Studies of GGBS and Metakaolin-Based Geopolymer Composite

Chungkham Jinita Devi, Nirmala Thoudam, Tanya Lairikyengbam, and Suresh Thokchom

1 Introduction

The rapid increase in industrialization, urbanization, innovations, and development in many fields in almost every part of the world has brought an appalling condition of the environment. With the growth of the human population, the demand for construction industries has increased to fulfil all human necessities [1, 2]. The main ingredient that is used as a binding agent in concrete is cement. Since the demand for concrete is increasing, the production of cement will also increase simultaneously. Cement manufacturing plants are accountable for about 5–7% of all CO_2 emissions since the production of one tonne of cement emits approximately 0.9 tonnes of CO_2 into the atmosphere [3, 4]. In recent decades, researchers have been pursuing in many fields to decrease the level of CO_2 emissions into the atmosphere. The exertion use of supplementary cementitious materials such as GGBS, rice husk ash, fly ash, metakaolin, red mud, silica fume which are of natural origin or industrial waste by-products have been encouraged in construction industries in order to reduce the global CO_2 emission into the atmosphere [5, 6].

One of the efforts that have been put up to reduce cement production is the development of geopolymer, which has emerged as a new environment-friendly construction material in order to reduce environmental carbon footprints. Geopolymer is a new emanating cementless binder first synthesized by Joseph Davidovitis in the 1970s. Geopolymers are formed by materials that are rich in silica and alumina like metakaolin, fly ash, GGBS, silica fume, etc. with a solution of alkali or alkali

C. J. Devi (✉) · N. Thoudam · T. Lairikyengbam · S. Thokchom
Department of Civil Engineering, Manipur Institute of Technology, Imphal 795001, Manipur, India
e-mail: jinitachungkham95@gmail.com

C. J. Devi · T. Lairikyengbam · S. Thokchom
Department of Civil Engineering, Manipur Institute of Technology, Manipur University, Imphal, India

B. P. Swain (ed.), *Recent Advances in Materials*, Springer Proceedings in Materials 25,
https://doi.org/10.1007/978-981-99-3844-5_2

salts [7]. The source materials are required to undergo polymerization which is an aluminosilicate material and alkaline activator solution [8–11].

The present study investigates the characteristics and properties of the effect of blending percentage of GGBS and metakaolin activated with sodium silicate (Na_2SiO_3) and sodium hydroxide (NaOH) solution. Three series of geopolymer mix were studied with the ratio of GGBS: metakaolin (MK) as 100:0, 95:5, 90:10. The microstructural and chemical composition of the resulting geopolymer has been characterized by SEM, EDAX, and XRD.

2 Experimental

2.1 Materials Used in the Study

GGBS, metakaolin (MK), sodium silicate (Na_2SiO_3), sodium hydroxide (NaOH), and water are the materials used in the study. GGBS sourced from JSW cement, Quality Polytech, and Mangalore was used for the study. Metakaolin was collected from AJ Corporation, Mumbai. The alkaline activator solution that was utilized in the study was an amalgamation of Na_2SiO_3 and 9 M NaOH solution, and they were collected from Bharath corporation, Guwahati.

2.2 Specimen Preparation

In this study, different amounts of GGBS and metakaolin (MK) were blended to prepare the specimen. The required quantities of GGBS and MK were taken according to the proportion, i.e. 100:0, 95:5, 90:10. The activator solution of 9 M NaOH and Na_2SiO_3 was prepared and was kept for one day prior in ambient temperature. On the next day, the activator solution was then mixed with the required quantities of GGBS and MK to get the desired geopolymer mix.

The mix was prepared in a nonabsorptive container and was then transferred into the cube mould of size 50 × 50 × 50 mm and was vibrated in a vibrating table for about 1–3 min to get rid of the entrapped air from the specimen. The specimens were demoulded after 24 h and were kept at ambient temperature for 28 days or until tested. Details of the sample mix proportion are given below in Table 1.

Table 1 Details of the specimen mix

Specimen Code	GGBS %	Metakaolin %
MG0	100	0
MG5	95	5
MG10	90	10

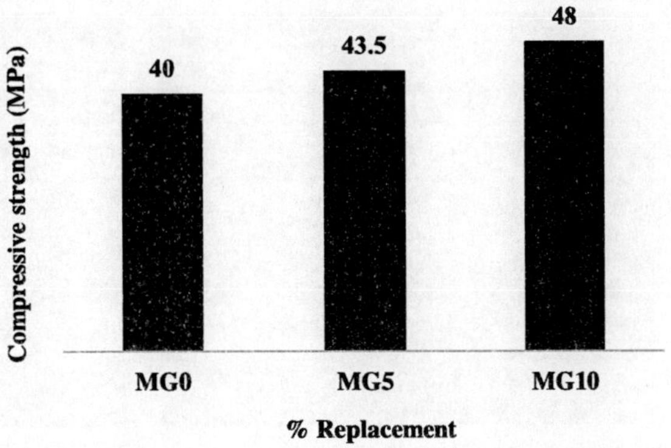

Fig. 1 Compressive strength of MG0, MG5, and MG10 paste specimen

2.3 *Test Procedure*

The direct compressive strength of the geopolymer samples was determined at the age of 28 days of curing as per ASTM C109 [12]. After the compressive strength test, the broken pieces of the specimen were collected. The broken specimens were used to study the microstructure of the sample and the quantification of elements using SEM and EDAX. Further, the broken specimens were ground to a fine powder and sieved through 45 microns mesh sieve to study the mineralogical composition of the specimen using XRD.

3 Result and Discussion

3.1 *Compressive Strength*

The values given in the Fig. 1 were the average of three samples. Specimen MG0, MG5, and MG10 indicate compressive strength of 40, 43.5, and 48 MPa. The test result shows that the strength of the samples increases with the increasing percentage of metakaolin dosage.

3.2 *Scanning Electron Microscope (SEM)*

The microstructure of the geopolymer paste specimen was studied through a scanning electron microscope while elemental quantification of the specimens was done by energy dispersive X-ray (EDAX) analysis. The microstructure of 100% GGBS and blended GGBS with metakaolin are shown below in Fig. 2. Images of the specimens indicate the formation of the gel. However, the incorporation of metakaolin

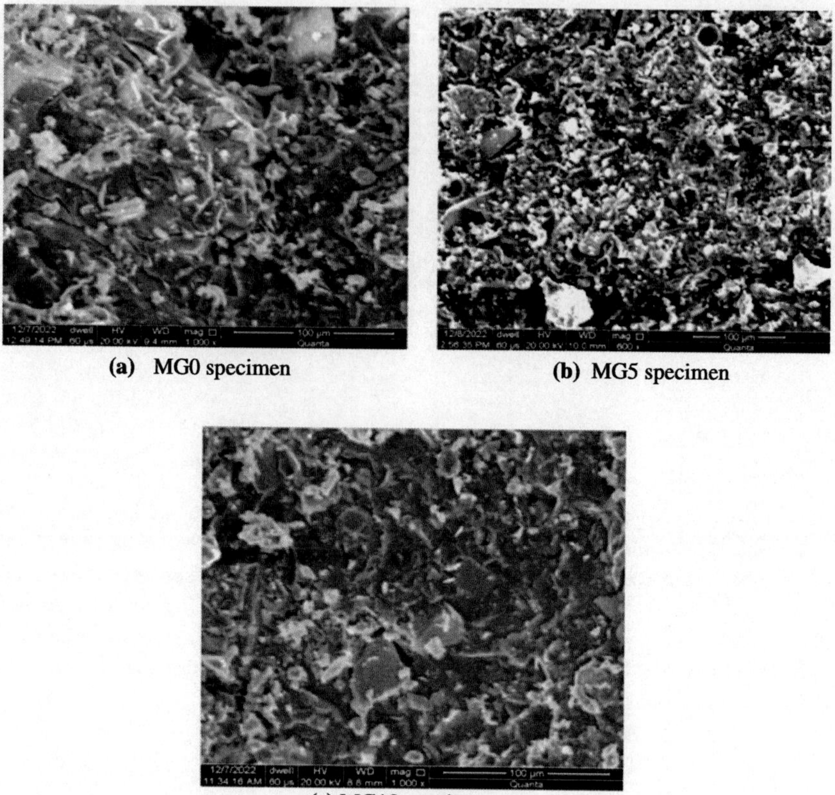

(a) MG0 specimen

(b) MG5 specimen

(c) MG10 specimen

Fig. 2 SEM image of the geopolymer paste specimens

into GGBS results in more dense intermolecular face bonding as compared to the GGBS100% (MG0) specimen. It is observed that MG10 specimens form better gel formation as compared to the other two specimens.

3.3 Energy Dispersive X-ray (EDX)/(EDAX)analysis

Quantification of the elements was studied through EDAX. Figure 3 and the tables given below shows the EDAX spectra of all the geopolymer paste specimen presenting major elements such as Calcium-Ca, Silicon-Si, Aluminium-Al, and Sodium-Na are shown in Tables 2, 3 and 4 while other elements such as Carbon-C, Oxygen-O, Magnesium-Mg, Phosphorus-P, Sulphur-S, Chlorine-Cl, Potassium-K, Titanium-Ti, Manganese-Mn, Iron-Fe, Cobalt-Co, Copper-Cu were noticed as traces. EDAX spectra of unblended paste specimen MG0 indicated a high weight percentage of calcium. As the metakaolin content in the blended specimen MG5 and MG10 increases, the intensity of calcium is observed to decrease while that of silicon where noticed to increase.

3.4 X-ray Diffraction Analysis

Figure 4 indicates 2θ(degree) in the X-axis horizontal and intensity counts in the Y-axis vertical. It is observed that GGBS unblended paste specimen peaks of quartz observed at around 27 and 30° 2θ and calcite is observed at 32° 2θ. In GGBS blended with metakaolin paste specimen peaks of quartz and kaolinite were noticed at around 27 and 30° 2θ for quartz and 33° 2θ for kaolinite. All the XRD pattern indicates amorphous nature. Blending GGBS with metakaolin produces peaks of kaolinite. However, the calcite present in the unblended GGBS paste specimen was not observed in the blended geopolymer specimen near the point where it was observed.

4 Conclusion

The effect of blending different percentages of metakaolin to GGBS was experimentally investigated. It is noticed that the compressive strength of the geopolymer was improved by adding 10% metakaolin to GGBS. The microstructure of the addition of metakaolin to GGBS is more homogenous and densified than 100% GGBS geopolymer.

The EDAX spectra of the unblended geopolymer paste specimen of GGBS-MG0 indicated a high weight percentage of calcium. As the metakaolin content in the blended specimen MG5 and MG10 increases, the intensity of calcium is observed to decrease while that of silicon where noticed to increase. A 10% addition of metakaolin to GGBS-based geopolymer is considered to be the optimum percentage of blending.

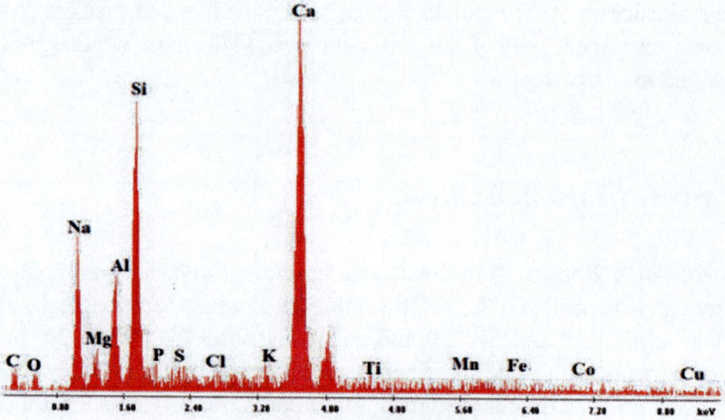

(a) 100% GGBS(MG0) specimen

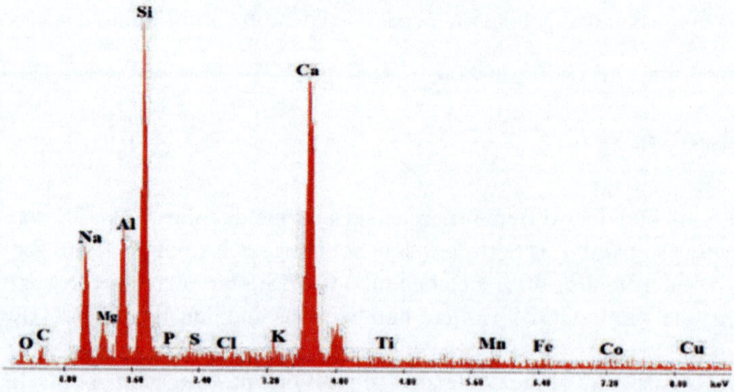

(b) MG5 specimen

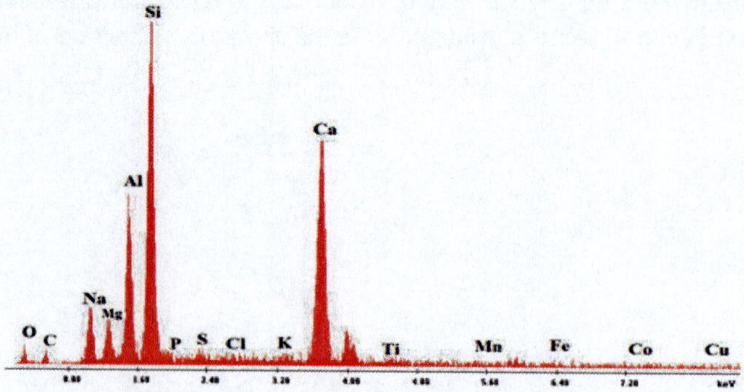

(c) MG10 specimen

Fig. 3 EDAX spectra result of the geopolymer paste specimens (a), (b), (c)

Table 2 The major element of the MG0 specimen

Elements	Wt %	At %
Si	20.08	18.78
Ca	31.14	20.41
Na	13.02	14.87
Al	7.20	7.01
Mg	2.76	2.98

Table 3 Major elements of MG5 specimen

Elements	Wt %	At %
Si	24.85	23.93
Ca	28.45	19.20
Na	11.99	14.10
Al	10.34	10.36
Mg	3.11	3.46

Table 4 Major elements of MG10 specimen

Elements	Wt %	At %
Si	28.72	26.18
Ca	25.28	16.15
Na	6.30	7.02
Al	11.83	11.23
Mg	3.72	3.91

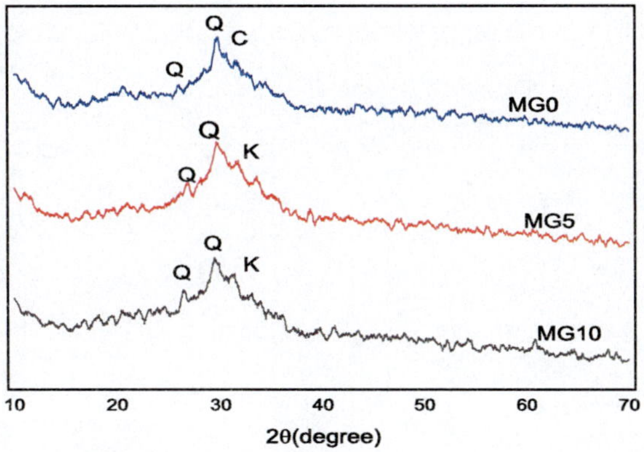

Fig. 4 XRD patterns of MG0, MG5, and MG10 (Q-quartz, C-calcite, K-kaolinite)

References

1. Mahmood A, Noman MT, Amor N (2019) Geopolymer and fibre reinforced concrete composites in civil engineering. Poly, 13:2099

2. Sharma S, Sood H (2016) Abrasion resistance of geopolymer concrete at varying temperature. J Mech Civil Eng 12:22–26
3. Jayshankar TN, Nagaraja PS (2018) Evaluation of water absorption and sorptivity properties of fly ash, GGBS, M-sand based glass fiber reinforced geopolymer concrete. Int Res J Eng Tech (IRJET) 5:11
4. Memon FA, Nuruddin MF, Shafiq N (2013) Effect of sodium hydroxide concentration on fresh properties and compressive strength of self-compacting geopolymer concrete. J Eng Sci Tech 8:44–56
5. Kumar BSC, Ramesh K (2016) Experimental study on strength properties of Metakaolin and GGBS based Geopolymer concrete. ARPN J Eng Appl Sci 11
6. Nazeer M, Kumar RA (2014) Strength studies on Metakaolin blended high volume fly ash concrete: Int J Eng Adv Tech, 176–179
7. Arun R, Shajahan S (2018) Feasibility studies on ternary blends in gepolymer concrete. IJCRT 6:2320–2882
8. Davidovits J (1991) Geopolymers: inorganic polymeric new materials. J Thermal Anal 37:1633–1656
9. Faris MA, Al Bakari MM, Muniandy A (2022) Review on mechanical properties of Metakaolin geopolymer concrete by inclusion of steel fibres. Arch Metall Mater 67:261–267
10. Kumar BSC, Ramesh K (2017) Durability studies of GGBS and metakaolin based geopolymer concrete. Int J Civil Eng Tech (IJCIET) 8:17–28
11. Thakur P, Singh S (2018) An experimental study on geopolymer concrete with bagasse ash and metakaolin: a green concrete. Int J Civil Eng Tech (IJCIET) 9(7)
12. ASTM. ASTM-C109 (2002) Standard test method for compressive strength of hydraulic cement mortars (using 2-in or [50mm] cube specimens), Vol 04.01

Study on a GGBS-Based Geopolymer Composites

Nirmala Thoudam, Tanya Lairikyengbam, Chungkham Jinita Devi, and Suresh Thokchom

1 Introduction

Due to industrialization and rapid growth of population, there has been increasing use of cement which cause a rapid rise in the level of carbon dioxide (CO_2) into the atmosphere [1]. According to reports, cement production accounts for 5–7% of all worldwide carbon dioxide emissions into the environment. Hence, to protect the environment and to conserve limestone ore to some extent, we must come up with a sustainable alternative for ordinary Portland cement [2, 3]. Geopolymer concrete is a new innovative concreting technology which emits lesser carbon dioxide and is considered as a green concrete [4]. It not only reduces emission of carbon dioxide but also utilizes large amount of industrial wastes such as ground granulated blast furnace slag (GGBS), granite waste powder, fly ash, silica fume [5]. Davidovits, a French academic, introduced "geopolymer" in 1978 [6]. The source of aluminosilicate minerals and an alkaline activator are the two essential elements of geopolymer concrete. By activating the alumina and silica-rich components with an alkaline activator solution, geopolymer binder is created. The widely used alkaline activator solution is a combine solution of sodium silicate (Na_2SiO_3) and sodium hydroxide (NaOH) [7–9]. Combination of potassium hydroxide and potassium silicate solution can also be used. In previous studies, fly ash has been the most often employed ingredient in the production of geopolymers [10, 11]. GGBS as a source material is also found in some literature. However, blending of silica fume with GGBS in manufacturing geopolymer composites has not been reported yet by any authors.

N. Thoudam (✉) · T. Lairikyengbam · C. J. Devi
Research Scholar, Department of Civil Engineering, Manipur Institute of Technology, Takyelpat, Manipur 795001, India
e-mail: thnirmala12@gmail.com

S. Thokchom
Department of Civil Engineering, Manipur Institute of Technology, Manipur University, Imphal, India

B. P. Swain (ed.), *Recent Advances in Materials*, Springer Proceedings in Materials 25,
https://doi.org/10.1007/978-981-99-3844-5_3

Table 1 Physical properties of materials

Property	GGBS	Silica fume
Colour	Whitish Grey	Black
Form	Fine powder	Fine powder
Particle size	<75 microns	<75 microns

Through the partial blending or replacement of GGBS with silica fume up to 10% in equal increments of 5%, this work aims to investigate the impact of silica fume on various characteristics of GGBS-based geopolymers. The tests carried out included various destructive and non-destructive tests on the geopolymer specimens such as compressive strength, bulk density, water absorption, rebound hammer, ultrasonic pulse velocity. The outcome of the present work might serve as a useful information in the field of geopolymer for suitable applications in construction activities.

2 Experimental

2.1 Materials

GGBS, silica fume (SF), NaOH, Na_2SiO_3, and water were the ingredients utilized to make the geopolymer composites. Alkaline activator solutions of sodium hydroxide and sodium silicate are employed to activate the binders. Table 1 displays the aluminosilicate materials' physical characteristics.

2.2 Specimen Preparation and Details

For the preparation of geopolymer composites, initially required quantities of sodium hydroxide (NaOH) pellets and water were mixed in a glass beaker to make sodium hydroxide solution of desired molarity. For this study, 9 M NaOH solution was used. The prepared sodium hydroxide solution was then mixed with required amount of sodium silicate solution in the proportion of 1NaOH: $2Na_2SiO_3$ to make activator solution. Before utilizing the resulting activator solution to prepare the geopolymer mix, it was allowed to sit at room temperature for 24 h. GGBS and SF were mixed in dry condition before the addition of alkaline solution to get the homogenous slurry. It was then transferred into the mould of size 50 × 50 × 50 mm and vibrated for around 2–3 min in a vibrating machine to remove entrapped air. The specimens were removed after 24 h and then kept in room temperature for 28 days for curing. Table 2 shows the details of geopolymer specimens mix casted for the experimental programme.

Table 2 Details of paste specimens

Specimen ID	GGBS (%)	SF (%)
SO	100	0
S5	95	5
S10	90	10

Table 3 Rebound hammer test results

Specimen	Compressive strength (N/mm^2)
SO	44
S5	49
S10	54

3 Results and Discussion

3.1 *Rebound Hammer Test of the Specimens*

Compressive strength of the specimens in N/mm^2 can be determined by calibration from the graphical chart attached in the rebound hammer with the rebound number obtained from the test. Table 3 presents the quality grading of the specimens with respect to the rebound number as per IS-13311 part-2, 1992. Up to 10% replacement of GGBS by SF, the quality grading of the geopolymer specimen is very good. It has been found that compressive strength rises as silica fume dosage is increased.

3.2 *Ultrasonic Pulse Velocity (UPV) Test*

The quality of the specimen as per IS-13311(1992)-part-2 are given in the Table 4. Similar to the results obtained in rebound hammer test, geopolymer specimens made by replacing GGBS with SF up to 10% yields good quality. The information from rebound hammer and ultrasonic pulse tests could be related to other properties including the compressive strength of the specimens.

Table 4 UPV test results

Specimens	Pulse velocity by cross probing (km/s)	Quality grading
SO	4.386	Good
S5	4.032	Good
S10	3.87	Good

Fig. 1 Water absorption of specimens

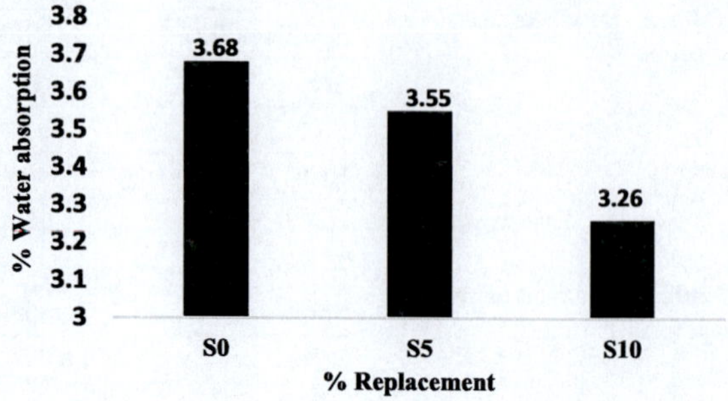

3.3 Water Absorption

The geopolymer composites durability is tested using a water absorption test. The geopolymer specimens are dried in an oven at a specific temperature and time for the water absorption test, after which they are put in desiccators to cool. According to the test results, the specimen's percentage of water absorption reduces as silica fume dosage is increased. The reduced water absorption was caused by the particle's poor porosity. The amount of pore space or porosity in a hardened geopolymer that is occupied by water when it is saturated is measured by the water absorption percentage (Fig. 1).

3.4 Bulk Density

Figure 2 displays the specimens' results for bulk density. It is obvious from the graphic that adding silica fume to GGBS-based geopolymer enhances bulk density. This might be as a result of the specimens getting denser as replacement goes up. The S10 specimen has the largest bulk density, measuring 20.26 kN/m^3. These numbers might be connected to other physical characteristics like porosity and water absorption.

3.5 Compressive Strength

Figure 3 displays the geopolymer specimens' compressive strength data, which was conducted in compression testing machine. As anticipated, the replacement of GGBS by silica fume increased as seen in the geopolymer specimens. The greatest compressive strength of 47.5 N/mm^2 was achieved by a geopolymer specimen containing 10% silica fume. The specimen with the lowest strength was S0, a geopolymer made entirely of GGBS with no substitutes.

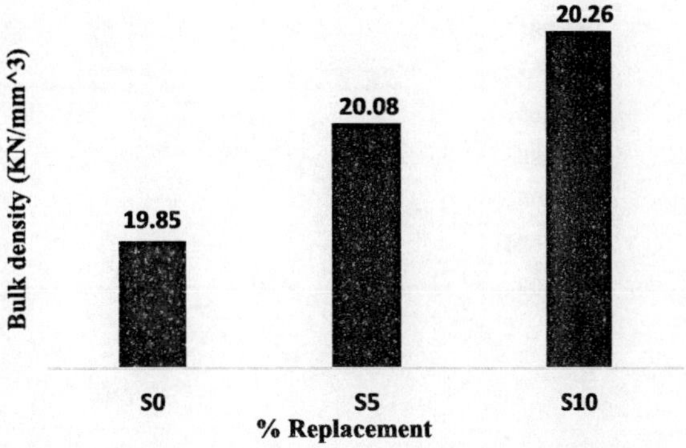

Fig. 2 Bulk density of geopolymer specimens

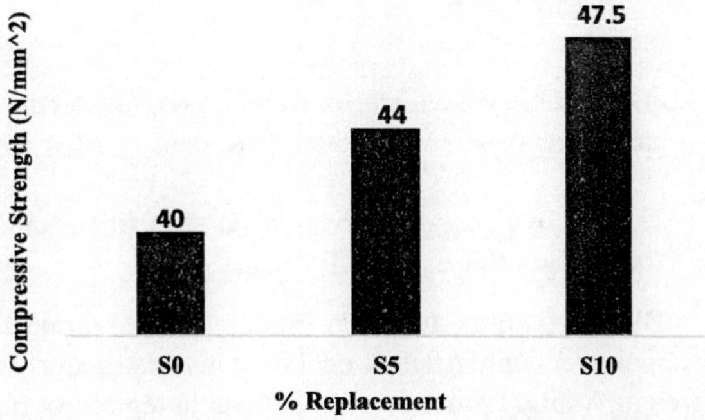

Fig. 3 Compressive strength of geopolymer specimens

3.6 Water Sorptivity

Sorptivity is an important property with respect to durability. Sorptivity test has been performed on the specimens. Figure 4 show the plot of increase in mass per unit area vs $\sqrt{}$time. It is observed that specimen S0 gives highest sorptivity among the specimen whereas specimen S10 results in least sorptivity. This could be due to higher porosity in specimen with 100% GGBS-based geopolymer.

4 Conclusions

The conclusions that are made from the current experimental study are as follows:

1. GGBS-based specimens exhibit very good properties in terms of strength and physical properties.
2. With the increase in dosage of silica fume, the compressive strength of the geopolymer specimen increases.

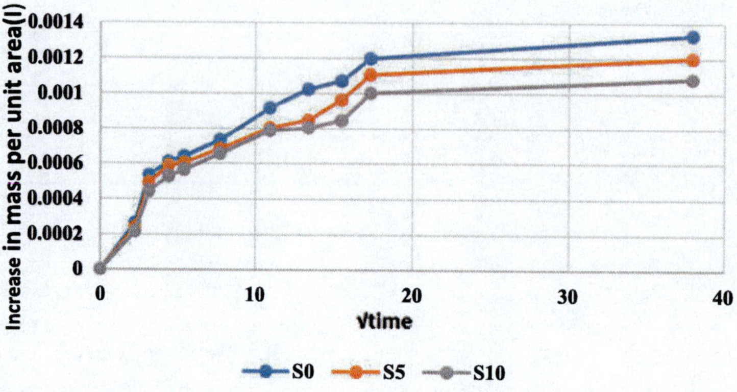

Fig. 4 Water sorptivity

3. Because they are lower in weight, geopolymer composites may be used to build towering structures as their bulk density rises with the amount of silica fume used.
4. The GGBS geopolymer replaced with silica fume up to 10% resulted in good quality grading as per UPV tests.

Blending silica fume in some suitable proportion with GGBS in making geopolymer could result in economy in construction works. Moreover, due to lighter weight, it might prove to be beneficial in regions of high seismic zones.

References

1. Saravanan R, Kumar MP, Elavenil S (2019) Study on mechanical properties of geopolymer concrete. Int J Inno Tech Expl Eng (IJITEE) 8(7), May. ISSN: 2278-3075
2. Mclellan BC, Williams RP, Lay J, Van Riessen A, Corder GD (2011) Costs and carbon emissions for geopolymer pastes in comparison to ordinary Portland cement. J Clean Prod 19(9):1080–1090
3. Hadid A, Farhan MNS, Sheikh NA, Neaz M (2017) Design of geopolymer concrete with GGBS at ambient curing condition using Taguchi method. Const Building Mat140:424–431
4. Hardjito D, Wallah SE, Sumajouw DMJ, Rangan BV (2004) On the development of fly ash-based geopolymer concrete. ACI Mat J, Nov and Dec
5. Anusha D (2017) Study on geopolymer concrete using GGBS. Int Res J Eng Tech 4(2), Feb
6. Aleem MIA, Arumairaj PD (2012) Geopolymer concrete-a review. Int J Eng Sci Emerg Techn 1(2), Feb
7. Rao GS, Kumar BSC (2019) Experimental investigation of GGBS based geopolymer concrete with steel fibers. Int J Recent Tech Eng 7(6C2), April. ISSN:2277-3878
8. IS-13311(1992), Part-1, Method of non-destructive testing of Concrete- Ultrasonic Pulse velocity, Bureau of Indian Standards, New Delhi
9. IS-13311(1992), Part-2, Method of non-destructive testing of Concrete- Rebound Hammer, Bureau of Indian Standards, New Delhi

10. Padmanaban MS, Sreerambabu J (2018) Geo polymer concrete with GGBS(Ground granulated blast furnance slag). Int J Eng Sci Res Tech, Feb. ISSN: 2277-9655
11. Yahya Z, Abdullah MMAB, Ramli NM, Burduhos-Nergis DD, Razak RA (2018) Influence of Kaolin in flyash based geopolymer concrete: destructive and non-destructive testing. In: IOP conference Series: Materials Science and Engineering, 374–2018

Characterization of GGBS-Based Geopolymer Blended with RED MUD Using EDAX and XRD

Tanya Lairikyengbam, Chungkham Jinita Devi, Nirmala Thoudam, and Suresh Thokchom

1 Introduction

The solid waste produced during the manufacture of alumina is known as red mud. It is the most highly desirable industrial waste require for utilization [1, 2]. Iron oxides, which give red mud its colour, are among the many oxides that make up the substance. Red mud is a form of industrial waste created when alumina is generated from bauxite ore using Bayer's process [3]. Because of its high alkalinity of red mud [4], it poses a risk to both the environment and human health [5]. The largest difficulty in reducing its impact is finding an efficient, inexpensive disposal method for alumina. Currently, hazardous waste is dumped into deep ocean or into waterbodies like rivers or the sea via pipelines or barrages [6–8]. However, red mud waste discharge into aquatic bodies was discontinued starting in 2016. Disposal of red mud residue into the environmental needs to be control. Massive efforts have been undertaken by today's researchers to treat, recycle, and use red mud [9–13]. Geopolymer was first proposed by davidovites in the year 1978. Geopolymer is becoming more popular as an alternative binder in experimentation and development. Geopolymer is a highly cementitious material prepared by rich aluminosilicates [1, 3] raw material like ground granulated blast furnace slag, fly ash, coal gangue, etc. [14, 15], which can replace ordinary Portland cement.

The current paper presents the ground granulated blast furnace-based geopolymer incorporated partially by red mud at a ratio of 5% and 10% and then compared with the control specimen. The geopolymeric specimen was measured microscopically using X-ray diffraction (XRD), Scanning electron microscope (SEM), and Energy dispersive X-ray analysis (EDAX/EDX).

Present Address:
T. Lairikyengbam (✉) · C. J. Devi · N. Thoudam · S. Thokchom
Department of Civil Engineering, Manipur Institute of Technology, Manipur University, Imphal, India
e-mail: tanya.lairikyengbam@gmail.com

B. P. Swain (ed.), *Recent Advances in Materials*, Springer Proceedings in Materials 25,
https://doi.org/10.1007/978-981-99-3844-5_4

2 Materials and Methods

2.1 Materials

Ground granulated blast furnace slag (GGBS), Red mud (RM), Sodium hydroxide, Sodium silicate, and tap water were the main constituent of the geopolymer paste specimen. GGBS is a by-product of iron and steel making from blast furnace. It mainly consists of silicates, aluminosilicates, and calcium-alumina-silicates. It exhibits hydraulic cementitious properties in finely ground form GGBS was collected from Quality Polytech, Mangalore, India. Bauxite residue is termed as red mud. It is a waste generated in production of alumina from bauxite in Bayer process. It contains minerals of bauxite residue. Red mud was collected from nature and greens, Gujarat. Combination of sodium hydroxide solution and sodium silicates solution is used as an alkaline solution. Both sodium hydroxide and sodium silicate were collected from Bharat Trading Cooperation, Guwahati, Assam.

2.2 Samples Preparation

By dissolving a NaOH flask in distilled water and letting it cool to room temperature, sodium hydroxide solution (9M) was prepared. In order to create the final alkaline solution, Na_2SiO_3 and NaOH solutions were combined and stirred until the solution was obtained uniformly. The solution is used after one day to allow the exothermically heated solution to cool down at ambient temperature.

Geopolymer paste specimen were prepared by mixing GGBS and RM and alkaline solution until a homogenous slurry is obtained. A vibrating table was used to remove entrapped air from slurry before it was casted. The cube mould size is 50 mm × 50 mm × 50 mm. The samples are cured at ambient temperature. The sample were maintained at ambient temperature after casting until they were mechanically tested. One mixture was produced using only GGBS as control specimen name as 'G' while other two mixtures were produced by replacing GGBS by weight in 5% (G + 5R) and 10%(G + 10R) with red mud as in Table 1.

Table 1 Details of specimen

Specimen ID	GGBS (%)	RM (%)
G	100	–
G + 5R	95	5
G + 10R	90	10

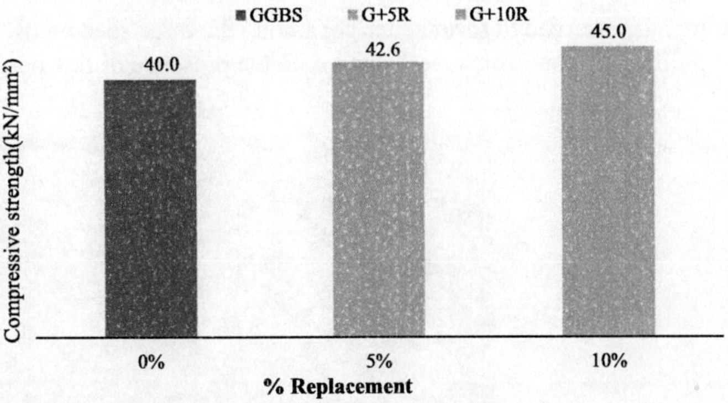

Fig. 1 Compressive strength of G, G + 5, and G + 10 paste specimen

2.3 *Methods*

The samples' compressive strength is evaluated. After the compressive strength test, the samples are crushed. A part of crushed samples is ground until it passes through 45 microns mesh sieve for XRD. The other part of the crushed samples was collected for SEM and EDAX.

3 Results and Discussion

3.1 *Compressive Strength*

The compressive strength of paste specimen sample prepared from unblended GGBS and blended GGBS with red mud is given in Fig. 1. The strength of paste specimens is tested at 28th days curing. The strength for geopolymer blended with red mud increases with increasing dosages of red mud.

3.2 *Microscopic Measurements*

3.2.1 Scanning Electron Microscope(SEM)

The microstructure of the specimen was examined using SEM. The microstructure of unblended GGBS paste specimen and blended GGBS with red mud are shown in Fig. 1(G), Fig. 2(G + 5R), and Fig. 3(G + 10R), respectively. The microstructure shows formation of gel along with one partially reacted particles. The microstructure indicates improvement with increasing blend with RM upto 10%, i.e. G + 10R paste specimen is observed to form better gel among the three specimens. However, unreacted GGBS particles are noticed along with the presence of few pores.

Fig. 2 G paste specimen

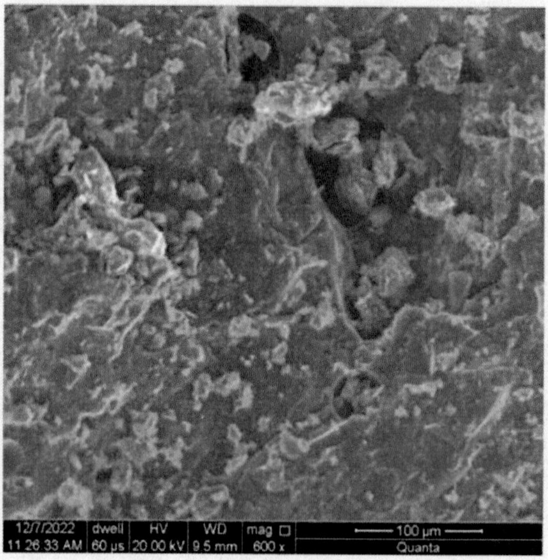

Fig. 3 G + 5R paste specimen

3.2.2 Energy Dispersive X-Ray Analysis (EDAX)

EDAX was used to evaluate the chemical composition of the specimens. Figures 4, 5 and 6 shows the EDAX spectra of G, G + 5R, and G + 10R specimen, respectively. In all the spectra, the major elements identified are Oxygen, Sodium, Aluminium, Silica, Calcium, etc. while traces of other elements like carbon, magnesium, phosphorous, sulphur, chlorine, potassium, titanium, manganese, iron, cobalt, copper were also observed. The weight % of the major elements identified in EDAX spectra are presented in Tables 2, 3, 4. The weight % of silica increased with blending RM with GGBS. However, weight % of calcium remain approximately equal in all the geopolymer blended specimen (Fig. 7).

Fig. 4 G + 10R paste specimen

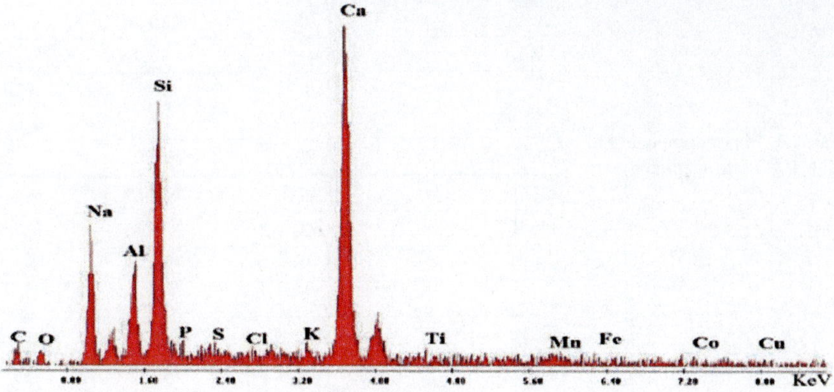

Fig. 5 EDAX diagram of G paste geopolymer specimen

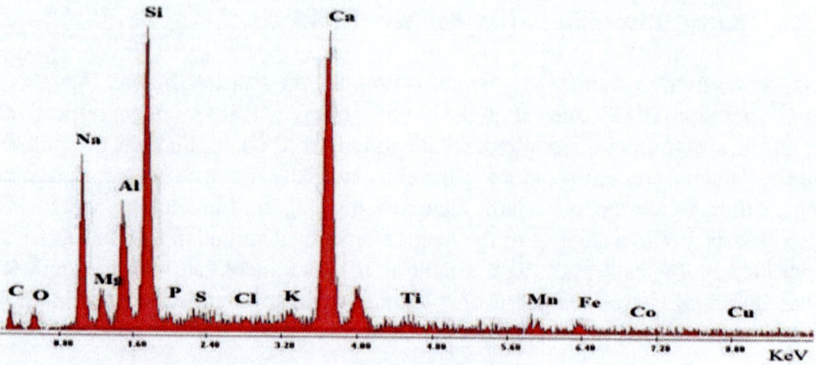

Fig. 6 EDAX diagram of G + 5R paste geopolymer specimen

Table 2 Major component of G paste specimen

Elements	Wt %	At %
O K	4.26	6.99
Na K	13.02	14.87
Al K	7.20	7.01
Si K	20.08	18.78
Ca K	31.14	20.41

Table 3 Major component of G + 5R paste specimen

Elements	Wt %	At %
O K	4.97	8.00
Na K	14.22	15.93
Al K	8.49	8.11
Si K	21.70	19.90
Ca K	26.83	17.25

Table 4 Major component of G + 10R paste specimen

Elements	Wt %	At %
O K	3.96	7.03
Na K	5.52	6.81
Al K	10.39	10.93
Si K	26.55	26.82
Ca K	30.52	21.61

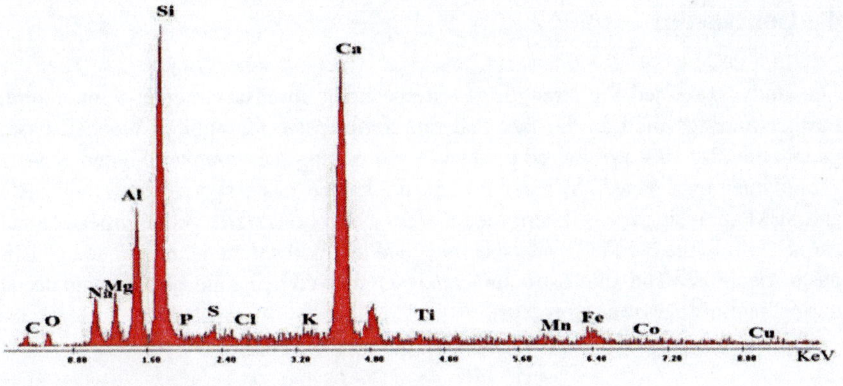

Fig. 7 EDAX diagram of G + 10R paste geopolymer specimen

3.2.3 X-Ray Diffraction

Figure 8 shows the XRD patterns of GGBS unblended paste specimen and GGBS blended specimen with red mud. Patterns in the range of 10–70° are shown in Fig. 8. It is observed that GGBS unblended paste specimen peaks of quartz and calcite, respectively, at 2Θ. In GGBS blended with red mud paste specimen peaks of quartz, hematite, and calcite were noticed at 2Θ approximately. All the XRD patterns indicates amorphous nature. Blending GGBS with red mud produces peaks of hematite peaks of quartz and calcite. However, the hematite present in blended geopolymer paste specimen was not observed in unblended GGBS paste specimen.

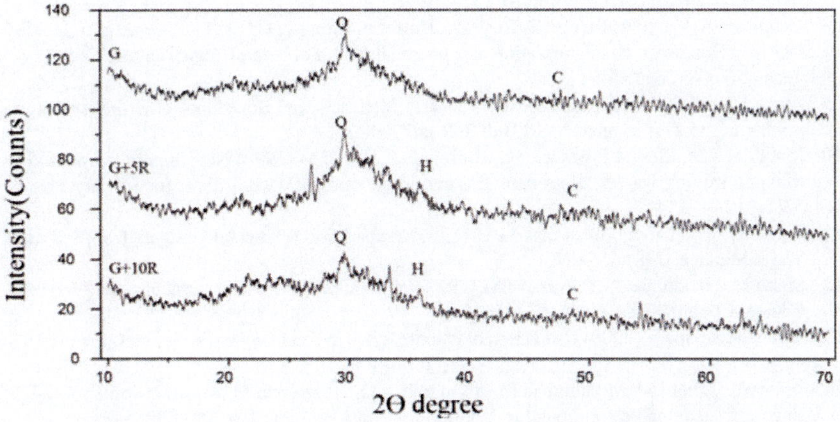

Fig. 8 XRD patterns of ambient cured specimens (Q-Quartz; HHematite; C- Calcite)

4 Conclusion

The study examined the strength of the specimen cured at ambient temperature. Further findings include the fact that the compressive strength of blended paste geopolymer by 10% weight red mud has higher strength as compared blended paste geopolymer by 5% weight red mud and unblended paste specimen. Using XRD and SEM analysis, it was determined that an amorphous paste geopolymer gel had formed, while the blended paste specimen had less pores than the unblended GGBS paste specimen. The microstructure analysis showed red mud as filling material during geopolymerization process.

References

1. Davidovits J (1994) Properties of geopolymer cements. In: First International Conference on Alkaline Cement and Concrete
2. Davidovits J (1994) Chemistry of geopolymeric systems, Terminology Geopolymer, International Conference
3. Zhang G, He JA, Gambrell RP (2010) Synthesis, characterization and mechanical properties of red mud-based geopolymer. Trans Res Record 2167:1–9
4. Xu B, Smith P, Wingate C, Siva L De (2010) The effect of calcium and temperature on the transformation of sodalite to cancrinite in Bayer digestion. Hydrometallurgy 105:75–81
5. Joseph CG, Taufiq-Yap YH, Krishnan V, Puma GL (2019) Application of modified red mud in environmentally-benign applications-a review paper. Environ Eng Res 25:795–806
6. Alp A, Goral MS (2003) The influence of soda additive on the thermal properties of red mud. J Therm Anal Calorim 73:201
7. Atasoy A (2005) An investigation of characterization and thermal analysis of the Aughinish red mud. J Therm Anal Calorim 81:357–361
8. Gelencser A, Kovts N, Turoczi B, Rostsi A, Hoffer A, Imre K, et al (2011) The red mud accident in Ajka (Hungary): characterization and potential health effects of fugitive dust. Environ Sci Tech 45:1608–1615
9. Ayres RU, Holmberg J, Andersson B (2001) Materials and the global environment: waste mining in the 21st century. MRS Bull 26(6):477–480
10. Marabini AM, Plescia P, Maccari D, Burragato F, Pelino M (1998) New materials from industrial and mining wastes: glass-ceramics and glass- and rock-wool fibre. Int J Miner Process 53(1–2):121–134
11. Singh M, Upadhayay SN, Prasad PM (1996) Preparation of special cements from red mud. Waste Manage 16(8):665–670
12. Singh M, Upadhayay SN, Prasad PM (1997) Preparation of iron rich cements using red mud. Cement Concrete Res 27(7):1037–1046
13. Yalcin N, Sevinc V (2000) Utilization of bauxite waste in ceramic glazes. Ceram Int 26(5):485–493
14. Asim N, Alghoul M, Mohammad M, Amin MH, Akhtaruzzaman M, Amin N, Sopian K (2019) Emerging sustainable solutions for depollution. Const Building Mat 199:540–548
15. Chen S, Wu C, Yan D (2019) Binder-scale creep behaviour of metakaoline based geopolymer. Cement Const Res, 12

Metal Additive Manufacturing Technique in Construction Industry: A Review Paper

Laiphrakpam Indrajit Singh and Khwairakpam Sachidananda

1 Introduction

Additive manufacturing or 3D printing is the technique of manufacturing which employed the concept in which finite section of the element are attaching to each other in a well-organized pattern so as to develop a full structure. With this concept of manufacturing at the finite level of an element, properties of the structure can be altered and modified accordingly from the ground level. AM offers great benefits in terms of much efficient structure, freedom in adopting geometrical shapes, utilization of lesser material and reduction of material wastage, opportunity for developing new functionally graded material and prestressing, additionally for repair and strengthening benefits.

Traditional method of manufacturing is subtractive by nature meaning when any structure is developed such as either concrete or steel block, extra materials are removed during shaping or machining which leads to reduction in the overall integrated mass. Additive manufacturing adds material progressively [1], the concept of building layer over layer from 3D model data. The technical concept has been observed in 1860s from a three-dimensional sculpture developed from a two-dimensional sculpture portrait [2].

Comparing with concrete, steel can be molded and shaped depending on our desirable aesthetical views and purposes. Applying the concept of additive manufacturing in steel manufacturing will be great boom in terms of freedom of choosing shapes and sizes, generating more optimized structure, utilizing the undesired residual stress in a beneficial manner, etc. Wire and Arc Additive Manufacturing (WAAM) is suitable for steel construction. Here wire electrode serves as printing material to produce large

L. I. Singh (✉) · K. Sachidananda
Department of Civil Engineering, National Institute of Technology Manipur, Imphal, India
e-mail: indrajitlpk@gmail.com

B. P. Swain (ed.), *Recent Advances in Materials*, Springer Proceedings in Materials 25,
https://doi.org/10.1007/978-981-99-3844-5_5

components layer by layers. This level of manufacturing rate can be applied successfully in commercial field for making connections and joining technology rather than fabrication.

In milling process, materials are removed from the structure, on the other hand additive manufacturing creates structure by adding materials. With additive manufacturing in metal, the concept of thermal and mechanical behaviors of the material used to play an important role. Traditional welding has a limited number of deposition tracks whereas AM has thousands of such weld tracks overlapping each other. In normal welding a large temperature variation is developed around the melt pool due to high-energy concentric heat. The thermal effect due to deposition of material and non-uniform cooling is associating with residual stress which is responsible for unexpected failure, strength of the build structure, fatigue life, and dimensional inaccuracy. The undesirable residual stress can be converted into desirable pre-stress by controlling the moving heat source manually in additive manufacturing. The heat is introduced layer after layer which is termed as 'thermal cycle'. This method helps to control cooling effect rather than non-uniform cooling as observed in traditional welding.

2 Additive Manufacturing in Construction

2.1 3D Printing in Concrete

Concrete is the very basic and widely used raw material in construction and it is of low cost and locally available easily. Concreting requires huge expenditure for the arrangement of scaffolding and formwork accounting half of the construction cost.

WinSun, a Chinese advanced materials supplier turned architectural business, originally began investigating additive manufacturing in 2005. By 2013, WinSun had constructed the first residential building made using additive printing. The first AM office structure, depicted in Fig. 1, was constructed in China and shipped to Dubai in 2016. [3].

In 2016, concrete pedestrian bridge [4] spanning 12 m was also completed in urban park of Castilla-La Manicha near Madrid, Span (Fig. 2). In 2017, a bicycle bridge was also built in Gemert, Netherland (Fig. 3), printed using concrete layer of 1 cm thick from an AM nozzle with zero formwork [5, 6].

2.2 Polymer Printing

Using polymer as printing material a çanal house of 6 m tall designed in 2014 by DUS Architects developed blocks and built in Amsterdam [7], Netherland resembling the traditional Dutch gabled canal house (Fig. 4).

Fig. 1 Winsun Building, Dubai [3]

Fig. 2 Castilla-La Mancha Pedestrian bridge [4]

Fig. 3 Gemert Bicycle Bridge [6]

Fig. 4 3D print canal house in Amsterdam [7]

2.3 Metal Printing

Initially metal printing was used for smaller components such as connections and facade nodes. The Nematox facade node [8] (Fig. 5) gives an example of how metal AM helps to optimized the façade geometries by giving connection outside the nodes, reducing drawbacks due to the effect of sealing condition. The Arup Lightning node [9] which was redesigned by Arup by topology optimizing the previous structure is shown in Fig. 6. In the optimized structure materials are reduced or removed wherever lower stresses occur. This reduces the total weight to 75% of the previous model.

The MX3D bridge (Fig. 7) of Amsterdam, Netherland [10] is the most recently developed full size metal additive manufacturing masterpiece completed in the year 2019 built using WAAM technique. The 2.5 m wide bridge with 10 m spanning was designed with engineers in partnership with Arup in collaboration with researchers from Imperial College London. Numerical simulation and load test are also completed.

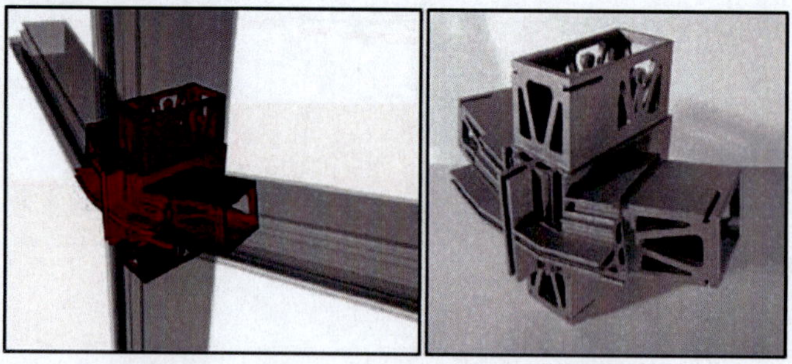

Fig. 5 Nematox node [8]

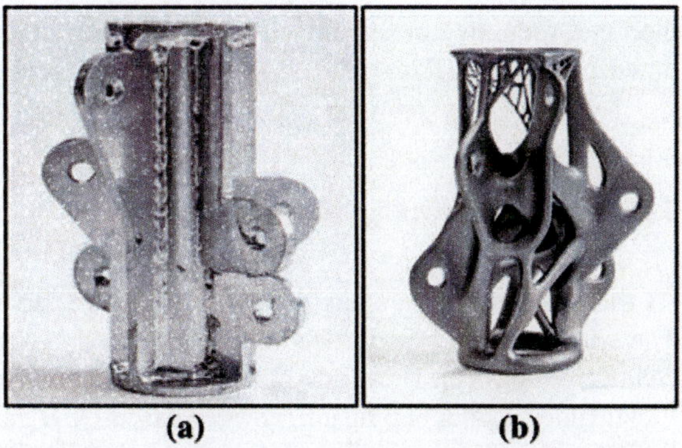

Fig. 6 a Conventional lightning node b Arup optimized node [9]

Fig. 7 MX3D bridge, Amsterdam [10]

3 Technologies in Metal Additive Manufacturing

Based on the standard of ISO/ASTM 52900 [11] powder bed fusion and directed energy deposition are the popular and easily available methods for metal additive manufacturing.

3.1 Powder Bed Fusion (PBF)

Laser or electron beam source heat energy is incorporated and material contained in a powder bed is selectively fused and create molten pool of metal and bonded together. Pieces with complicated geometries are a good fit for this technique [12]. The existing build time for items may be tens of hours, and the building rate is roughly 50 g/hour because they are constructed in tens of micron-thick individual layers. It can give a very smooth surface finish with roughness lesser than 20 μm [13]. A very inert and vacuum atmospheric condition is required to prevent oxidation of metallic powder [2]. So PBF is solely adopted in Lab rather than using in-situ construction.

3.2 *Directed Energy Deposition (DED)*

As the name suggests, energy deposition as directed by the laser or electron beam. The wire or metal powder are deposited directly at the focal point of the beam of electron or laser [2]. They form molten pool of metal and form mass as it cooled down. Identical to powder bed fusion, powder-based DED has a maximum component size limitation, requires an inert atmosphere to avoid oxidation, and has lengthy build periods (currently depositing at a rate of roughly 1 kg/hour). Typically, it gives not a smooth surface with values ranging from 20 to 100 μm [13].

4 Researches in Metal Additive Manufacturing

Additive manufacturing products have been mostly in the researching stage. In the past 5–6 years significant researches have been conducted based on metal additive manufacturing building components and parts. Numerical finite element modeling of the scale model, optimization, and load testing are also carried out. Comparative studies between machined components and printed components with varying cross-sectional dimensions such as square hollow section (SHS) and circular hollow section (CHS).

4.1 *Research Based on Powder Bed Fusion (PDF)*

Using powder bed fusion method for manufacturing Grade 316L stainless steel material with different orientation of building, the stress–stain properties were evaluated in many previous studies [14]. Buchanan et al. [14] investigated the stress–strain curve of the AM PBF shown as in Fig. 8. Three building orientations were investigated. The angle between the building's horizontal surface and the coupon's center line is known as the building orientation. A significant degree of porosity in PDF produced material than in ordinary stainless steel was shown to be related with a lower elastic modulus. Yet, the fabrication process's quick cooling effects and crystals result in a higher strength.

Marouene Zouaou et al. [15] developed a finite element model of a fused filament fabricated (FFF) specimen made up of new polymeric pre-structured material using Abaqus software. Three angles of building orientation (0°, 45°, and 90°) were adopted. Tensile strength as predicted by the software matches very closely to the experimental results. The longitudinally oriented specimen (0°) attains the maximum

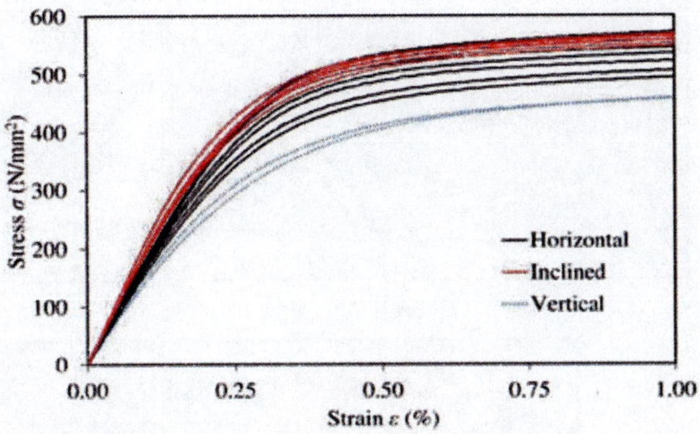

Fig. 8 Measured stress–strain curves by Buchanan et al. [14]

tensile yield strength with good ductile behavior while the transversal (90°) one has the weakest and exhibit fragile behavior at the yielding point.

Grade 316L stainless steel open cellular lattice structures [16], circular hollow section as well as square hollow section compression elements [14] have all been studied in the structural application of powder bed fusion metallic materials. Generally stockier AM cross-sections had stronger axial resistance compared with conventional formed stainless steel SHS when same vertical material properties are used in both cases. However, in the case of slender section, AM specimens have slightly lower resistance than the conventionally built section which may due to residual stress effect.

4.2 *Research Based on Directed Energy Deposition (DED)*

Prior to the work of MX3D bridge, researches based on the use of wire and arc additive manufacturing (WAAM) applications were carried out. Two perpendicular orientations of tensile coupons for mild steel were developed, with no notable change in yield strength detected, although stainless steel coupons were constructed in the same way. Full sized load test experiment as well as the numerical simulation of the elements of the MX3D were also conducted.

Research based on AM metal components such as beam hook, stiffener, clamping elements for diagonal bracing, end plate replacement, node for space frame were also carried out following the concept of optimization [17] (Fig. 9).

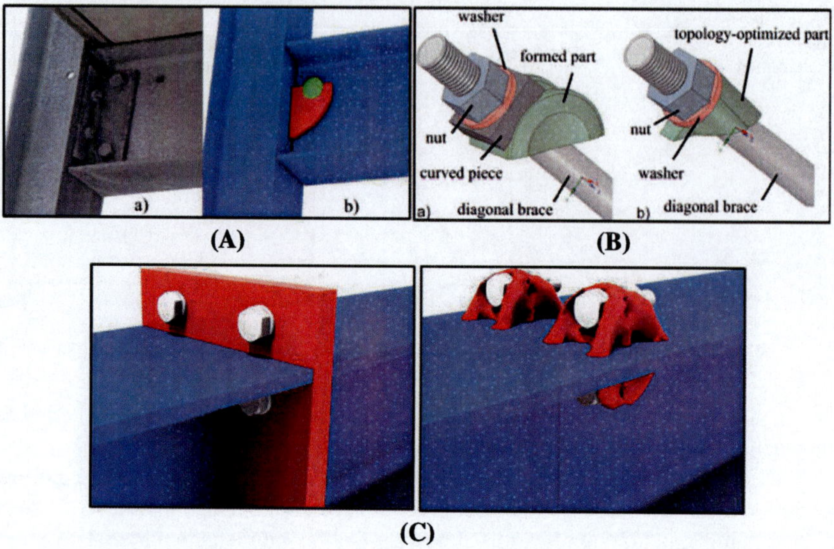

Fig. 9 **A** Conventional connection versus beam hook, **B** conventional versus optimized clamp **C** conventional versus optimized T-stub end plate [17]

5 Opportunities and Challenges

5.1 Flexibility in Geometry and Material Optimization

With the introduction of additive manufacturing, a new system of structural components such as hollow cross-sectional elements are readily fabricated. These elements are designed to obtained sections with unregular thicknesses at the same time without compromising the strength. Additionally, these hybrid type of elements when coupled with the conventional type of material can produced much higher strength.

The idea of additive manufacturing implies not only for the freedom of customization but also the concept of selective deposition of material to the region subjected to higher stresses. This concept drastically reduces the material consumption and hence the cost. Materials are supplied to only wherever needed to fulfill the bending moment diagram's and hence produces a geometrically more ductile cross-section.

By using high strength material in areas with a large stresses and moments, like the middle of beams, and higher ductility in areas with higher ductility needs, such connections, by controlled cooling system a good ductile and strengthen material can be produced [18]. The porosity nature of the biomedical human implant in order to reduce material stiffness within the neighboring human bone [19] is extracted and applied to the concept of addictively manufactured elements. AM elements have greater advantages in terms of forces and moment distribution with lower stiffness to attract lower forces in specific locations. The porosity nature gives tremendous benefit in energy dissipation capacity for seismically subjected structures.

5.2 *Degree of Customization*

Additive manufacturing allows for levels of customization previously unattainable in a low-margin business like construction. Structural engineers might make each structural component distinctive without incurring printing expenses; the cost of producing two identical or different variations of the same product is practically the same [20]. Design modifications might also be implemented quickly. This customization is anticipated to increase demand for additive manufacturing, which will assist to reduce prices.

5.3 *Construction Time*

Additive manufacturing provides various potential to reduce building time. With additive manufacturing in construction, the duration of project can be reduced to 65% when compared with the conventional method of building [21].

5.4 *Hybridization and Strengthening*

Additive manufacturing is thought to enhance rather than replace traditional construction approaches. Additive manufacturing techniques may also be used to repair damaged or corroded structural parts or to reinforce a structure in-situ, lowering the cost of repairs or strengthening. Any repair procedure gives a chance to upgrade the structural element's design.

5.5 *Environmental Effects*

Construction accounts about 30% of global greenhouse gas emissions [18]. Elements such as walls, columns, beams, and floors with definite proportions, traditional building procedures employ standardized components such as steel structural cross-sections or reinforcing bar. By producing specific, optimal structural components for each project, additive manufacturing minimizes the amount of raw materials used and waste that has to be disposed of. In general, additive manufacturing processes may reduce waste by 40% when compared to subtractive approaches [20].

5.6 *Human Factors*

Labor expenditures are predicted to account for 15–50% of overall building costs [22]. Some claim that by increasing automation and working in adverse weather and at night, many worksite operations may be completed more safely, correctly, and quickly [22]. Building procedures in additive manufacturing are often highly automated, lowering labor costs and lowering the danger of human mistake throughout the production process. However, because there are fewer possibilities for human intervention during manufacturing, 3D CAD models must be of high quality and mistake free [1]. This growth in automation will represent a significant departure from traditional, more manual system.

5.7 *Challenges*

The initial cost of machineries and installation would be very expensive with this current technology. And there is limitation of manufacturing process such as PBF is completely confined to laboratory only which requires very inert environment. With regard to design methodology and digital workflow, advanced computational tools are required and as these advances further, the cost of production will decrease. Besides there is a fear in the general people mind regarding this robotic technology will give unemployment also.

6 Conclusion

This article examined ongoing metal AM techniques, current research, early applications in construction, and application in many technical industries, as well as the numerous potential and difficulties that lie ahead for widespread metal AM adoption in the construction industry. For use in constructing, powder bed fusion and directed energy deposition are the two best metal additive manufacturing technologies (DED). Quality construction is possible using PBF and DED technologies, though there are drawbacks on expense, lead time, and maximum size. Practically unbounded item sizes are possible using wire and arc additive manufacturing DED, but dimensional precision and surface polish are restricted.

Using the technique of metal additive manufacturing (AM), massively scalable structural that are conventionally not possible, costly and time-consuming are now coupled with traditional structural parts to produce hybrid structures. Modifications in the microstructure can alter the basic mechanical behavior, and functionally graded materials can be used to regulate the internal dispersion of forces and moments.

Initially AM is to cost more than conventionally developed section. It will be necessary to develop new digital design processes, focus more toward functionality rather than manufacturing and adapt codified structural methodological approaches to factor in different material properties and more variable shape.

References

1. Kathryn M, Moroni G, Vaneker T, Fadel G, Campbell RI, Gibson I, Bernard A, Schulz J, Graf P, Ahuja B, Martina F (2016) Design for additive manufacturing: trends, opportunities, considerations, and constraints. CIRP Ann Manuf Technol 65:737–760
2. Gao W, Zhang Y, Ramanujan D, Ramani K, Chen Y, Williams CB, Wang CCL, Shin YC, Zhang S, Zavattieri PD (2015) The status, challenges, and future of additive manufacturing in engineering. Comput Aided Des 69:65–89
3. Dubai Future Foundation (2016) The world's first 3D printed office. http://officeofthefuture.ae/ , October 10, 2017
4. IAAC (2017) Large scale 3D printing. https://iaac.net/research-projects/large-scale3d-printing/3d-printed-bridge/ accessed August 8, 2017
5. Clarke C (2017) Royal BAM Group use 3D printer to make concrete bicycle bridge with TU Eindhoven. 3D Printing Industry. https://3dprintingindustry.com/news/royalbam-group-concrete-3d-printer-concrete-bicycle-bridge-with-tu-eindhoven-116359/, accessed August 8, 2017
6. Scott C (2017) TU/e and BAM Infra get to work on 3D printed concrete bicycle bridge. 3D Print. https://3dprint.com/178462/eindhoven-3d-printed-bridge/, accessed August 8, 2017
7. Frearson A (2016) DUS architects builds 3D-printed micro home in Amsterdam. Dezeen. https://www.dezeen.com/2016/08/30/dus-architects-3d-printed-micro-homeAmsterdam-cabin-bathtub/ accessed October 10, 2016
8. Strauss H, Partner AGEP, Knaack U (2015) Additive manufacturing for future facades: the potential of 3D printed parts for the building envelope. J Facade Design Eng 3:225–35
9. Arup (2015) 3D makeover for hyper-efficient metalwork. Arup News. http://www.arup.com/news/2015_05_may/11_may_3d_makeover_for_hyperefficient_metalwork, accessed October 10, 2016
10. Goehrke SA (2017) 3D printed steel pedestrian bridge will soon span an Amsterdam canal. 3D Print. https://3dprint.com/72682/amsterdam-3d-printedbridge/, accessed August 8, 2017
11. ASTM International. ISO/ASTM 52900:2015(E) Standard terminology for additive manufacturing technologies—general principles—terminology, 2015
12. Williams SW, Martina F, Addison AC, Ding J, Pardal G, Colegrove P (2015) Wire + arc additive manufacturing. Mater Sci Technol 32:641–647
13. Gu DD, Meiners W, Wissenbach K, Poprawe R (2012) Laser additive manufacturing of metallic components: materials, processes and mechanisms. Int Mater Rev 57:133–164
14. Buchanan C, Matilainen V-P, Salminen A, Gardner L (2017) Structural performance of additive manufactured metallic material and cross-sections. J Constr Steel Res 136:35–48
15. Zouaoui M, Gardan J, Lafon P, Makke A, Labergere C, Recho N (2021) A finite element method to predict the mechanical behavior of a pre-structured material manufactured by fused filament fabrication in 3D printing. Appl Sci
16. Shen Y, McKown S, Tsopanos S, Sutcliffe CJ, Mines RAW, Cantwell WJ (2010) The mechanical properties of sandwich structures based on metal lattice architectures. J Sandwich Struct Mater 12:159–180
17. Lange J, Feucht T, Erven M (2020) 3D printing with steel—additive manufacturing for connections and structures. Steel Const 13(3):144–153
18. Labonnote N, Rønnquist A, Manum B, Rüther P (2016) Additive construction: state-of the-art, challenges and opportunities. Autom Constr 72:347–366
19. Krishna BV, Xue W, Bose S, Bandyopadhyay A (2008) Engineered porous metals for implants. JOM 60:45–48

20. Berman B (2012) 3-D printing: the new industrial revolution. Bus Horiz 55:155–162
21. Buswell RA, Soar RC, Gibb AGF, Thorpe A (2007) Freeform construction: mega-scale rapid manufacturing for construction. Autom Constr 16:224–231
22. Richardson V (2014) 3D printing becomes concrete: exploring the structural potential of concrete 3D printing
23. Tolosa I, Garciandía F, Zubiri F, Zapirain F, Esnaola A (2010) Study of mechanical properties of AISI 316 stainless steel processed by "selective laser melting", following different manufacturing strategies. Int J Adv Manuf Technol 51:639–647
24. Guan K, Wang Z, Gao M, Li X, Zeng X (2013) Effects of processing parameters on tensile properties of selective laser melted 304 stainless steel. Mater Des 50:581–586
25. Niendorf T, Leuders S, Riemer A, Richard HA, Tröster T, Schwarze D (2013) Highly anisotropic steel processed by selective laser melting. Metall Mater Trans B 44:794–796
26. Spierings AB, Herres N, Levy G (2011) Influence of the particle size distribution on surface quality and mechanical properties in AM steel parts. Rapid Proto J 17:195–202
27. Rehme O (2009) Selective laser melting of honeycombs with negative Poisson's ratio. J Laser Micro/Nanoeng 4:128–134
28. Kahlen F-J, Kar A (1999) Tensile strengths for laser-fabricated parts and similarity parameters for rapid manufacturing. J Manuf Sci Eng 123:38–44
29. Wu P, Wang J, Wang X (2016) A critical review of the use of 3-D printing in the construction industry. Autom Constr 68:21–31
30. Buchanan C, Gardner L (2019) Metal 3D printing in construction: A review of methods, research, applications, opportunities and challenges. Eng Struct 180:332–348. ISSN 0141-0296
31. Petrovic V, Gonzalez JVC, Ferrando OJ, Gordillo JD, Puchades JRB, Griñan LP (2011) Additive layered manufacturing: sectors of industrial application shown through case studies. Int J Product Res 49:1061–1079

A Study on the Load Carrying Capacity of Shallow Foundation on Reinforced Sandy Soil

Kangujam Monika and Th. Kiranbala Devi

1 Introduction

The foundation transfers the load from the structure to the soil. To stand the vertical load transfer, the bearing capacity of the soil needs to be strong enough. To increase the bearing capacity and to minimize the settlement, reinforcement is given to the soil either by mixing the soil with additives like cement, natural fibers or synthetic fibers, etc. or with geotextile, geogrid, geocell, etc.

Some researchers have shown the effect of reinforcement on soil and also illustrated the influence of reinforcement like geotextile or geogrid with the embedded depth-to-breadth ratio and the numbers of the reinforcement layer. Kolay et al. [1] concluded that the bearing capacity of the soil increased when reinforcement was provided. They showed the bearing capacity for two-layered soil using one geogrid layer at the interface of soils in which the ratio embedded depth (u) to the width of footing (B) is 0.667 and had an average increase of 16.67%, and for one geogrid in the middle of the sand layer with u/B equal to 0.33, the bearing capacity increased with an average of 33.33%. Chakraborty and Kumar [2] also illustrated an increase in bearing capacity after the soil was reinforced and also showed that the critical position of reinforcements lay between 0.29 B and 0.57 B for single layer reinforcement, which would give the maximum bearing capacity of a strip footing placed over granular and cohesive-frictional soils. Chakraborty and Kumar [3] also stated that for circular footing on the sand, the embedment depth of the circular reinforcement sheet within 0.15 diameter of footing (D) to 0.43 D had maximum bearing capacity. Shirazi et al. [4] reviewed the effectiveness of the ratio of first geotextile depth to footing width (d/B), the ratio of geotextile spacing to footing width (S/B), ratio of geotextile length to footing width (L/B), and reinforcement layers number (N) on the bearing capacity.

K. Monika · Th. K. Devi (✉)
Manipur Institute of Technology, Manipur University, Imphal, Manipur, India
e-mail: kiranbala_th@gmail.com

B. P. Swain (ed.), *Recent Advances in Materials*, Springer Proceedings in Materials 25,
https://doi.org/10.1007/978-981-99-3844-5_6

Roy [5] reviewed different reinforcement types, like natural and synthetic fibers, and concluded that the reinforcement fiber increased the bearing capacity and reduced settlement. Omar et al. [6] concluded that in the case of geogrid, the effect of reinforcement lay within 1.4B for square foundations on sand. From the literature, it could be concluded that the optimum effect of the reinforcement could be achieved if the reinforcement lay within 0.6B. However, the influence of reinforcement with the embedded depth differs with different types of soil and type of reinforcement used.

In this paper, two different embedment depths were considered, and fiberglass mats were used as the reinforcement. The load test of shallow foundation models was performed in different soil bedding conditions, i.e., plain soil, fiberglass mat embedded at 7.5 cm (embedment depth/ breadth of the foundation ratio is 0.5) depth soil below the ground level and fiberglass mat embedded at 15 cm (embedment depth/ breadth of the foundation ratio is 1) depth below the ground level. The influence of the fiberglass mat on the soil bed and foundation model was also studied.

2 Experimental Process

The experimental process is categorized, namely (i) materials used and (ii) test box set up along with soil bedding.

2.1 Materials

The soil sample was collected from Dhansiri River bank, Dimapur, Nagaland. The detailed characteristic properties of the soil sample are shown in Table 1. Direct shear test was performed to find the shear strength parameters as per IS:2720 (Part 13)-1986 [7]. The particle grain size distribution and normal-shear stress graph are given in Figs. 1 and 2, respectively. The soil sample is classified as well-graded sand (SW), and effective diameter is 0.2 mm. The fiberglass mat used was a double-layer 200 gsm non-woven fabric and had a cross section of 60 cm × 50 cm.

The concrete of M25 grade, which has a ratio of 1 cement:1 fine aggregate:2 coarse aggregate, was used for the preparation of the foundation model. Since the models have a 2.5 cm thickness, the coarse aggregate of size ranging from 4.75 mm to 10 mm is employed. The fine aggregate has been classified as Zone-II as per IS:383–1970 [8]. The cement is 53-grade ordinary Portland cement having a consistency of 32%, an initial setting time of 2 h 9 min, and a final setting time of 4 h 30 min. The compressive strength of the concrete at 7 and 28 days is 20 MPa and 33 MPa, respectively.

Table 1 Characteristic properties of the soil sample

Characteristic property	Value
Gravel (greater than 4.75 mm)	2.8%
Sand (ranging between 4.75 and 0.75 mm)	97.2%
Silt (ranging between 0.75 and 0.002 mm)	0.8%
Clay (smaller than 0.002 mm)	Nil
Specific gravity	2.63
Liquid limit (LL)	N/A
Plastic limit (PL)	Non-Plastic
Co-efficient of uniformity (Cu) Co-efficient of curvature (Cc) Bulk density	4 1 1.6 g/cm^3
The angle of internal friction Cohesion	29.73° 2.336 kPa

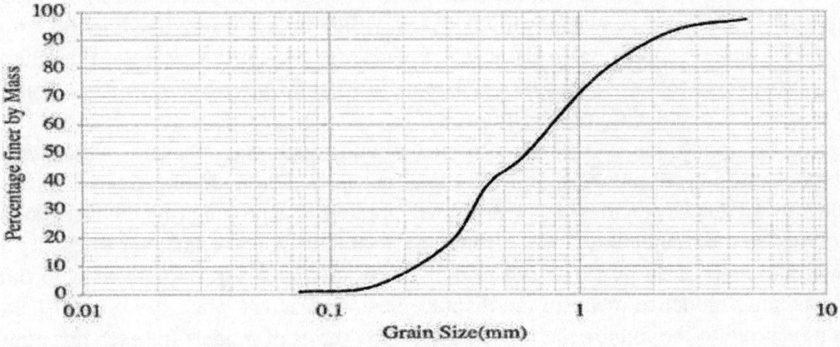

Fig. 1 Particle size distribution of the soil sample is shown

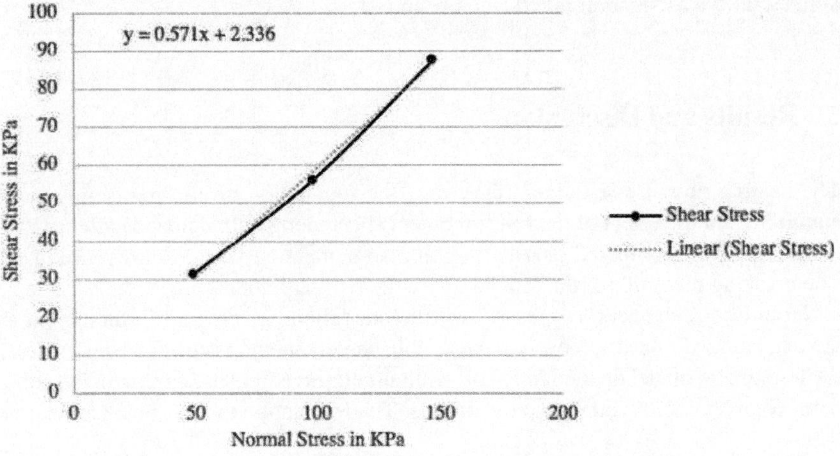

Fig. 2 Normal stress–shear stress graph gives the shear strength parameter values of the soil sample

2.2 *Test Box Along with Soil Bed Preparation*

The thickness of the foundation model was made at 2.5 cm, and the ratio of the thickness of the foundation model to the test box was 1:40. The length and breadth of the model were 150 mm (15 cm) × 150 mm (15 cm). The test box was made of 4 mm thick steel. The internal dimensions of the box were 1000 mm in length, 1000 mm in breadth, and 1000 mm in height. The test box was made rigid in such a way that there would be less plain displacement in all directions. The soil was poured uniformly by pluviation method as in Vaid and Negussey [9] in five layers from constantly changing the pouring trip to have constant the height of drop 0.5 m into the testing box so that the relative density of 22.22% was maintained. The soil bed was again compacted with a plyboard placed over the surface of the bed to give a uniform surface. The procedures were repeated when the soil was disturbed for fiberglass embedment. The soil bed was arranged for three series. The first series is only plain sand without any reinforcement. For the second series, the fiberglass double-layered mat was embedded at a depth of 7.5 cm, in which the ratio of embedment depth/ breadth of the foundation (*u/B*) is 0.5. Finally, for the third series, the fiberglass double-layered mat was embedded at a depth of 15 cm, where the embedment depth/ breadth of the foundation ratio (*u/b)* is 1.

At the top frame of the test box, a hydraulic jack was attached, which was connected to a proving ring of 25 kN load capacity. The hydraulic jack produced downward displacement when pressure was applied. The proving ring reads the load applied to the foundation model through the hydraulic jack. A dial gauge was fixed with the help of an iron beam. The dial gauge was later kept in contact with the foundation model to measure the displacement/ settlement when the vertical load was applied to the foundation model. Three test series of models for each different soil bedding were carried out. When the load was applied through the hydraulic jack, the readings of both the proving ring and the dial gauge were taken. The load was applied till the foundation failed (Figs. 3 and 4).

3 Results and Discussion

The bearing pressure-settlement curves of the foundation of all three series were analyzed, and the effect of the reinforcement sheet along with embedded depth was considered. The compared bearing pressure-settlement curves are shown in Fig. 5. The breaking pressures of the foundation models vs settlements are given in Table 2.

From Fig. 5, the bearing pressure of the foundation, which has reinforcement at 7.5 cm, has the highest value when the double tangent method is used. From Table 2, the foundation model on the sandy soil without reinforcement has a maximum settlement compared to the other two conditions. The foundation models break at nearby

Fig. 3 Experimental test box set up for the load-settlement test

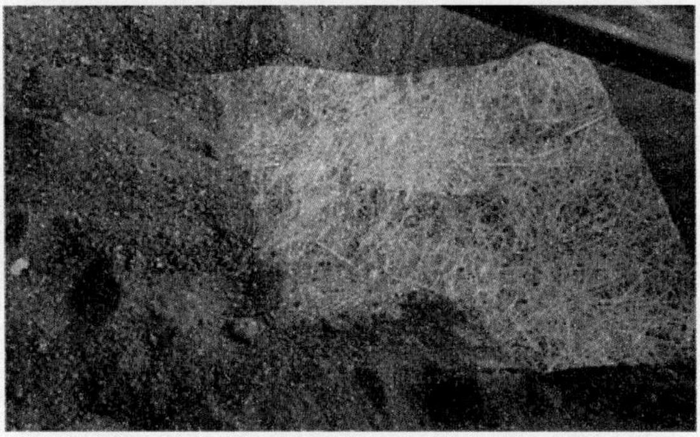

Fig. 4 Fiberglass mat embedded inside the soil

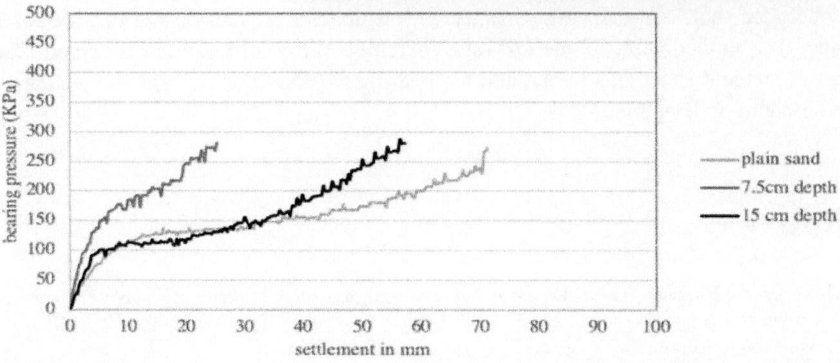

Fig. 5 Comparison of bearing pressure-settlement curves of the shallow foundation models

Table 2 Breaking pressure of the shallow model foundation in different soil bedding conditions

Soil bed condition	Breaking pressure in KPa	Settlement in mm
Plain soil	273.78	71.1
7.5 cm depth fiberglass mat embedded	281.24	25.2
15 cm depth fiberglass mat embedded	280	57.3

values of breaking pressure. However, there are significant differences in the settlement values. The results show that the reinforcement increases the ultimate bearing capacities of the foundation model and decreases the settlement.

Moreover, the soil with reinforcement at 0.5 B embedded depth has less settlement than the other two conditions. This shows the influence of the reinforcement's embedded depth. Similar findings were found in [1–3] and [6]. As the reinforcement's embedded depth increases, the bearing capacity decreases and settlement increases.

4 Conclusion

The study concludes that the fiberglass mat as reinforcement to the soil improves bearing capacity and strengthens the soil. The foundation model, placed on the soil with the fiberglass mat at 7.5 cm, has the highest bearing pressure/ load carrying capacity with less settlement. In both conditions of reinforcement with fiberglass mat, the values of settlement of the foundation model decrease as compared to the unreinforced soil. The settlement of the foundation model, which was embedded at a depth of 7.5 cm below the ground level (ratio of embedment depth to width of the model as 0.5), has 64.55% less settlement than the plain soil and 56% less settlement than the plain soil which was embedded at a depth of 15 cm below the ground level (the ratio of embedment depth to width of the model as 1). Moreover, it can be concluded that load carrying capacity and settlement depend on the depth of the fiberglass embedment. As the *u/B* ratio increases, the contribution of the fiberglass reinforcement sheet gets lesser, and the bearing capacity decreases, which leads to an increase in settlement.

References

1. Kolay PK, Kumar S, Tiwari D (2013) Improvement of bearing capacity of shallow foundation on geogrid reinforced silty clay and sand. J Const Eng 201:Article ID 293809, 10 p. Hindawi Publishing Corporation
2. Chakraborty D, Kumar J (2014) Bearing capacity of strip foundations in reinforced soils. Int J Geomech 14(1):45–58. ASCE
3. Chakraborty D, Kumar J (2015) Bearing capacity of circular footings on reinforced soils. Int J Geomech 15(1):04014034. ASCE

4. Shirazi MG, Rashid ASBA, Nazir RB, Rashid AHBA, Moayedi H, Horpibulsuk S, Samingthong W (2020) Sustainable soil bearing capacity improvement using natural limited life geotextile reinforcement-a review. Minerals 10:479
5. Roy S (2020) Effects of reinforcement elements in soil reinforcement. Int J Geol Geotech Eng 6(2):22–33. ISSN: 2581-5598
6. Omar MT, Das BM, Puri VK, Yen SC (1993) Ultimate bearing capacity of shallow foundations on sand with geogrid reinforcement. Can Geotech J 30(3):545–549
7. Indian Standard: Methods of Test for Soils, Part 13 Direct Shear Test (Second Revision). IS: 2720 (Part 13)-1986 (Second Revision)
8. Indian Standard: Specification for Coarse and Fine Aggregates from Natural Sources for Concrete (Second Revision). IS: 383-1970. (Reaffirmed 2002)
9. Vaid YP, Negussey D (1984) Relative density of pluviated sand samples. Soils Found 24(2):101–105. Japanese Society of Soil Mechanics and Foundation Engineering

Assessment of Dimapur Sand with Local Sand Found in Manipur

Khwairakpam Selija, Kosygin Leishangthem, Koko Karbia, and Devasis Laishram

1 Introduction

Sand is a granular substance comprised of tiny fragments of rock and mineral. It can also be described as a mixture of rock and tiny, fine granular elements. Sand is also frequently described by its size, i.e. that it is finer than gravel and coarser than silt, and that its size ranges from 0.06 mm to 2 mm. Sand is a widely utilized topographical material that can be employed for a variety of Civil Engineering projects. Typically, quarry rocks and stones are blasted, followed by repeated crushing cycles to reduce the sand's particle size. Once more, sifting is done on the sand that was produced by blasting and crushing quarry stones. The sand is first sieved, then cleaned to get rid of the small impurities and pollutants. Providing an affordable and environmentally friendly alternative to natural sand is the primary goal of synthetic sand. Because the grains are typically sharp, angular, rough, and coarse, they can form solid connections on surfaces. Pit sand often has little salt, which reduces the possibility of excessive moisture absorption when utilized in a building. Checking the degree of coarseness is necessary to make sure that it isn't above the allowed limits since, as we have learnt, pit sand is of course type and that the size of the grains is larger than those of sand from other sources. The degree of coarseness can be checked by using the sieve analysis testing method. Sand from rivers is also known as natural sand. It is classified as fine sand and is frequently discovered next to streams and river banks. River sand can be used in masonry and concrete projects due to its softness. For plastering, RCC construction, and other types of block work, river sand is perfect.

River sand has high silica content. The silica concentration must therefore be examined before utilizing it in any form of construction job. The river sand should have a silica level of less than 5%. When utilized in buildings, river sand doesn't need a lot of water because it has a high moisture content. Manufactured sand is also

K. Selija (✉) · K. Leishangthem · K. Karbia · D. Laishram
Department of Civil Engineering, Manipur Technical University, Imphal, India
e-mail: selijakhwairakpam@gmail.com

B. P. Swain (ed.), *Recent Advances in Materials*, Springer Proceedings in Materials 25,
https://doi.org/10.1007/978-981-99-3844-5_7

known as artificial sand, since it is created by humans by crushing rocks and other natural raw materials. The alternative to river sand is more environmentally friendly. Crushing granite, stones, etc. produces manufactured sand at factories or facilities. To reduce the size of the resulting sand, quarry stones and boulders are often blasted and then put through many crushing cycles. The sand produced by blasting and crushing quarry stones goes through another process. Normally, quarry stones and rocks are blasted and then crushed several times for size reduction. The crushed sand is then sieved and washed after it has been crushed to remove the impurities and fine particles. As manufactured sand is an environmentally friendly and economical alternative to natural sand, it provides a faster and more efficient way to build. Typically, quarry boulders and stones are blasted before undergoing many crushing cycles to reduce the size of the sand that results. The sand that is produced by blasting and crushing quarry stones is sieved once more. After sieving, the sand is rinsed. To get rid of the contaminants and tiny particles, the sand is sieved and washed. The primary goal of synthetic sand is to offer a more affordable and environmentally friendly alternative to natural sand is finely graded, and the resulting sand is then tested before using it in construction. The sand should pass easily through the 3/8 sieve after production.

Fine sands are those sand that passes through 16 number sieves, i.e. 1.18 mm^2 openings. Fine sands are most commonly used in plastering works. Sand is one of the main materials used in building, and how it is utilized frequently affects the strength and longevity of the finished product. The general rule is to steer clear of contaminated materials combined with clay, topsoil or surface soil, or vegetation when utilizing sand in construction. Using impure sand or sand with high clay and silt content affects the formation of proper bonds between the cement and the sand giving weak bonding which thus results in weak bonded building or structure. Poor bonds usually result in a structure that is not as strong as it would be if the right type of sand was used. So, there is a need for testing of sand to determine whether the sand to be used in the construction will be pure enough to give a good cement bonding. Sand is available freely and plenty in nature but not all the sand which are available in nature can be used for construction. The characteristics of sand vary depending on its source, which may either help or prevent it from having the optimal qualities needed for effective construction sand. Let's take the example of desert sand and sea sand. The overwhelming bulk of desert sand has no use in construction as the grains are too smooth and fine to bind together for making concrete or building material. The desert sand is eroded by wind rather than water, this is the reason why they have smooth grains and its surface chemistry is not able to offer a sufficient number of multidirectional chemical linkages. Sand can also be found in the sea, but because of its high salt content, which encourages corrosion and also results in efflorescence, sea sand is not recommended for use in building. As part of this effort, we also investigate the chemical composition of sand to better understand its characteristics.

In this project, we work on testing Dimapur sand and some commonly used local sand in Manipur. The main objective of the project is to compare the properties and find out which sand has the best quality based on the result of different testing conducted and give a credible reason to society why it is the best so far. To study the

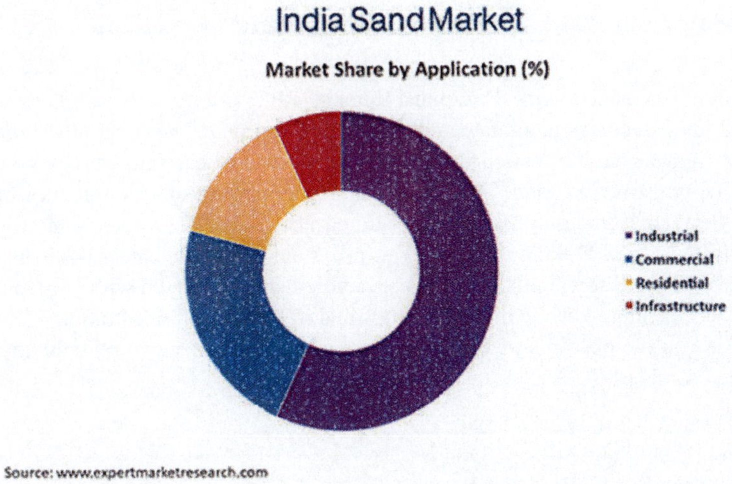

Fig. 1 India sand market [1]

properties of local sand commonly used in the state and utilize the use of local sand. To minimize the cost of construction by using local sand found in the state.

In 2020, the Indian sand market reached 833 million tonnes. In India, the market for the sand sector is expanding at a yearly pace of 6–7% [1]. Application-wise, the sand market is segmented into (Fig. 1):

i Industrial
ii Commercial
iii Residential
iv Infrastructure.

In the India sand market, more than 50% of the sand is used up for industrial purposes which marks the highest, commercial purpose marks the second, residential buildings marks the third and infrastructure with the least used percentage. Whereas in the context of Manipur residential marks are the highest. The price range of the sample collected are shown in Table 1.

Table 1 Price range of the local sand which we used in this project

S. no	Name of sand	Price range/truck load
1	Dimapur	₹ 45,000
2	Hundung	₹ 17,000
3	Yairipok	₹ 13,000
4	Dolaithabi	₹ 15,000

2 Materials and Experimental Procedure

For this experimental work, Pozzolana Cement (PPC) is used in this study according to IS 1489 [2], coarse aggregate and sand. Sand testing is the process used to determine if the sand has the desired properties that favour the ideal characteristics required for good construction sand. This project is mainly about testing Dimapur sand and some local sand found in Manipur. For the project, we select three commonly used sand in the State: Dolaithabi sand, Yairipok sand, and Hundung sand. The sands collected from different places are tested, studied, and compared with Dimapur sand which is considered as the higher quality sand in the context of Manipur.

Based on the classification of the source of sand in the above classification, the selected sand falls under.

i Dimapur Sand—River Sand
ii Dolaithabi Sand—River Sand
iii Yairipok Sand—River Sand
iv Hundung Sand—Manufactured Sand

The sand sample collected (Fig. 2) were tested for the properties such as pH test, sand equivalent test, moisture content, specific gravity, and sieve analysis. The mechanical properties of concrete casted with four different soil samples collected were tested.

3 Results and Discussion

3.1 PH Test

pH test is to measure acidic/basic in the sand. The pH scale ranges from 0 to 14 and 7 is considered neutral. pH scales less than 7 indicate acidic and greater than indicate basic. Testing of pH in the sand is to examine sand quality related to the impurities found in the sand. Silica sand is pure sand, free from impurities that are close to neutral pH of 7. The experimental data proves that the Dimapur sand and Yairipok sand are almost similar and neutral in nature (Table 2).

3.2 Sand Equivalent Test

The idea that most soils, gravel bases, etc., are combinations of preferable coarse particles, sand, and typically unfavourable fine particles, or clay, is expressed by the term *sand equivalent*. The sand equivalent test offers a quick way to distinguish between the smaller clay-like particles and the larger grains or sand sizes, and the relative proportions are compared on an arbitrary volume basis by a straightforward

Fig. 2 Sand

Table 2 PH test of sand

Sample	pH value
Hundung	5.5 (slightly acidic)
Dimapur	7 (neutral)
Yairipok	7.5 (slightly alkaline)
Dolaithabi	9 (highly alkaline)

process that tends to magnify or expand the volume of clay somewhat in proportion to its unfavourable or objectionable effects [3]. The sand equivalent test assesses the relative content of sand against clay in the soil as per IS: 2720 [4]. Table 3 gives the experimental data of sand equivalent.

Table 3 Sand equivalent

S. no	Types of sand	Sand equivalent in %
1	Dimapur	98
2	Hundung	96
3	Yairipok	90
4	Dolaithabi	85

Table 4 Result of moisture content

Sand type	W_1 (g)	W_2 (g)	W_3 (g)	Moisture content %
Hungdung	41	200	99	1.73
Dimapur	41	200	99	1.73
Yairipok	41	200	97.35	1.81
Dolaithabi	41	200	97.27	1.82

Table 5 Specific gravity of the sand samples

Sand sample	Specific gravity
Hundung	3.3
Dimapur	2.2
Yairipok	2.86
Dolaithabi	2.5

3.3 Moisture Content

Moisture content percentage is the ratio of the weight of wet sand to the weight of dry sand as per IS 2720 (Part II)—1973 [5]. The table 4 provides the experimental date of moisture content. The four different sand has almost similar of moisture content.

3.4 Specific Gravity

The specific gravity of solid particles (G) is defined as the ratio of the mass of a given volume of solids to the mass of an equal volume of water at 4 °C (IS 2386 (Part III)—1963) [6]. The specific gravity results are tabulated in Table 5.

3.5 Sieve Analysis

This technique describes how to use sieving or screening to determine the distribution of particle sizes for fine, coarse, and all-in-aggregates (IS 2386 (Part I)—1963) [7]. Sieve analysis has been done for the four different type of sand and has been plotted in Fig. 3.

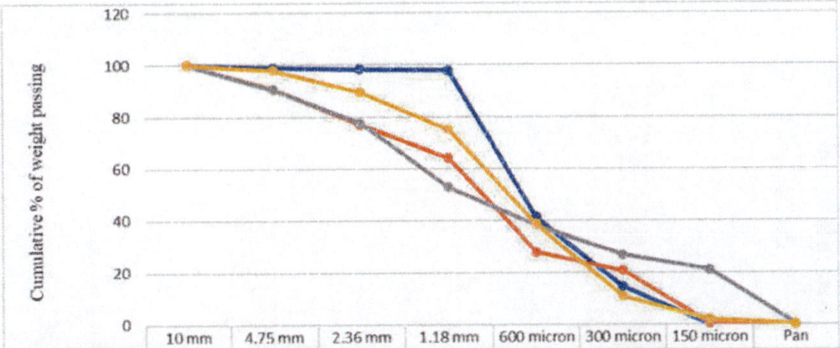

	10 mm	4.75 mm	2.36 mm	1.18 mm	600 micron	300 micron	150 micron	Pan
Dimapur	100	98.95	98.45	97.9	41.2	14.4	0.2	0
Yairipok	100	90.8	76.8	63.8	27.7	20.4	0.35	0
Dolaithabi	100	90.6	77.6	52.6	38.1	26.8	21	0
Hundung	100	98	89.65	74.95	38.35	10.7	1.85	0

Fig. 3 Sieve analysis of sand

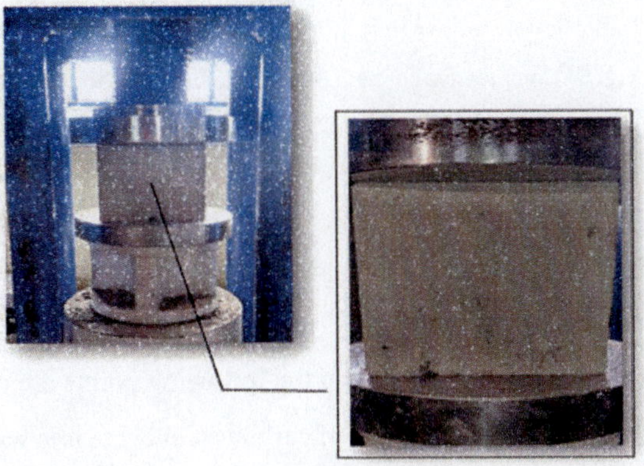

Fig. 4 Test specimen sample of compressive strength

3.6 Compressive Strength

Compressive strength can be defined as the capability of concrete to resist loads before it is compromised (IS: 516) [8]. Three samples were casted for each mix design. The test specimen of compressive strength is shown in Fig. 4. The Dimapur sand and Dolaithabi have a similar compressive strength of 21.33kN (Fig. 5).

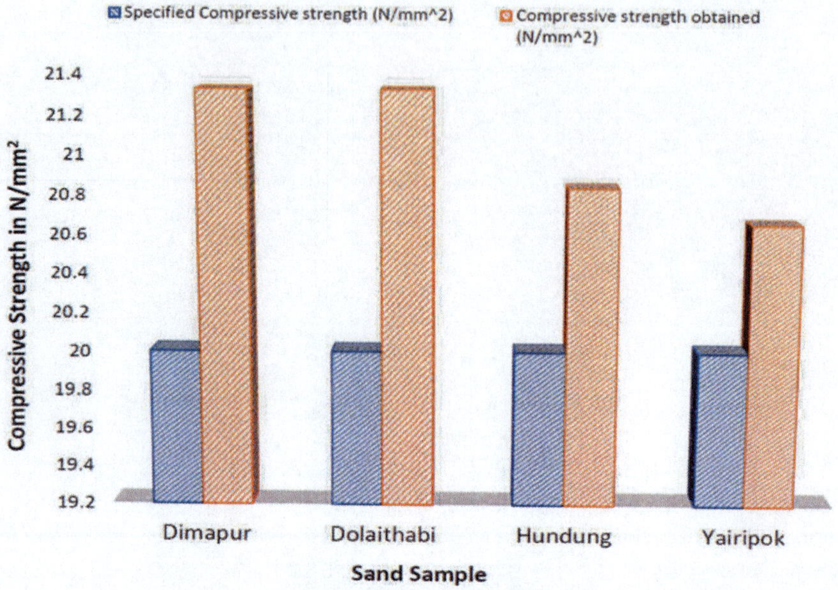

Fig. 5 Compressive strength of concrete cast with four different sand

4 Conclusion

The main aim of this project has been achieved. The study of this project provided knowledge about the sand properties of Manipur along with the help of Dimapur sand. The study has been done on the properties of the four selected sand by conducting different types of tests and the following points are observed from the study.

- The rate of corrosion of carbon steel decreases with increasing pH value towards the alkaline. Among the tested sample, Hundung sand is found to have the least pH value of 5.5 and is slightly acidic. Dolaithabi sand has the highest pH of value 9 and is basic. Dimapur sand is neither acidic nor basic, it is found to be neutral. Sand with high clay content affects the formation of proper bonds between cement and sand. A high clay content percentage in the sand decreases the compressive strength of the concrete.
- Excessive moisture content in sand affects the strength and durability of the concrete.
- Based on the result of sieve analysis, Yairipok sand and Dolaithabi sand is found to be Zone I sand, i.e. coarse sand. Hundung sand is Zone III, medium sand, and Dimapur sand of Zone IV (fine sand).
- Zone I–II sand is highly recommended for Reinforced concrete work so we can conclude that Yairipok, Dolaithabi, and Hundung sand are good RCC sand.
- The specific gravity of sand is considered a measurement of strength.

- Sand with high clay content affects the formation of proper bonds between cement and sand.
- A high clay content percentage in the sand decreases the compressive strength of the concrete.
- Among the tested sample Dimapur sand has the least clay content and Dolaithabi sand have the highest.
- The compressive strength of unwashed Dimapur sand and properly washed (impurities like clay content removed) Dolaithabi sand is same. This shows that the quality of local sand in terms of strength is the same provided the sand is properly washed.

References

1. Expert market Research report, India, Sand Market Outlook (2022) https://www.expertmarketresearch.com/reports/india-sand-market, Retrieved January 28, 2023
2. IS 1489-1 (1991) Specification for Portland pozzolana cement, Part 1: flyash based. Bureau of Indian Standards, New Delhi, India
3. IS 2720-37 (1976) Methods of test for soils, Part 37: determination of sand equivalent values of soils and fine aggregates. Bureau of Indian Standards, New Delhi, India
4. IS 2720-2 (1973) Methods of test for soils, Part 2: determination of water content. Bureau of Indian Standards, New Delhi, India
5. IS 2386-3 (1963) Methods of test for aggregates for concrete, Part 3: specific gravity, density, voids, absorption and bulking. Bureau of Indian Standards, New Delhi, India
6. IS 2386-1 (1963) Methods of test for aggregates for concrete, Part I: particle size and shape. Bureau of Indian Standards, New Delhi, India
7. IS 516 (1959) Method of tests for strength of concrete. Bureau of Indian Standards, New Delhi, India
8. Hveem FN (1953) Sand-equivalent test for control of materials during construction. National Academy of Sciences-National Research Council, 238–250. https://onlinepubs.trb.org/Onlinepubs/hrbproceedings/32/32-015.pdf

Effect on Structural and Optical Properties of NiO Thin Film on Ag Incorporation

Laishram Thoibileima Chanu, Shagolsem Romeo Meitei, and Naorem Khelchand Singh

1 Introduction

In the past decades, nickel oxide (NiO)-based thin film (TF) has been employed due to the efficiency of the materials utilization and extensive improvement in suitability for a wide range of applications. NiO (p-type) is a transition metal oxide with a wide bandgap of 3.2–3.8 eV [1] and has promising properties like highly stable chemically and thermally [2, 3], and tunable optical properties [4]. NiO has been extensively studied for various applications such as photonic [5], photocatalytic [1], battery [6], sensor [7], etc. To extend the range of applicability of NiO over wide fields, tuning the properties of NiO by incorporating it with other metal oxide or metal may be effective [7–10]. Moreover, there have been reports investigating the enhanced properties by doping or incorporating with metals like lanthanum [7], lithium [6], gold [11], platinum [12], etc. Among them, silver (Ag) is the cheapest with excellent properties like good thermal and electrical conductivity, highly reflective, etc. Moreover, it is a non-reactive material, which makes it favorable to incorporate [13, 14].

In this work, we fabricate bare NiO TF and Ag incorporation NiO (Ag-NiO) TF using electron-beam (e-beam) evaporation. Comparing the deposited films, an in-depth analysis of the enhancement effect of Ag incorporation is studied with different characterizations. For structural and morphology, X-ray diffraction (XRD), as well as field-emission scanning electron microscope (FESEM) analysis, to investigate the elemental composition, energy-dispersive spectroscopy (EDS) measurement, for

L. T. Chanu (✉) · S. R. Meitei · N. K. Singh
Department of Electronics and Communication Engineering, National Institute of Technology, Chumukedima, Nagaland 797103, India
e-mail: thoibileimalaish@gmail.com

L. T. Chanu
Department of Computer Science, Tetso College Nagaland, Dimapur 797103, India

S. R. Meitei
Department of Physics, IIT (ISM Dhanbad), Dhanbad 826004, India

B. P. Swain (ed.), *Recent Advances in Materials*, Springer Proceedings in Materials 25,
https://doi.org/10.1007/978-981-99-3844-5_8

optical properties photoluminescence (PL) analysis, and absorption analysis was performed. Each characterization study was performed based on as-deposited bare NiO TF and as-deposited Ag-NiO TF.

2 Experimental Details

An e-beam evaporation technique (model—BC 300, HHV India) was employed to deposit NiO TF and Ag-NiO TF onto silicon (Si) substrate (p-type single crystal <100>) each of which cut in a dimension of 1 × 1 cm. Before the deposition, the sequential steps of cleaning the substrate were carried out using a solution of acetone, methanol, and deionized water (purity of 99.99% each) in the ultrasonic cleaner consecutively for 3 min each. After cleaning, the cleaned substrates were loaded inside the vacuum coating chamber using a substrate holder. The target source material (maintained purity 99.99%) for evaporation, NiO granules for NiO TF deposition, and Ag-NiO composite (ratio of $1_{Ag:}$ 4_{NiO}) granules for Ag-NiO TF were utilized. Between the target source and the substrate holder, a distance of 22 cm was maintained perpendicularly. The films (bare NiO TF and Ag-NiO TF) were deposited with a consistent deposition rate of 1 Å/s and a base pressure at ~5 × 10^{-6} mbar. The depositing density of the films was monitored using digitized quartz crystal during the deposition. An in-depth study on the properties of the films has been performed with different characterizations, i.e., Rigaku Ultima IV emitting Cu Kα radiation of monochromatic X-ray (XRD) to examine the crystallinity behavior, FESEM to deduce morphology, EDS to analyze and elemental composition using ZEISS-SIGMA, 5 kV. For the optical properties analysis, PL (Hitachi F-7000 spectrophotometer), and UV-Vis spectrophotometer (Hitachi-UH-4150) were employed. All the characterization studies were performed using two samples, i.e., as-deposited bare NiO TF and as-deposited Ag-NiO TF.

3 Results and Discussion

3.1 Structure and Morphology Analysis

Figure 1 shows the XRD pattern of the bare NiO TF and Ag-NiO TF deposited on Si substrate, obtaining pure crystallographic properties. Three peaks for bare NiO TF orientations at (222), (400), and (440) phases and three peaks at (222), (400), and (440) with two peaks for Ag at (111) and (200) were observed. Both the XRD pattern shows a polycrystalline nature, and with Ag incorporation, the XRD pattern does not find any reaction phases. Moreover, (100) plane observed on both the patterns corresponds to the Si substrate [15]. For each diffraction peaks, the respective FWHM

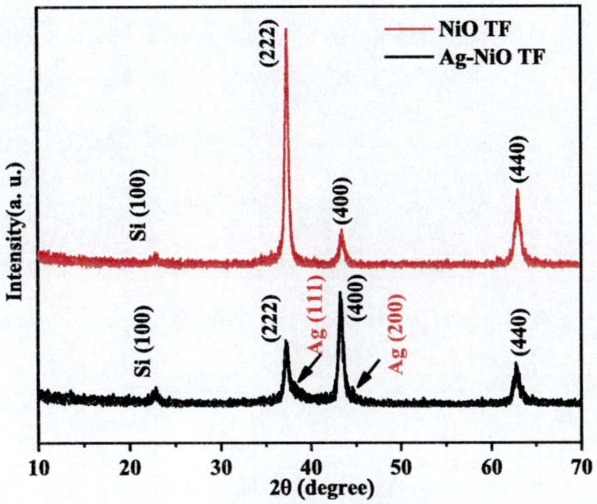

Fig. 1 Bare NiO TF and Ag-NiO TF XRD pattern

values were applied to calculate the average crystalline size 'D' implementing the Debye–Scherrer [16].

$$D = \frac{0.9 \cdot \lambda}{\beta \cdot \mathrm{Cos}\theta}, \tag{1}$$

where the X-ray radiated wavelength (1.54056 Å) denoted by 'λ', corresponding FWHM of the obtained peaks is denoted by 'β,' and Bragg's angle by 'θ.' From the evaluation of Debye–Scherrer formula, the average crystallite size was found to be 14.44 nm for bare NiO TF and 13.44 nm for Ag-NiO TF. Here, crystallite size reduction for Ag-NiO TF is observed, which may be due to integration of Ag atoms in grain boundaries as well as surface of the film. Using the Stock's Wilson relation [17], the lattice strain for both the sample was evaluated. The lattice strain seems increased from 0.00125 for NiO TF to 0.00134 for Ag-NiO TF.

Figure 2 shows the morphology of both the deposited samples using FESEM analysis. Figure 2a, b display the top-view and cross-sectional view for bare NiO TF and Ag-NiO TF, respectively. From the images, we observe the well-growth of both samples with densely packed. The acquired thickness of both samples was found to be ~375 nm. Moreover, with the Ag incorporation, the surfaces increase lattice strain as a rough appearance of the surfaces can also be seen in Fig. 2b. This result also correlates with the reduction in crystallite size in XRD analysis, which in turn arouse a surface tension and increase of stress with incorporation of Ag [18].

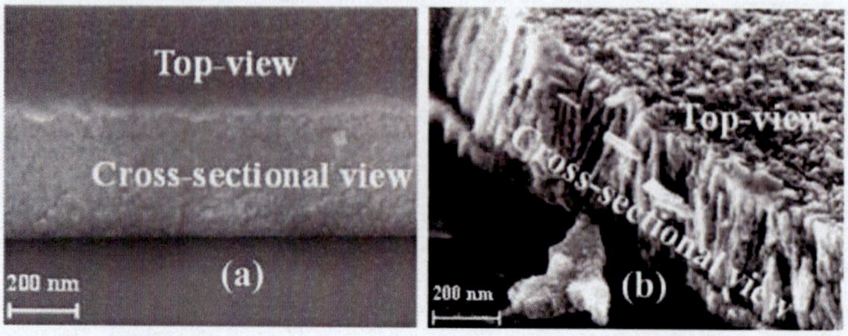

Fig. 2 Top and cross-sectional view FESEM image of **a** bare NiO TF and **b** Ag-NiO TF

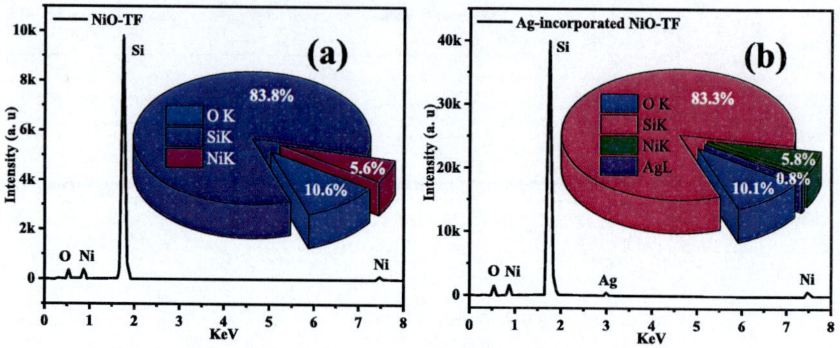

Fig. 3 EDS spectrum and atomic % analysis for **a** bare NiO TF **b** Ag-NiO TF

3.2 EDS Analysis

The elemental composition investigation using EDS analysis is shown in Fig. 3a, b. Figure 3a displays the composition of Si, Ni, and O of bare NiO TF, whereas Fig. 3b indicates the existence of Si, Ni, O, and Ag. Both the deposited films show that there are no free unwanted impurities present in them. Moreover, in Fig. 3b, the appearance of Ag supports the XRD analysis revealing separate peaks for Ag.

3.3 Optical Properties

The PL spectra of bare NiO TF and Ag-NiO TF excited at 250 nm are shown in Fig. 4. Both spectra show a wide range of UV–Vis spectral emission from 320 to 600 nm. Both films show emissions at around ~330 nm (3.75 eV), ~400 nm (3.1 eV), ~436 nm (2.84 eV), and ~467 nm (2.65 eV). The strong UV peaks at 3.75 eV and 3.1 eV can be accredited to the electronic transition of Ni^{2+} ions between the conduction band (CB) and valence band (VB) and 3.1 eV as near band edge emission [19, 20]. Further, blue emissions at 436 nm and 467 nm may be due to the trapped-electron transitions of

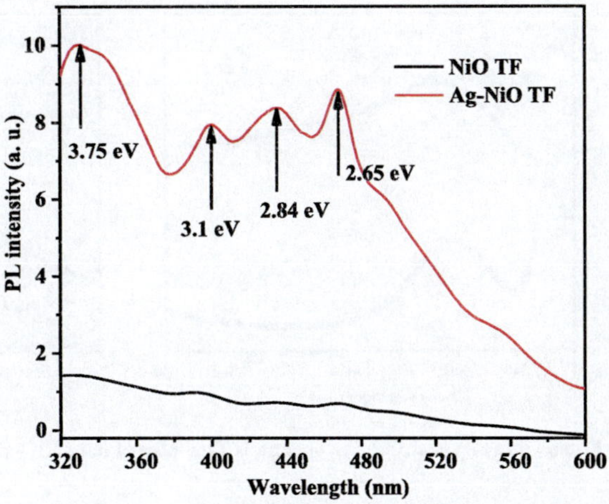

Fig. 4 PL spectra for bare NiO TF and Ag-NiO TF

nickel vacancies [21]. Comparing the emissions intensity of bare NiO and Ag-NiO TF, we can observe an increase in intensity, which may be due to the incorporated Ag behaving as a charge recombination center and the low intensity of the bare NiO reveals small amount of photogenerated carriers [8]. Related research was also reported by Romeo et al. on β-Ga_2O_3 nanowires by integrating Ag, Prasenjit et al. reported the study of Ag incorporation on TiO2 nanowires/reduced graphene oxide, employing e-beam evaporation [18, 22].

UV–Vis absorption spectra for bare NiO and Ag-NiO TF are displayed in Fig. 5a. Both spectra show the absorption edge in the UV-region and visible-region wide absorption. Comparing both spectra, the absorption intensity increased for Ag-NiO film, which indicates the improvement in absorption due to photon absorption by incorporating Ag. This can also be attributed to the numerous scatterings of light inside the nanostructure due to plasmonic effect of Ag incorporation. Figure 5b shows the bandgap calculation by plotting Tauc relation [8]. From the plot, the evaluated bandgap is found to be 3.54 eV for bare NiO TF and 3.12 eV for Ag-NiO TF. The bandgap of the films seems to decrease for Ag-incorporated films, which may be due to the shifting of free electrons between the CB and VB [18]. With the introduction of Ag, more excitation of electron occurs, that ascribe to the surface plasmon resonance of Ag. A similar study was reported by Romeo et al. for β-Ga_2O_3, Henam et al., and Suresh *et* for NiO [8, 18, 23]. However, Khedkar et al. obtained the similar results reporting that the lowering of bandgap is due to electronic interaction of Ag and NiO at the interface.

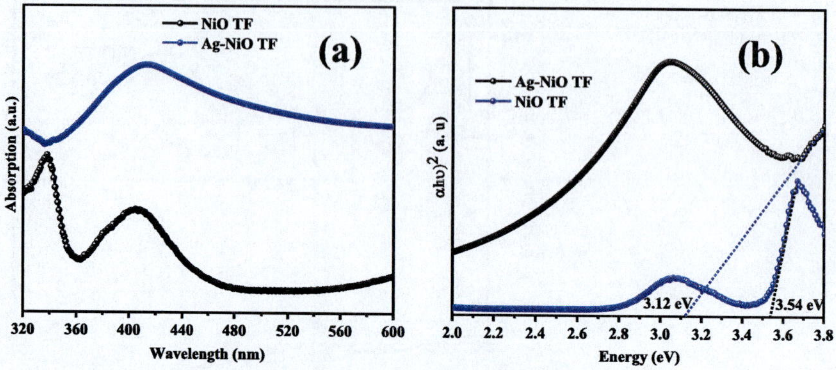

Fig. 5 **a** UV–Vis absorption spectra **b** Tauc plot of bare NiO and Ag-NiO TF

4 Conclusion

NiO TF and Ag-NiO TF have been successfully fabricated employing the e-beam evaporation method. The XRD study promotes the crystallinity of the fabricated films with an evaluated average crystallite size of 14.44 nm for bare NiO TF and 13.44 nm for Ag-NiO TF. The elemental composition for Ag incorporation is confirmed with energy-dispersive X-ray spectroscopy (EDS) study as well as XRD analysis. The benefit of Ag incorporation can be observed by PL and absorption analysis displaying enhanced intensity and improvement in absorption due to the collaborative action of Ag. The obtained results demonstrated the positive impact of Ag incorporation which can be a good candidate for various applications such as a photodetector, gas sensor, photocatalytic.

Acknowledgements We authors would like to show gratitude to the CoE in Advanced Material, and NIT Durgapur for FESEM analysis, NIT Manipur for PL analysis. We also thankful to Science & Humanities department and Electronics and Communication department, NIT Nagaland for providing XRD measurement, and for providing financial support.

References

1. Al-Ghamdi AA, Abdel-wahab MS, Farghali AA, Hasan PMZ (2016) Structural, optical and photo-catalytic activity of nanocrystalline NiO thin films. Mater Res Bull 75:71–77. https://doi.org/10.1016/J.MATERRESBULL.2015.11.027
2. Zhang X, Zhang Y, Zhao B, Lu S, Wang H, Liu J, Yan H (2015) Improvement on optical modulation and stability of the NiO based electrochromic devices by nanocrystalline modified nanocomb hybrid structure. RSC Adv 5:101487–101493. https://doi.org/10.1039/c5ra16876g
3. Palmer DA, Bénézeth P, Xiao C, Wesolowski DJ, Anovitz LM (2011) Solubility measurements of crystalline NiO in aqueous solution as a function of temperature and pH. J Solution Chem 40:680–702. https://doi.org/10.1007/s10953-011-9670-x

4. Al-Kuhaili MF, Ahmad SHA, Durrani SMA, Faiz MM, Ul-Hamid A (2015) Application of nickel oxide thin films in NiO/Ag multilayer energy-efficient coatings. Mater Sci Semicond Process 39:84–89. https://doi.org/10.1016/j.mssp.2015.04.049
5. Saha B, Sarkar K, Bera A, Deb K, Thapa R (2017) Schottky diode behaviour with excellent photoresponse in NiO/FTO heterostructure. Appl Surf Sci 418:328–334. https://doi.org/10.1016/j.apsusc.2017.01.142
6. Liu H, Wang G, Liu J, Qiao S, Ahn H (2011) Highly ordered mesoporous NiO anode material for lithium ion batteries with an excellent electrochemical performance. J Mater Chem 21:3046–3052. https://doi.org/10.1039/c0jm03132a
7. Mrabet C, ben Amor M, Boukhachem A, Amlouk M, Manoubi T (2016) Physical properties of La-doped NiO sprayed thin films for optoelectronic and sensor applications. Ceram Int 42:5963–5978. https://doi.org/10.1016/j.ceramint.2015.12.144
8. Pandey SK, Tripathi MK, Ramanathan V, Mishra PK, Tiwary D (2021) Highly facile Ag/NiO nanocomposite synthesized by sol-gel method for mineralization of rhodamine B. J Phys Chem Solid 159. https://doi.org/10.1016/j.jpcs.2021.110287
9. Na D-M, Satyanarayana L, Choi G-P, Shin Y-J, Park JS (2005) Surface morphology and sensing property of NiO-WO 3 thin films prepared by thermal evaporation. Sensors 5:519–528. http://www.mdpi.org/sensors
10. Karimi- M, Behpour M, Babaheidari AK, Saberi Z (2017) Efficiently enhancing photocatalytic activity of NiO-ZnO doped onto nanozeoliteX by synergistic effects of p-n heterojunction, supporting and zeolite nanoparticles in photo-degradation of Eriochrome Black T and Methyl Orange. J Photochem Photobiol A Chem 346:133–143. https://doi.org/10.1016/j.jphotochem.2017.05.038
11. Mattei G, Mazzoldi P, Post ML, Buso D, Guglielmi M, Martucci A (2007) Cookie-like Au/NiO nanoparticles with optical gas-sensing properties. Adv Mater 19:561–564. https://doi.org/10.1002/adma.200600930
12. Kim SS, Park KW, Yum JH, Sung YE (2006) Pt-NiO nanophase electrodes for dye-sensitized solar cells. Sol Energy Mater Sol Cells 90:283–290. https://doi.org/10.1016/j.solmat.2005.03.015
13. Hameed MA, Ali OA, Al-Awadi SSM (2020) Optical properties of Ag-doped nickel oxide thin films prepared by pulsed-laser deposition technique. Optik (Stuttg), 206. https://doi.org/10.1016/j.ijleo.2020.164352
14. Reddy YAK, Reddy AS, Reddy PS (2014) Effect of oxygen partial pressure on the properties of NiO-Ag composite films grown by DC reactive magnetron sputtering. J Alloys Compd 583:396–403. https://doi.org/10.1016/j.jallcom.2013.08.180
15. Miyata H, Kuroda K (1999) Preferred alignment of mesochannels in a mesoporous silica film grown on a silicon (110) surface. J Am Chem Soc 121:7618–7624. https://doi.org/10.1021/ja990758m
16. El-Nahass MM, Emam-Ismail M, El-Hagary M (2015) Structural, optical and dispersion energy parameters of nickel oxide nanocrystalline thin films prepared by electron beam deposition technique. J Alloys Compd 646:937–945. https://doi.org/10.1016/j.jallcom.2015.05.217
17. Stokes AR, Wilson AJC (1944) The diffraction of X-rays by distorted crystal aggregates-I. In: Proceedings of Physical Society, 56, 174. https://doi.org/10.1088/0959-5309/56/3/303/meta
18. Meitei SR, Ngangbam C, Singh NK (2021) Microstructural and optical properties of Ag assisted β-Ga_2O_3 nanowires on silicon substrate. Opt Mater (Amst) 117. https://doi.org/10.1016/j.optmat.2021.111190
19. Chakrabarty S, Chatterjee K (2011) Synthesis and optical manifestation of NiO-silica nanocomposite. ISRN Nanotech 2011:1–6. https://doi.org/10.5402/2011/719027
20. Karthikeyan B, Pandiyarajan T, Hariharan S, Ollakkan MS (2016) Wet chemical synthesis of diameter tuned NiO microrods: microstructural, optical and optical power limiting applications. Cryst Eng Comm 18:601–607. https://doi.org/10.1039/c5ce02232k
21. Jayalakshmi G, Saravanan K, Navas J, Arun T, Panigrahi BK (2019) Fabrication of p-NiO nanoflakes/n-Si(100) heterojunction architecture for high sensitive photodetectors. J Mater Sci: Mater Electron 30:6811–6819. https://doi.org/10.1007/s10854-019-00993-y

22. Deb P, Dhar JC (2020) Boosted photoresponsivity using silver nanoparticle decorated TiO2 nanowire/reduced graphene oxide thin-film heterostructure. Nanotechnology 31. https://doi.org/10.1088/1361-6528/ab8084
23. Devi HS, Singh TD, Singh NR (2017) Green synthesis and catalytic activity of composite NiO-Ag nanoparticles for photocatalytic degradation of dyes

Chevronic TiO_2 Thin Film Fabrication Using E-Beam Evaporation for UV Photodetection Applications

Pinky Khundrakpam, Biraj Shougaijam, and Ashish Ranjan

1 Introduction

Photodetector devices with boosted sensitivity and speed have been promising area of investigation in the recent past. Ultraviolet (UV) photodetectors, with its varied spectrum of potential applications in the areas of purification of water, optical communication, chemical analysis, remote control, diagnosis of medical conditions, and detection of missile plume have been a potential research interest [1–3]. Unlike transition metal oxide semiconductors, conventional photodetectors fabricated from Ge, Si, and GaAs exhibit bandgaps which are narrow and thereby absorbing a wide wavelength range. Titanium dioxide (TiO_2) being a semiconductor with wide bandgap of 3.2 eV is suited for UV detection [4–6], besides its other applications such as biosensors [7], optoelectronics [8], cancer therapy [9], photocatalysis [10], owing to its outstanding electrical, optical, and catalytic properties [11, 12]. A varied of techniques, including sol–gel [13], hydrothermal [14], glancing angle deposition [15], and chemical vapour deposition [16] have been employed by many researchers to grow TiO_2 nanostructures. Nevertheless, these above-mentioned methodologies necessitate either a complicated fabrication process or pose difficulty in controlling the nanostructure's growth direction. These problems may be overcome through the introduction of a simple oblique angle deposition methodology (OAD) to grow chevronic TiO_2 thin films (TFs) possessing high surface area on a silicon substrate. TiO_2 TFs grown on silicon substrate at slanted angles with OAD technique has been theoretically and experimentally characterized in [17]. Another paper reported the

P. Khundrakpam (✉) · A. Ranjan
Department of Electronics and Communication Engineering, National Institute of Technology Manipur, Imphal, Manipur, India
e-mail: pinks.kh@gmail.com

P. Khundrakpam · B. Shougaijam
Department of Electronics and Communication Engineering, Manipur Technical University, Imphal, Manipur, India

B. P. Swain (ed.), *Recent Advances in Materials*, Springer Proceedings in Materials 25,
https://doi.org/10.1007/978-981-99-3844-5_9

analysis of fabricated TiO_2 TFs on silicon substrate using OAD method [18]. The OAD of TiO_2 on glass [19] and silicon [20] substrates forming zig-zag TiO_2 TFs has also been reported. Nevertheless, these literatures deliberate on the growth mechanisms and morphology of the cylindrical TFs but exclude the structural analysis of the TiO_2 TFs grown through the OAD.

OAD is a prominent physical vapour deposition methodology for creating remarkably efficient TFs with configurable cylindrical morphologies such as S-shapes, slanted columns, chevron/zig-zag structures, C-shapes and helices, through variation of the two fundamental axes of substrate rotation [21, 22]. These cylindrical microstructure morphologies result from the blocking of the incoming deposition flux by the shadowing effect of the tall surface facets thereby initiating a combative growth and hence formation of slanted structures towards the approaching flux [23–25]. High nanoporosity and anisotropic growth dependent on the material, angle of deposition, and deposition conditions, due to the limited diffusion of adatom and self-shadowing effect result to lower refractive index in TFs grown by OAD methodology as compared to other growth techniques. Moreover, these TFs microstructures formed by the OAD technique cause optical anisotropy [26] that reveals birefringence property [27]. These structures also exhibit greater surface area then conventional thin films. Out of all the potential quantities that represent a cylindrical film microstructure, the relationship of the vapour flux incident angle (α) with the column tilt angle (β), for substrates that do not rotate, is of immense pertinence and has been a topic of deliberation amongst the researchers. The famous tangent rule proposed by Nieuwenhuizen and Haanstra in [28] is represented by

$$\tan\alpha = 2\tan\beta. \tag{1}$$

In another extension of OAD technique, known as glancing angle deposition (GLAD) technique, the manipulation of the substrate incline is done during the deposition of the film and thereby additionally requires active manipulation of both the angle of rotation and substrate incline angle while fabrication of columnar structure growth.

In this work, an attempt has been made to fabricate chevronic TiO_2 thin film-based UV photodetectors having chevronic/zig-zag nanostructures and analyse their optical, morphological, structural, and photodetection properties using the OAD technique in comparison the conventional TiO_2 thin film-based UV photodetector.

2 Experimental Details

Chevronic/zig-zag TiO_2 thin films were deposited on silicon substrates (1 cm by 1 cm) within the electron beam evaporation system (HHV India's Smart Coat 3.0,) with the OAD methodology. 99.999% pure TiO_2 (acquired from Tecnisco Advanced Materials Private Limited, Singapore) was utilized as the source material to fabricate the chevronic TiO_2 TFs. The integrated OAD unit in the electron beam chamber ensues the change of the desirable angle with the help of the shaft of the OAD unit. The schematic of simple OAD system is shown in Fig. 1. The e-beam chamber is set up for deposition by cleaning at the outset with acetone preceding to vacuuming the chamber. The pressure of the e-beam chamber falls to ~2×10^{-5} mbar from the originally maintained ~6×10^{-6} mbar in the course of the deposition which may owe to the evaporation of the source material releasing oxygen gas. A quartz crystal was utilized to track the rate of deposition which was sustained at 1 Å s^{-1}. A 20 cm separation was maintained between the source material to be evaporated and the substrate holder. Initially, a thin film of TiO_2 of 50 nm was grown on the Si substrate prior to placing the substrate holder at an 85° inclination with the normal axis between the substrate holder and the source material to be evaporated. The fabrication of the 450 nm-thick chevronic TiO_2 TF was achieved by deposition of 150 nm of thick film of the TiO_2 interchangeably thrice by rotating the substrate 180° azimuthally twice. The schematic depiction of the conventional TiO_2 TF and the chevronic TiO_2 TF can be observed from Fig. 2. A normal TiO_2 TF of thickness 200 nm was also fabricated separately by keeping the substrate holder at normal with evaporation material for comparative analysis with the chevronic TiO_2 TF.

The analysis of the chevronic TiO_2 TF morphology was achieved with the FEG-SEM (field emission gun-scanning electron microscopes) (JEOL, JSM-7600F). Additionally, the X-ray diffraction (XRD) characterization was also accomplished with the help of the X-Pert Pro Pan analytical having Cu K-alpha radiation (k = 1.54060 Å) for both the conventional and chevronic TFs. The UV–vis spectrophotometer (Shimadzu's UV-1800) and the UV–Vis spectrophotometer (AN-UV-6500N ANTech) were utilized to characterize the absorption and photoluminescence spectra with the use of a 370 nm filter under a wavelength of excitation of 340 nm for both the conventional and chevronic TFs.

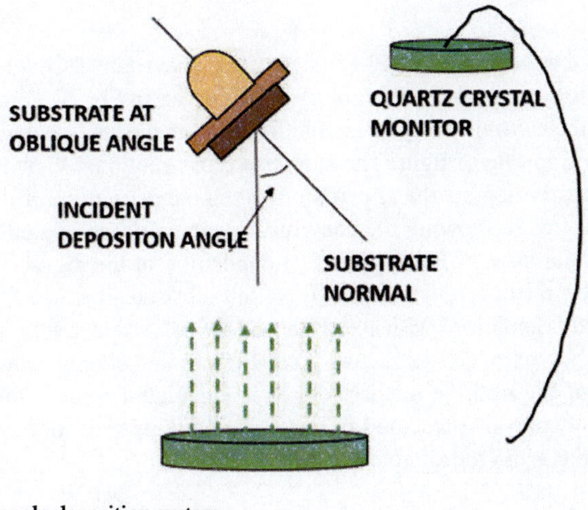

Fig. 1 Oblique angle deposition system

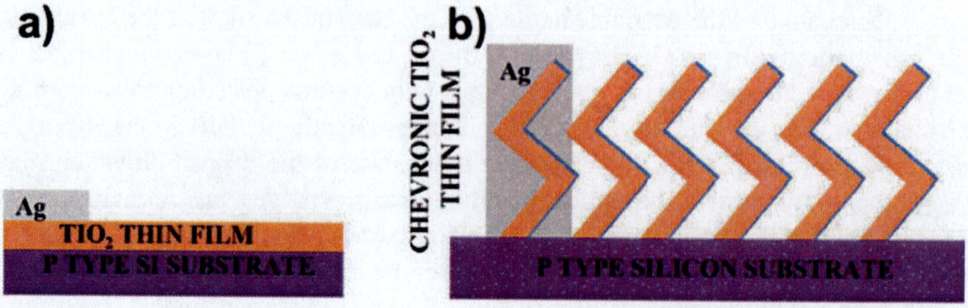

Fig. 2 Schematic representations of UV detectors based on **a** TiO_2 thin film and **b** chevronic TiO_2 thin film

Two types of photodetector devices were fabricated. Thermal evaporation was utilized to deposit metal (Ag) electrode contacts of 200 nm thickness on the surfaces of Si/TiO_2-TF and Si/Chevronic-TiO_2-TF with the help of a shadow mask having holes of circular shape with area ~7.8×10^{-3} cm^2. The I–V measurement of Si/TiO_2-TF and Si/Chevronic-TiO_2-TF-based UV photodetectors were measured with the help of semiconductor parameter analyser (Keithley's 4200-SCS) beneath low-intensity white light illumination and dark conditions. The analysis of response time of the photodetectors was conducted with ON/OFF switching of illumination of UV light.

3 Results and Discussion

3.1 Morphological Analysis

The sectional view of the FEG-SEM image of the chevronic TiO_2 TF fabricated by OAD methodology at an inclination of 85° is observed in Fig. 3. The porous nature of the fabricated chevronic TiO_2 TF is established from the sectional view which will facilitate better trapping of light. The sample's cross-sectional view is presented in Fig. 4 which clearly depicts the chevronic/zig-zag nanostructure of the TiO_2 TF. A more magnified image showing the chevron/zig-zag nature indicated by red arrows is illustrated in the inset of Fig. 4. The first deposition of the zig-zag nanostructure was deposited for the orientation of $\phi = 0°$ of the substrate position, followed by the second and third orientations with $\phi = 180°$ and $\phi = 0°$, respectively. The calculated height of the chevronic TiO_2 structure from the cross-sectional view is ~190 nm, and the angle of the oblique nanostructures is calculated to be ~46°. The energy dispersive X-ray analysis presented in Fig. 5 is depicting the existence of titanium (Ti) and oxygen (O_2) in the sample.

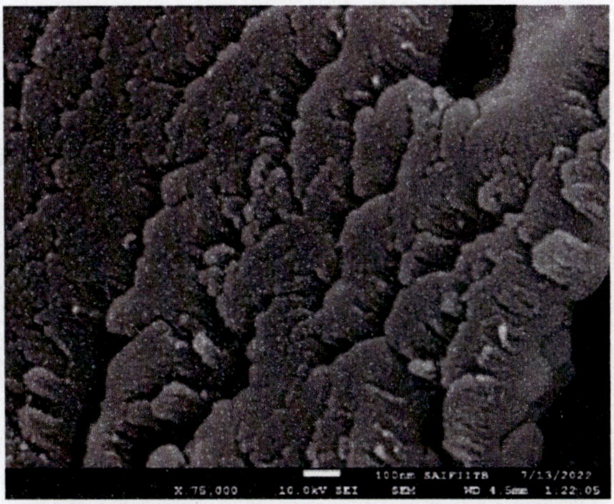

Fig. 3 Sectional view of chevronic TiO_2 thin film

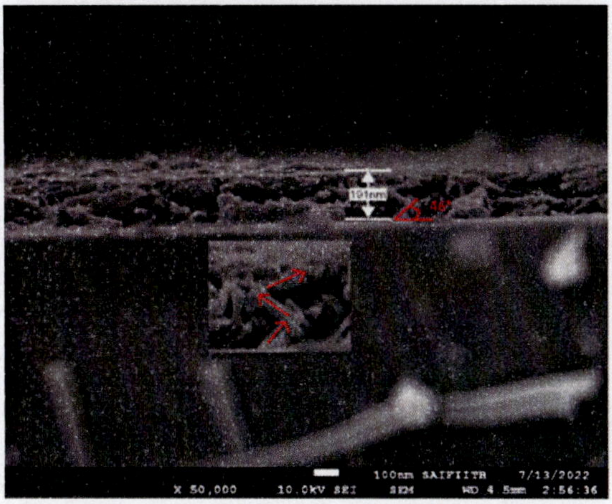

Fig. 4 Cross-sectional view of chevronic TiO_2 thin film

3.2 Structural Analysis

The analysis of the structural properties of the samples was achieved through X-ray diffraction (XRD) characterization. The XRD results of the conventional TiO_2 TF and chevronic TiO_2 TF grown by the OAD methodology are depicted in Fig. 6 whereby weak peaks at ~25° corresponding to the anatase TiO_2 crystal lattice of (101) (JCPDS No. 84–1286) were seen in both the samples. These confirm the amorphous nature of the samples. TiO_2 TFs grown through e-beam are reported to be usually amorphous in nature [29]. The structural characteristics of the amorphous TiO_2 TF can be improved by annealing the sample between 300 and 600 °C [30].

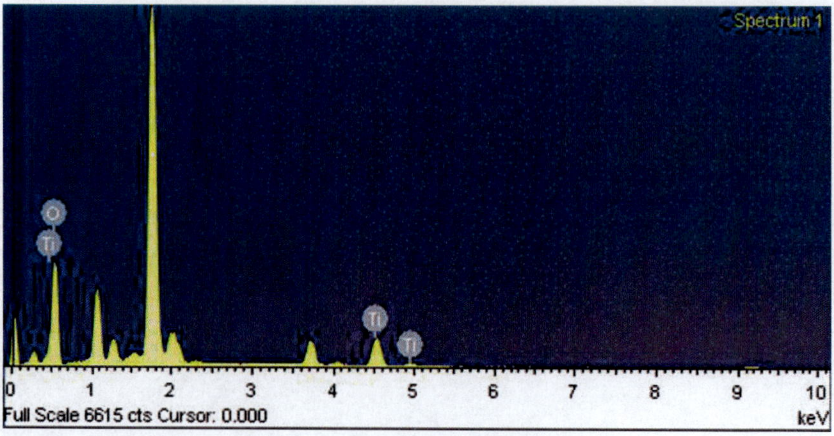

Fig. 5 Energy dispersive X-ray spectrum of chevronic TiO_2 thin film

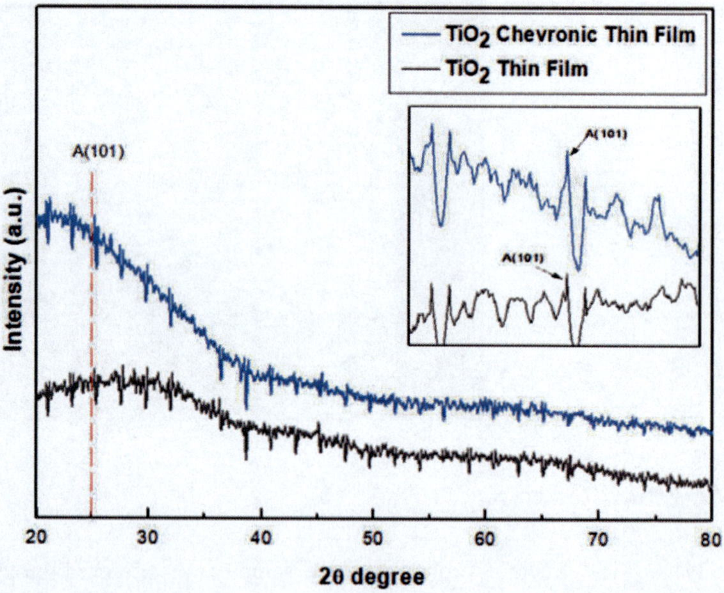

Fig. 6 X-ray diffraction measurement of TiO_2 thin film and chevronic TiO_2 thin film

3.3 *Optical Analysis*

To examine the absorption spectra of the TiO_2 TF and the chevronic TiO_2 TF samples, the room temperature optical absorption spectra were recorded from 310 nm to 800 nm wavelengths which are presented in Fig. 7. Enhanced absorption is shown

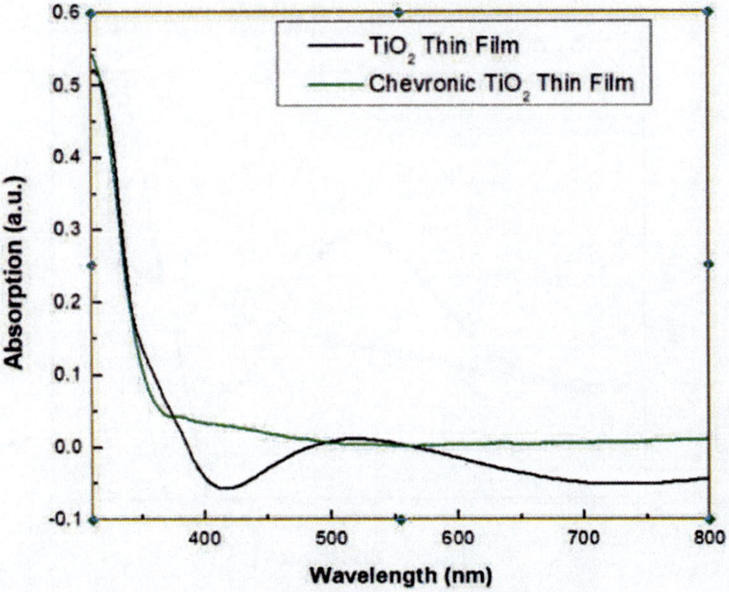

Fig. 7 Absorption spectra of TiO_2 thin film and chevronic TiO_2 thin film

in the ultraviolet (UV) range in comparison to the visible range due to the excited electrons transitioning to conduction band out of the valence band by both the TiO_2 TF and the chevronic TiO_2 TF samples. Further, slightly better absorption is shown by the latter. This may be due to better porosity of the chevronic TiO_2 TF sample as seen from the sectional view in Fig. 3.

The Tauc plots plotting $(\alpha h\nu)^2$ vs $h\nu$, where $h\nu$ is the incident photon energy on the TFs and α is the absorption coefficient, is presented in Fig. 8 for both the conventional TiO_2 TF and chevronic TiO_2 TFs. Calculation of the band gaps of both the samples was achieved through extrapolation of the linear portions of the curve to the energy axis (X axis) and observed to be ~3.23 eV and ~3.3 eV for the conventional TiO_2 TF and chevronic TiO_2 TFs, respectively, as shown in Fig. 8.

The photoluminescence intensity spectra of both the conventional TiO_2 TF and chevronic TiO_2 TFs which were excited with the use of a 370 nm filter at 340 nm wavelength are presented in Fig. 9. The plot shows two peaks 430 nm 413 nm for the chevronicTiO_2 TF and two peaks at 411 nm and 431 nm for the conventional TiO_2 TF. The photoluminescence spectrum of the chevronic TiO_2 TF fabricated through OAD technique is Gaussian fitted, and the fitted plot shows a peak at 423 nm corresponding to a bandgap of 2.93 eV in Fig. 10. The bandgap from the photoluminescence measurement is faintly smaller than the band gap calculated from the Tauc plot (~3.29 eV).

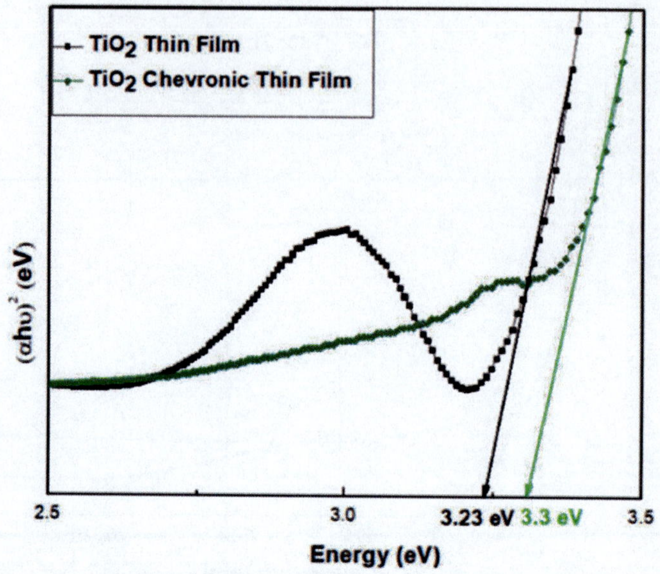

Fig. 8 Tauc plots of TiO_2 thin film and chevronic TiO_2 thin film

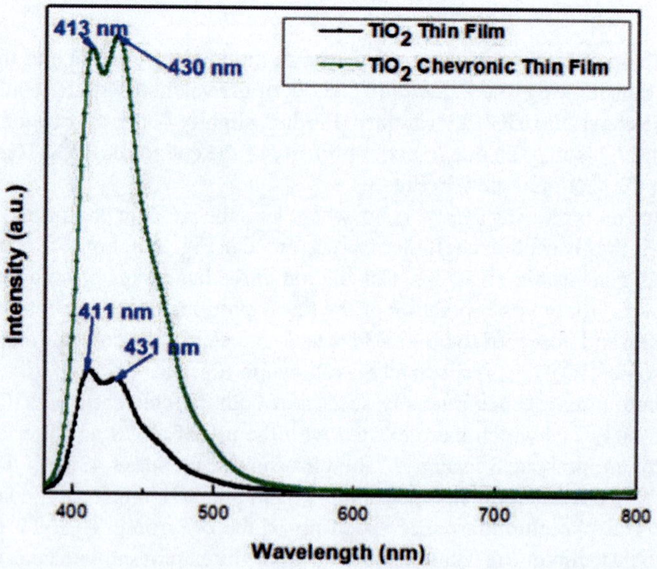

Fig. 9 Photoluminescence spectra of TiO_2 thin film and chevronic TiO_2 thin film

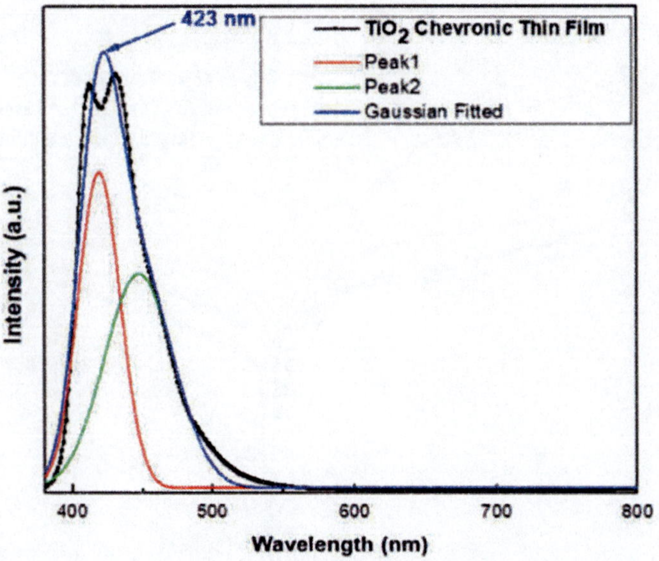

Fig. 10 Gaussian fitted photoluminescence spectra of TiO_2 chevronic thin film

3.4 Analysis of Photodetector Device

The investigation of the J-V characteristics of the Si/TiO_2-TF/Ag and Si/Chevronic-TiO_2-TF/Ag-based UV photodetectors thus fabricated is processed with biasing from −8 to 8 V in both forward and reverse bias in light and dark environments AS shown in Fig. 11. The devices were investigated under UV light illumination (~390 nm wavelength) with the UV LED light source having output optical power of 174 nW/cm^2 measured with optical power metre (OPHIR model no. PD300-1W, USA). The optical source was installed at a separation of 3 cm from the photodetectors during the measurements of UV lights. The rectification behaviour in both the photodetectors owes to development of p–n junction at the boundary between TiO_2 TFs and p-Si substrate. However, the Si/Chevronic-TiO_2-TF/Ag-based device shows stronger rectification in comparison to Si/TiO_2-TF/Ag-based device. This behaviour is due to the zig-zag nanostructures in the former device leading to greater surface area, better trapping of light, and greater porosity.

Moreover, the response time of both fabricated Si/TiO_2-TF/Ag and Si/Chevronic-TiO_2-TF/Ag-based UV photodetectors was also investigated by On and Off UV light switching with ~174 nW/cm^2 optical power at a biasing of −2 V, as presented in Fig. 12. The photosensitivity of photodetector is identified as the proportion of photocurrent to the dark current. The photosensitivity of both fabricated Si/TiO_2-TF/

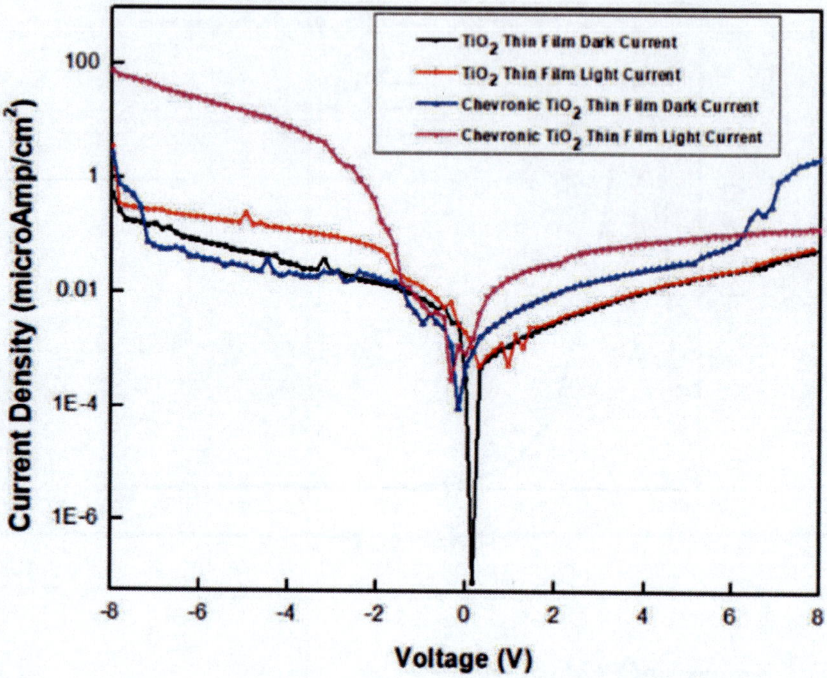

Fig. 11 J-V characteristics of TiO_2 thin film and chevronic TiO_2 thin film-based UV photodetectors

Ag and Si/Chevronic-TiO_2-TF/Ag-based photodetectors was calculated and found to be ~5.3 and ~7.4, respectively, at −2 V under UV light illumination showing better photosensitivity of the Si/Chevronic-TiO_2-TF/Ag-based photodetector owing to its porosity, greater surface area, and enhanced trapping of light. The fabricated Si/Chevronic-TiO_2-TF/Ag-based photodetector shows better photoresponsivity than the TiO_2-based UV photodetectors reported in [20, 31]. The rise time (t_r) and fall time (t_f) of the fabricated Si/Chevronic-TiO_2-TF/Ag-based photodetectors were computed from a switching cycle from Fig. 12 and presented in Fig. 13 revealing a rise time of 120.31 ms and a fall time of 120.31 ms. The fabricated Si/Chevronic-TiO_2-TF/Ag-based photodetector device shows better response time in comparison to the previous works as reported in [20, 31–33] with smaller biasing voltage of −2 V (Table 1).

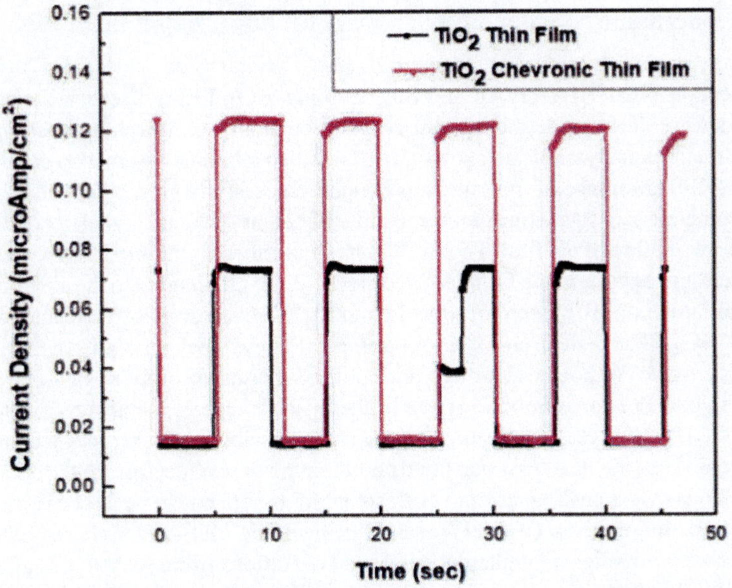

Fig. 12 On–off switching of TiO_2 thin film and chevronic TiO_2 thin film-based UV photodetectors at room temperature

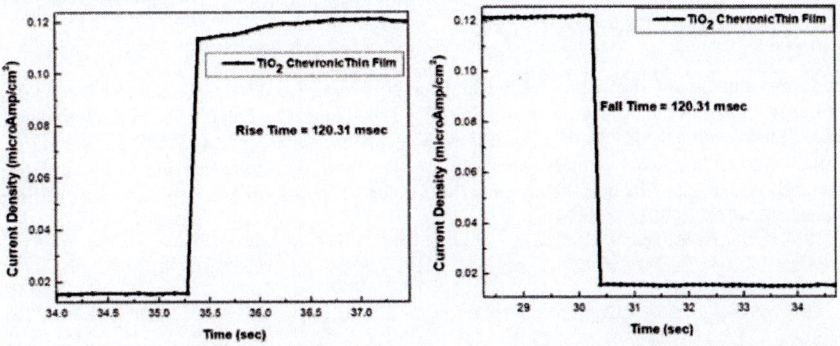

Fig. 13 Rise time and fall time of chevronic TiO_2 thin film UV photodetector

Table 1 Comparative table of rise time and fall time of photodetectors

Device	Rise time (s)	Fall time (s)
Annealed TiO_2/In_2O_3 NW [32]	0.26	0.13
Er doped TiO_2 NW [33]	1.23	0.6
TiO_2 chevronic thin film [present work]	0.1203	0.1203

4 Conclusion

Fabrication of Si/TiO_2-TF/Ag and Si/Chevronic-TiO_2-TF/Ag-based UV photodetectors were done inside the e-beam evaporation chamber. Morphological, optical, and structural analysis of the chevronic TiO_2 TF so fabricated were also conducted. The FEG-SEM image of the sample presented successfully grown TiO_2 thin film of chevronic/zig-zag shape on silicon substrate of height ~190 nm and also confirming the porous nature of the film. The EDX mapping further highlighted the existence of titanium and oxygen in the sample. Weak peaks at ~25° belonging to the anatase TiO_2 crystal lattice of (101) were observed from the XRD analysis of both the conventional TiO_2 TF and chevronic TiO_2 TF samples. These weak amorphous TiO_2 characteristics may be boosted through annealing. Superior absorption was seen in the ultraviolet (UV) range in comparison to the visible range in the absorption spectra from the UV–vis spectroscopy in both the samples with better absorption shown by chevronic TiO_2 TF. The Gaussian fitted photoluminescence spectrum of the chevronic TiO_2 TF shows a peak at 423 nm corresponding to a bandgap of 2.93 eV. The J-V characteristics of the Si/TiO_2-TF/Ag and Si/Chevronic-TiO_2-TF/Ag-based photodetectors were investigated under dark and light conditions from −8 V to 8 V and their ON/OFF switching was studied at −2 V. The Si/Chevronic-TiO_2-TF/Ag-based UV photodetectors showed better photoresponsivity and response time than the Si/TiO_2-TF/Ag-based UV photodetectors and also other reported UV photodetectors due to the porous nature of the chevronic TiO_2 thin film and larger surface area ensuing better harvesting of light.

Acknowledgements The authors extend their thankfulness to SAIF, Indian Institute of Technology Bombay, India, for the measurement of FEG-SEM and National Institute of Technology Manipur, India, for providing facility for the optical absorption, photoluminescence, and XRD measurements. Finally, the authors express their appreciation to the Dept. of Electronics and Communication Engineering, Manipur Technical University (MTU), for facilitating the fabrication funded by SERB under file no. ECR/ 2018/000834.

References

1. Muñoz E, Monroy E, Pau JL, Calle F, Omnès F, Gibart P (2001) III nitrides and UV detection. J Phys Condens Matter 13:7115–7137. https://doi.org/10.1088/0953-8984/13/32/316
2. Ates ES, Kucukyildiz S, Unalan HE (2012) Zinc oxide nanowire photodetectors with single-walled carbon nanotube thin-film electrodes. ACS Appl Mater Interfaces 4:5142–5146. https://doi.org/10.1021/am301402y
3. Zhu P, Nair AS, Shengjie P, Shengyuan Y, Ramakrishna S (2012) Facile fabrication of TiO_2-graphene composite with enhanced photovoltaic and photocatalytic properties by electrospinning. ACS Appl Mater Interfaces 4:581–585. https://doi.org/10.1021/am201448p
4. Zhang H, Ruan S, Li H, Zhang M, Lv K, Feng C, Chen W (2012) Schottky diode ultraviolet detector based on TiO_2 nanowire array. IEEE Electron Device Lett 33:83–85. https://doi.org/10.1109/LED.2011.2173896

5. Chen Z, Zhu Z, Huang L, Cheng C (2021) High sensitivity UV photodetectors based on low-cost TiO_2P25-graphene hybrids. Nanotechnology. 33:08LT01. https://doi.org/10.1088/1361-6528/ac3a37
6. Hoang Tran M, Bae J-S, Hur J (2022) Self-powered, transparent, flexible, and solar-blind deep-UV detector based on surface-modified TiO_2 nanoparticles. Appl Surf Sci 604:154528. https://doi.org/10.1016/j.apsusc.2022.154528
7. Ge C, Lu M, Zhang W, Cunningham BT (2010) Distributed feedback laser biosensor incorporating a titanium dioxide nanorod surface. Appl Phys Lett 96:163702. https://doi.org/10.1063/1.3394259
8. Cho K-S, Lee EK, Joo W-J, Jang E, Kim T-H, Lee SJ, Kwon S-J, Han JY, Kim B-K, Choi BL, Kim JM (2009) High-performance crosslinked colloidal quantum-dot light-emitting diodes. Nat Photonics 3:341–345. https://doi.org/10.1038/nphoton.2009.92
9. Kawamura G, Okuno T, Muto H, Matsuda A (2014) Visible-light-induced photocatalysis of 2D-hexagonal mesoporous SiO_2–TiO_2 deposited with Au nanoparticles. J Nanosci Nanotechnol 14:2225–2230. https://doi.org/10.1166/jnn.2014.8476
10. Nalwaed HS, Encyclopedia of nanoscience and nanotechnology. American Scientific Publishers, Los Angeles, CA
11. Deshmukh HP, Shinde PS, Patil PS (2006) Structural, optical and electrical characterization of spray-deposited TiO_2 thin films. Mater Sci Eng B 130:220–227. https://doi.org/10.1016/j.mseb.2006.03.016
12. Linsebigler AL, Lu G, Yates JT (1995) Photocatalysis on TiO_2 surfaces: principles, mechanisms, and selected results. Chem Rev 95:735–758. https://doi.org/10.1021/cr00035a013
13. Liang Y, Sun S, Deng T, Ding H, Chen W, Chen Y (2018) The preparation of TiO_2 film by the sol-gel method and evaluation of its self-cleaning property. Materials (Basel) 11:450. https://doi.org/10.3390/ma11030450
14. Keerthana GT, Solaiyammal T, Muniyappan S, Murugakoothan P (2018) Hydrothermal synthesis and characterization of TiO_2 nanostructures prepared using different solvents. Mater Lett 220:20–23. https://doi.org/10.1016/j.matlet.2018.02.119
15. Pandey AK, Tiwari AK, Paliwal HK (2022) Enhancement of mechanical, thermal and optical properties of TiO_2 thin films using glancing angle deposition technique. Opt Mater (Amst) 134:113054. https://doi.org/10.1016/j.optmat.2022.113054
16. Du J, Gu X, Guo H, Liu J, Wu Q, Zou J (2015) Self-induced preparation of TiO_2 nanowires by chemical vapor deposition. J Cryst Growth 427:54–59. https://doi.org/10.1016/j.jcrysgro.2015.07.004
17. Álvarez R, González-García L, Romero-Gómez P, Rico V, Cotrino J, González- AR, Palmero A (2011) Theoretical and experimental characterization of TiO_2 thin films deposited at oblique angles. J Phys D Appl Phys 44:385302. https://doi.org/10.1088/0022-3727/44/38/385302
18. Alvarez R, Lopez-Santos C, Parra-Barranco J, Rico V, Barranco A, Cotrino J, Gonzalez-Elipe AR, Palmero A (2014) Nanocolumnar growth of thin films deposited at oblique angles: beyond the tangent rule. J Vac Sci Technol B, Nanotechnol Microe-lectron Mater Process Meas Phenom 32:041802. https://doi.org/10.1116/1.4882877
19. Hwangbo CK, Park YJ, Sobahan K, Ahn J (2010) Optical properties of TiO_2 zigzag films prepared by oblique angle deposition. J Korean Phys Soc 56:1378–1381. https://doi.org/10.3938/jkps.56.1378
20. Tiwari AK, Mondal A, Mahajan BK, Choudhuri B, Goswami T, Sarkar MB, Chakrabartty S, Ngangbam C, Saha S (2015) Improved photo-detection using zigzag TiO_2 nanostructures as an active medium. J Nanosci Nanotechnol 15:5099–5104. https://doi.org/10.1166/jnn.2015.9821
21. Pursel S, Horn MW, Demirel MC, Lakhtakia A (2005) Growth of sculptured polymer submicron wire assemblies by vapor deposition. Polymer (Guildf). 46:9544–9548. https://doi.org/10.1016/j.polymer.2005.07.092
22. Robbie K, Brett MJ, Lakhtakia A (1995) First thin film realization of a helicoidal bianisotropic medium. J Vac Sci Technol A Vacuum, Surf, Film 13:2991–2993. https://doi.org/10.1116/1.579626

23. Meakin P (1998) Fractals, scaling, and growth far from equilibrium. Cambridge University Press
24. Yang H-Y, Lee M-F, Huang C-H, Lo Y-S, Chen Y-J, Wong M-S (2009) Glancing angle deposited titania films for dye-sensitized solar cells. Thin Solid Films 518:1590–1594. https://doi.org/10.1016/j.tsf.2009.09.026
25. Besnard A, Martin N, Carpentier L, Gallas B (2011) A theoretical model for the electrical properties of chromium thin films sputter deposited at oblique incidence. J Phys D Appl Phys 44:215301. https://doi.org/10.1088/0022-3727/44/21/215301
26. Woo S-H, and Hwangbo CK (2006) Optical anisotropy of TiO_2 and MgF_2 thin films prepared by glancing angle deposition. J Korean Phys Soc 49:2136
27. Sobahan KMA, Park YJ, Kim JJ, Shin YS, Kim JB, Hwangbo CK (2010) Nanostructured optical thin films fabricated by oblique angle deposition. Adv Nat Sci Nanosci Nanotechnol 1:045005. https://doi.org/10.1088/2043-6262/1/4/045005
28. Nieuwenhuizen JM, Haanstra HB (1966) Microfractography of thin films. Philips Tec Rev 27:87–91
29. Li Z, Zhu Y, Zhou Q, Ni J, Zhang Z (2012) Photocatalytic properties of TiO_2 thin films obtained by glancing angle deposition. Appl Surf Sci 258:2766–2770. https://doi.org/10.1016/j.apsusc.2011.10.129
30. Martin N, Rousselot C, Rondot D, Palmino F, Mercier R (1997) Microstructure modification of amorphous titanium oxide thin films during annealing treatment. Thin Solid Films 300:113–121. https://doi.org/10.1016/S0040-6090(96)09510-7
31. Sani SR (2014) Analysis of optoelectronic properties of TiO_2 nanowiers/Si heterojunction arrays. Chin Physics B 23:107302. https://doi.org/10.1088/1674-1056/23/10/107302
32. Pooja P, Chinnamuthu P (2021) Annealed n-TiO_2/In_2O_3 nanowire metal-insulator-semiconductor for highly photosensitive low-noise ultraviolet photodetector. J Alloys Compd 854:157229. https://doi.org/10.1016/j.jallcom.2020.157229
33. Lahiri R, Ghosh A, Choudhuri B, Mondal A (2018) Investigation on improved performance of Erbium doped TiO_2 nanowire based UV detector. Mater Res Bull 103:259–267. https://doi.org/10.1016/j.materresbull.2018.03.024

Paper-Based Capacitive Sensor for Detection of Arsenic in Drinking Water

Geetartha Sarma, Mrigendra Yadav, and Partha P. Sahu

1 Introduction

The essential component of life on earth is water, but it also contains a number of dangerous substances that can seriously harm a person's health. Among all of these dangerous substances, arsenic, which is most frequently found as arsenate (As(III)) and arsenite (As(V)), is known to cause skin cancer and other fatal diseases such cardio vascular disease, heart disease, and birth defects [1, 2]. According to the World Health Organization (WHO), arsenic in drinking water should not exceed 10 ppb [3]. However, there are relatively few trusted methods available for quickly testing for arsenic in drinking water. The traditional method to measure the amount of arsenic in drinking water needs skilled personnel, bulky, expensive apparatus, and is not environment-friendly. It also requires considerable sample pre-treatment [4].

To detect arsenic in drinking water, we have introduced an interdigitated electrode (IDE) paper-based capacitive sensor. By measuring capacitance or impedance for chemical sensing, IDE plays a crucial role in this direction [5]. On Whatman filter paper, a comb-based electrode was fabricated, and Fe@rGO nanocomposite was deposited on the electrode surface by using drop coating method. Reduced graphene oxide (rGO) has a sizable surface area, and when it is combined with iron to produce a nanocomposite, iron ions are deposited on rGO surface in the form of C-O-Fe. When arsenic contain water expose to this nanocomposite, a bond is form between As and iron [12]. The paper capacitor was calibrated with DI water, 100 ppb, 60 ppb As solution, and 100 ppb $FeCl_3$ sol.

G. Sarma · M. Yadav (✉) · P. P. Sahu
Department of Electronics and Communication Engineering, Tezpur University, Napaam, Tezpur, Assam 784028, India

G. Sarma
e-mail: geetarthasarma59@gmail.com

B. P. Swain (ed.), *Recent Advances in Materials*, Springer Proceedings in Materials 25,
https://doi.org/10.1007/978-981-99-3844-5_10

2 Experimental Section

2.1 Chemicals

All the chemical used for the preparation of sample such as graphite flake, sulfuric acid, potassium permanganate, hydrochloric acid, hydrogen peroxide, ferric chloride, ammonia (25%), ethanol, all chemicals are purchased from Merck and Sigma-Aldrich.

2.2 Synthesis of Graphene Oxide (GO)

The synthesis of GO using modified hummers method is followed as proposed by Abdolhosseinzadeh et al. [6].

2.3 Synthesis of Fe@rGO Nanocomposite

The Fe@rGO nanocomposite was prepared by a hydrothermal method as described by S. Radhakrishnan et al. [7]. 70 mL of ethanol was mixed with 50 mg of graphene oxide, then ultrasonically dissolved for 30 min. Following that, 100μL of 25% ammonia and 0.1 g of $FeCl_2$ were poured into the solution. The mixture was then placed in a Teflon-lined stainless steel autoclave and kept at 170 °C for 4 h before gradually cooling to ambient temperature. The precipitate was then filtered, repeatedly rinsed with distilled water and ethanol, and dried in a vacuum oven at 80 °C for 6 h.

2.4 Characterization

Using a Nicolet Impact 410 FTIR spectrophotometer, Fourier transform infrared spectroscopy (FTIR) was used to identify the functional groups contained in the sample. To evaluate the crystallite size and interlayer spacing, X-ray diffraction (XRD) was carried out using BRUKER AXS D8FOCUS, CuK radiation (k = 1.540598), 30 kV, 15 mA, and a scan rate of 1 min. In the JEOL JSM 6390LV, energy dispersive X-ray (EDX) experiments were conducted. TECNAI G2 20 S-TWIN underwent TEM investigation (200 kV).

2.5 Senor Design and Concept

The fabricated capacitive-based arsenic detection kit comprises of coupled comb electrode with finger width (n) of 'x' and gap of 'y' between the fingers printed on a Whatman filter paper with dimensions 3×2 cm^2 (Fig. 1d) and then it is covered with the sensing material. Considering eight fingers, two edge electrode capacitance and five interior capacitances are formed. Due to the application of +Ve and −Ve potential at adjacent fingers of electrode, there is a generation of traversing electric fields penetrating through sensing material and paper substrate and fringing fields traveling from one finger to other finger through air medium [8]. The total capacitance contributed by the above fields is the function of the dielectric constant of sensing material and the thickness of the sensing layer and gap between two adjacent finger and equated as:

$$C_{\text{Total}} = L(N - 3)C_1 + 2C_E, \tag{1}$$

where C_I is the initial capacitance of unit cell, and C_E is the edge capacitance which is neglected with respect to C_I as $C_I >> C_E$. For the design, we have taken four fingers in each electrode of comb structure ($N = 8$). Considering the variation of C_{Total} with the dielectric with dielectric constant (εm) of sensing material, w is obtained as 1.5 mm, and 'a' is found to be 1.5 mm.

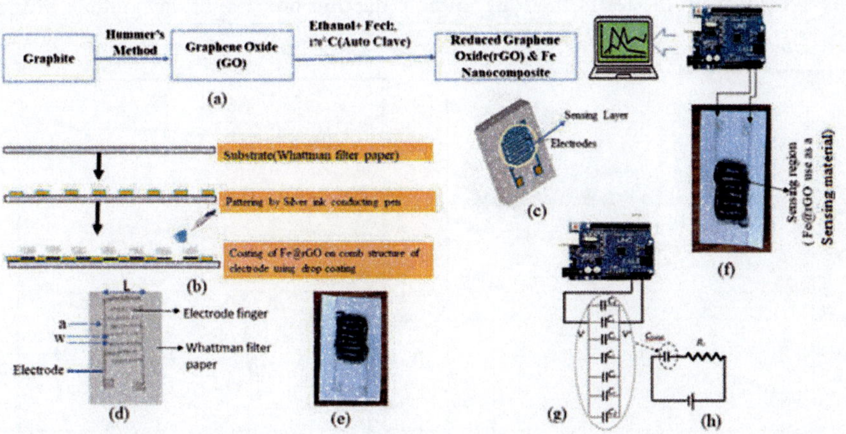

Fig. 1 Synthesis of Fe@rGO nanocomposite, fabrication of paper capacitive sensor with Fe@rGO sensing layer and its measurement set up **a** synthesis process of Fe@rGO nanocomposite, **b** fabrication steps of paper capacitor, **c** 3D view of paper capacitor, **d** paper capacitor before deposition of sensing layer **e** paper capacitor after deposition of sensing layer **f** capacitance measuring setup **g** by applying voltage through Arduino developed interior and edge capacitance **h** equivalent circuit having total capacitance (C_{Total}) and internal resistance of Arduino R1

3 Result and Discussion

Figure 2a shows the XRD spectra of GO rGO and Fe@rGO nanocomposite. The spectra for GO exhibited a characteristics peaks at a diffraction angle of 10.62 and 42.34 corresponding to (001) and (100) planes confirming the successful synthesis of GO. Upon reduction, the characteristics peak shifted to the higher diffraction angle attributed to the successful reduction of GO [9]. The spectrum for Fe@rGO nanocomposite exhibited characteristics peaks at $2\theta = 24.2$, 33.2, 35.6, 40.9, 49.4, 53.9, 57.4, 62.4, 63.9, and 71.8 corresponding to (012), (104), (110), (113), (024), (116), (122), (214), (300) planes of Fe@rGO and (002) plane of rGO, respectively [6, 10]. The characteristics peaks of rGO sheets are not seen in the composites (Fe@rGO), and all of the peaks are related to the Fe_2O_3 (JCPDS: 89-2810). This is because Fe_2O_3 nanoparticles are coated on each side of the two-dimensional, thin rGO sheets.

The FTIR spectra as prepared GO, rGO, and Fe@rGO are depicted in Fig. 2b, respectively. Broad peak at 3410 cm^{-1} corresponds to the stretching and bending vibrations of OH groups [11]. Carboxyl (C = O) absorption peaks were present at 1725 cm^{-1}, and the peak at 1620 cm^{-1}can be attributed to the stretching vibration of C = C. The FTIR spectra show the absorption peak caused by the stretching vibrations of C-O group at 1150 $cm^{-1}$and C–O–C group at1050 cm^{-1}. These oxygen-containing groups provide evidence that the graphite has been oxidized. The presence of the surface hydroxyl groups demonstrates how hydrophilic graphene is. It is evident from the result that due to the reduction process, the magnitude of the

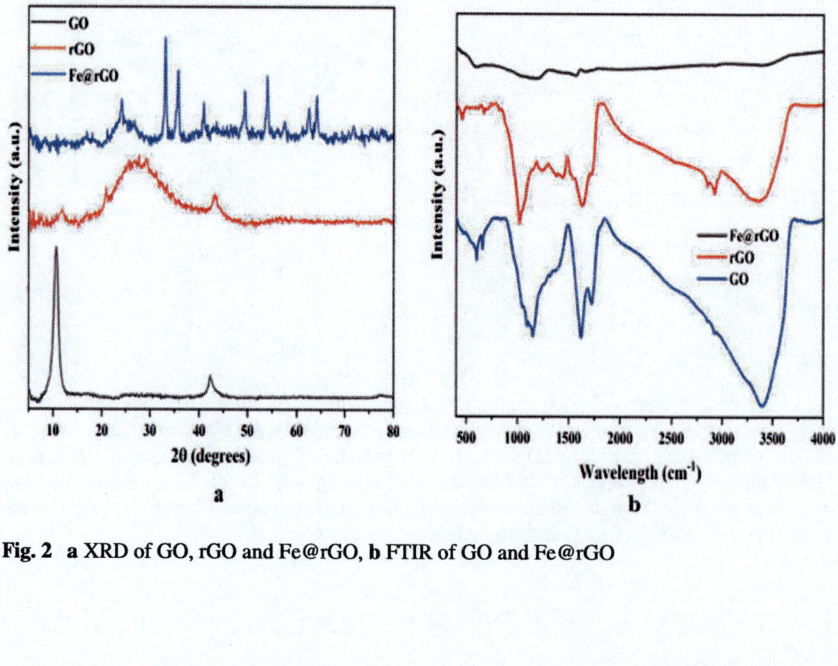

Fig. 2 **a** XRD of GO, rGO and Fe@rGO, **b** FTIR of GO and Fe@rGO

peaks corresponding to different functional group reduced confirming the reduction of GO. However, the spectrum of Fe@rGO nanocomposite shows very small deviations in its peaks because of complete reduction of rGO and then formation of Fe@rGO nanocomposite. None of the characteristics peaks of rGO are observed which confirms the removal of functional groups and addition of iron atoms over the surface of thin layered rGO.

The EDX spectra of GO, rGO and Fe@rGO presented in Fig. 3a–C, respectively. From the data, it can be observed that the percentage of the oxygen and carbon atom is reduced due to the reduction of GO to rGO via hydrothermal method and addition of the Fe atoms during the formation of Fe@rGO nanocomposite. Figure 3c, d shows the TEM images of rGO and Fe@rGO nanocomposite. From the figure, it can be clearly seen that the embedded Fe atoms over the 2D reduced graphene oxide [12].

The capacitance of paper capacitor was measured by experimental setup with charging and discharging through RC circuit having capacitance C_{Total} and internal resistance (R_I) of Arduino Uno (Fig. 1f). Before the exposure of As samples, the

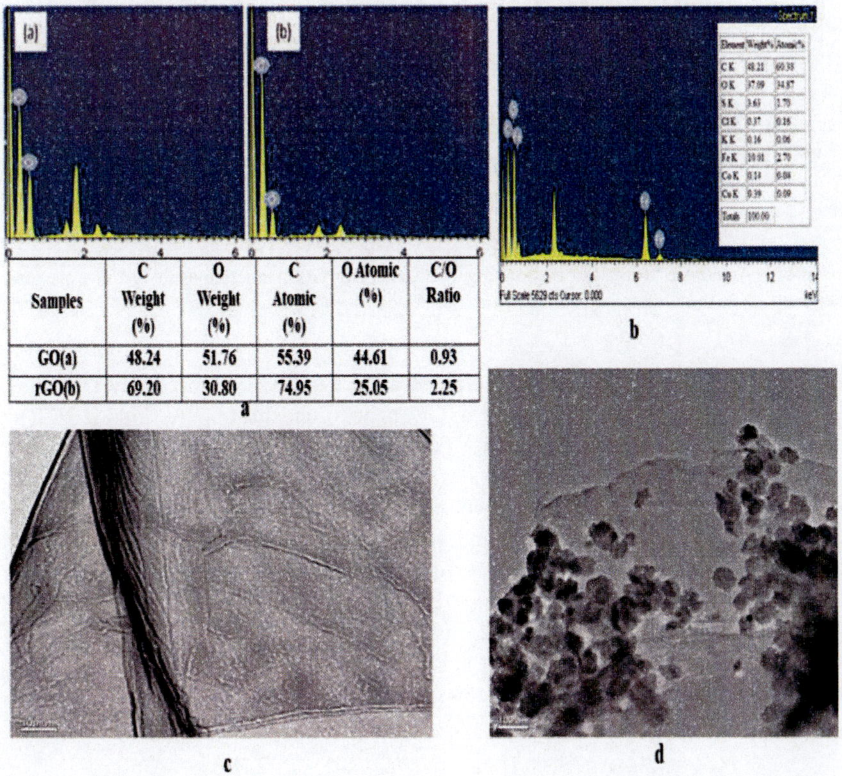

Samples	C Weight (%)	O Weight (%)	C Atomic (%)	O Atomic (%)	C/O Ratio
GO(a)	48.24	51.76	55.39	44.61	0.93
rGO(b)	69.20	30.80	74.95	25.05	2.25

Fig. 3 **a** EDX spectra of GO and rGO, **b** EDX spectra of Fe@rGO, **c** TEM images of rGO **d** TEM images of Fe@rGO

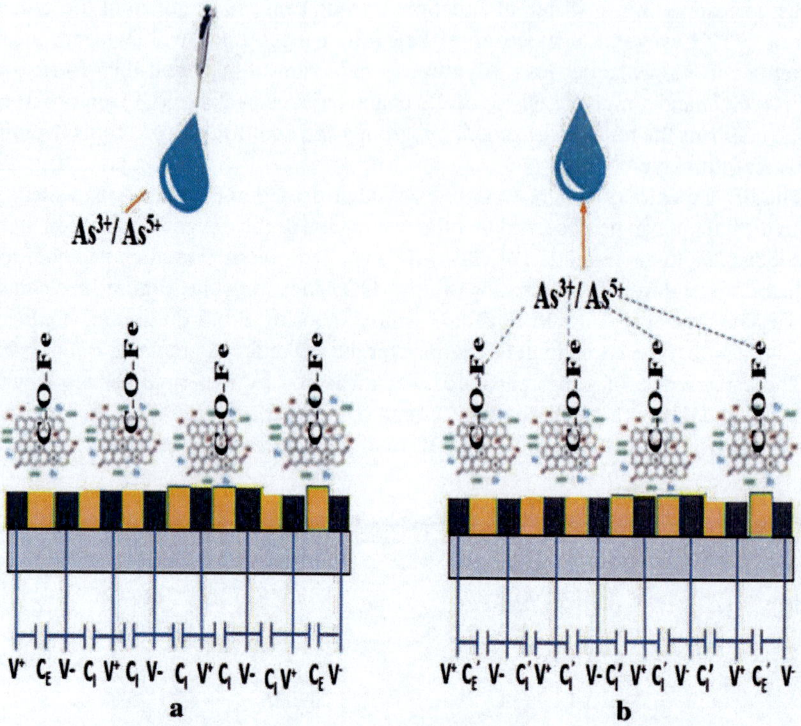

Fig. 4 Mechanism of interaction between arsenic and Fe@rGO nanocomposite a before exposure with Arsenic **b** formation of bonding between Fe@rGO nanocomposite and arsenic after exposure of Arsenic samples

interior capacitance and edge capacitance are CI and CE, respectively (Fig. 3a). During exposure, active sites of Fe@rGO nanocomposite trap As ions [13]. The mechanism of interaction of As with Fe@rGO nanocomposite over the electrode surface is shown in Fig. 4.

The calibration of paper capacitive sensor was carried out by DI water samples, 100 ppb arsenic, 60 ppb arsenic, and 100 ppb $FeCl_3$. In each case, ten numbers of samples are tested, and volume of each sample was 10 μl. Figure 5a shows the transient analysis of capacitance (C_{Total}) of paper capacitive sensor exposed to above sample. The C^s_{Total} 100 ppb As samples is in the range of 170–220μF. The 60 ppb As sample shows C^s_{Total} ~150μF almost at same detection time of DI water. The C^s_{Total} for DI water and $FeCl_3$ samples are 130 μF and 135μF, respectively, almost at same detection time. Figure 5b shows the C_{Total} values of different As samples.

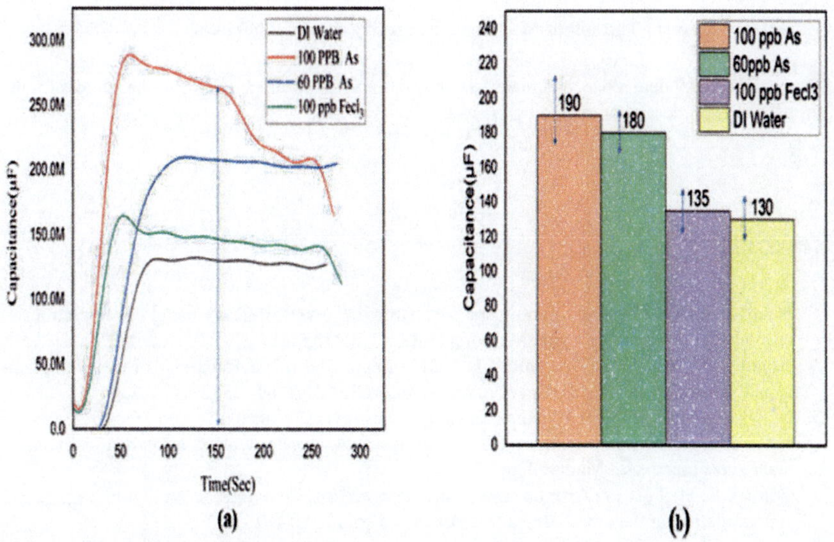

Fig. 5 **a** Time variation of capacitance after exposure of Fe@rGO surface of sensor with DI water, 100 ppb As, 60 ppb As, and 100 ppb FeCl3, **b** range of capacitance values of DI water, As, and $FeCl_3$

4 Conclusion

Here, capacitive-based sensors for detection of arsenic present in drinking water were designed and developed on degradable Whatman filter paper implying simple and efficient drop coating method. Initially, Ag com electrode was developed on the paper, followed by the deposition of Fe@rGO sensing material using simple drop coating technique. The fabricated sensor was calibrated using standard DI water. Upon exposure of arsenic samples, the fabricated sensor exhibited a capacitance value in the range of 170–220 μF and 170–190 μF for 100 ppb and 60 ppb, respectively. However, no variation is detected due to the exposure of Fe^{3+} ions, confirming the selectivity of the fabricated devices to arsenic. The sensor exhibited a detection time of 150 s. An additional advantage of the developed sensors is its biodegradable in nature.

Acknowledgements The authors acknowledge Dr. Rewrewa Narzary, Mr. Satyajit Das, Mr. Nithin Joseph Panicker research scholar, Microfabrication lab Dept. of ECE, Tezpur University, for helping in synthesis of sensing material, fabrication, and calibration of the sensor. The authors would like to thank Tezpur University's Sophisticated Analytical Instrument Center (SAIC) for providing characterization facilities.

Author Contribution GS and MY initiated and conceptualized the sensor and wrote the paper. GS fabricated the sensor. MY synthesis the sensing material. GS and MY testing the sensor. PPS supervised the work.

Competing Interest The authors declare no competing financial interests.

Data sharing All data, code, and materials used in the analysis are available to the authors for the purposes of reproducing or extending the analysis.

References

1. Ismail Z, et al (2013) Determining and comparing the levels of heavy metal concentrations in two selected urban river water. Measurement 46(10):4135–4144
2. Anjum NA, et al (2015) Too much is bad—an appraisal of phytotoxicity of elevated plant-beneficial heavy metal ions. Environ Sci Pollution Res 22:3361–3382
3. N. US EPA, OCSPP, OPPT, and chemical contaminant rules_drinking water requirements for states and public water systems_US EPA. https://www.epa.gov/dwreginfo/chemical-contaminant-rules (accessed: March2020)
4. Gupta DK, et al (2017) Arsenic contamination from historical aspects to the present. Arsenic contamination in the environment: the issues and solutions, 1–12
5. Gupta AK, et al (2022) Design and development of IDE sensor for naringin quantification in pomelo juice: an indicator of citrus maturity. Food Chem 377:131947
6. Abdolhosseinzadeh S, Asgharzadeh H, Kim HS (2015) Fast and fully-scalable synthesis of reduced graphene oxide. Scient Rep 5(1):1–7
7. Radhakrishnan S, et al (2014) A highly sensitive electrochemical sensor for nitrite detection based on Fe_2O_3 nanoparticles decorated reduced graphene oxide nanosheets. Appl Catalysis B: Environ 148:22–28
8. Das S, Sahu PP (2022) A novel electrochemical interdigitated electrodes sensor for limonin quantification and reduction in citrus limetta juice. Food Chem 381:132248
9. Narzary R, Phukan P, Sahu PP (2021) Efficiency enhancement of low-cost heterojunction solar cell by the incorporation of highly conducting rGO into ZnO nanostructure. IEEE Trans Elect Dev 68(7):3238–3245
10. Panicker NJ, Das J, Sahu PP (2021) Synthesis of highly oxidized graphene (HOG) by using HNO_3 and $KMnO_4$ as oxidizing agents. Mat Today: Proceed 46:6270–6274
11. Alam SN, Sharma N, Kumar L (2017) Synthesis of graphene oxide (GO) by modified hummers method and its thermal reduction to obtain reduced graphene oxide (rGO). Graphene 6(1):1–18
12. Basu AK, et al (2019) α-Fe2O3 loaded rGO nanosheets based fast response/recovery CO gas sensor at room temperature. Appl Surf Sci 465:56–66
13. Sui L-L, Peng L-N, Hong-Bo X (2021) Nanocomposites of Fe 2 O 3@ rGO for adsorptive removal of arsanilic acid from aqueous solution. Korean J Chem Eng 38:498–504

Investigation of Optical and Electrical Properties of $Au/SiO_x/ITO$ for Optoelectronic Applications

Rubila Laishram and Naorem Khelchand Singh

1 Introduction

Silicon Oxide (SiO_x) thin films (TFs) are among the most widely used nanomaterials for optical coatings, sensors and detectors, and storage applications [1, 2]. High-sensitivity detectors and sensors can be developed using the SiO_x coating on the Si surface [3, 4]. Suppression of dark current due to the SiO_x layer was also reported in the structure comprising SiO_x-$In_{2\text{-}x}O_{3\text{-}y}$-based Schottky detectors [5]. Meanwhile, Transparent Conducting Oxides (TCOs) are extensively used in transparent electronics, especially Indium Tin Oxide (ITO), as they possess a large band gap, making them suitable for UV absorption. It also has chemical inertness and hardness, substrate adhesion, and high visible region transmission. Due to this, ITO substrates are suitable for various optoelectronic applications, including photovoltaic cells, liquid crystal displays [6], and gas sensors [7]. Various researchers have reported using ITO TF for photodetector applications [8, 9].

In this study, SiO_x thin film was fabricated on an ITO substrate using a catalyst-free e-beam evaporation technique for photodetector application. The morphological analysis was characterized by Field Emission Scanning Electron Microscopy (FESEM-Carl Zeiss Sigma). Optical analysis was carried out using a UV-Vis-NIR spectrophotometer (Hitachi Model UH4150). Finally, the electrical study was conducted with the help of Keithley 4200-SCS.

R. Laishram (✉) · N. K. Singh
Department of Electronics and Communication Engineering, National Institute of Technology Nagaland, Dimapur 797103, India
e-mail: rubylaishram@gmail.com

B. P. Swain (ed.), *Recent Advances in Materials*, Springer Proceedings in Materials 25,
https://doi.org/10.1007/978-981-99-3844-5_11

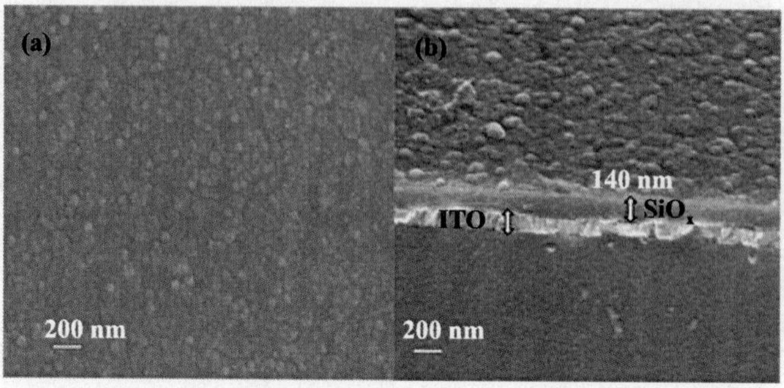

Fig. 1 FESEM images **a** top view of SiO_x TF **b** cross-sectional view showing thickness

2 Experimental Methods

SiO_x TF on ITO substrates was deposited with the e-beam evaporation method (BC-300, HHV, India). Initially, ITO substrates (~80% transmittance and 15 + 2.5 Ω resistance) were cleaned with isopropyl and Deionized (DI) water using ultrasonicate cleaner (KJI Group, MTI) for three minutes each sequentially. The cleaned ITO substrates were then fixed in a substrate holder kept 24 cm away from the crucible containing source materials, SiO_x granules (purity of 99.99%). Consistently, the deposition rate was monitored by a digital thickness monitor (DTM). During the deposition, an initial pressure of ~ 5×10^{-6} mbar and a constant evaporation rate of 1 Å/s were maintained. For metallization, Au contacts (top electrode) were evaporated above the SiO_x film by masking circular holes of a diameter of 1.5 mm.

3 Results and Analysis

3.1 FESEM

The FESEM analysis was conducted to study the morphology of the fabricated SiO_x TF. The top view of FESEM images, as shown in Fig. 1a, witnesses the successful growth of a uniform SiO_x TF. As shown in Fig. 1b, the average estimated thickness of the deposited SiO_x film was obtained as 140 nm.

3.2 Optical Absorbance

Optical absorption measurement was performed using a UV-Vis spectrophotometer for the wavelength up to 800 nm, as shown in Fig. 2a. A wide absorption peak was

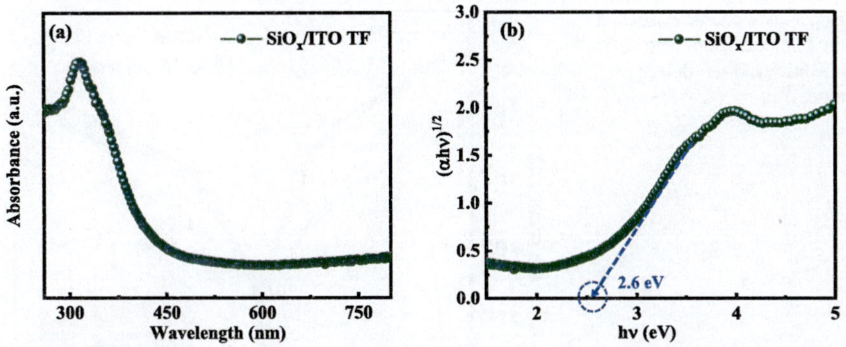

Fig. 2 **a** UV-Vis absorption spectrum of SiO_x/ITO **b** Tauc plot representing bandgap

observed in the UV-Vis region. Figure 2b depicts the optical bandgap of the SiO_x TF calculated using the Tauc's plot by plotting a graph between (hυ) and $(\alpha h\upsilon)^{1/2}$, where hυ is the energy of the photon and α represents the absorption coefficient [10]. A bandgap of ~ 2.6 eV was obtained by extrapolating a straight line in Tauc's plot. The obtained bandgap is well-matched with the SiO_x bandgap [11].

3.3 *Electrical Characteristics*

The electrical current–voltage (I–V) characteristics of Au/SiO_x/ITO TF under dark and light conditions are demonstrated in Fig. 3 with the device model (inset). Based on the rectifying I–V curve, it is evident that a Schottky junction has formed between the SiO_x TF and the Au contact. The conduction process of the device can be explained using the adsorption and desorption processes, which take place on the SiO_x surface sites. Under the dark condition, the proposed device exhibits low dark current due to the adsorption process where O_2 molecules from the atmosphere are attracted on the surface of SiO_x (O_2 (g) + e- → O^{2-}). The formation of oxygen ion (O^{2-}) leads to band bending, thereby increasing the depletion width. As a result, the conductivity of the device decreases.

During light illumination, carriers (holes and electrons) are generated due to the desorption process (O^{2-} + e- + h^+ → O_2 + e-), the ionized oxygen recombines with the holes releasing O_2 in the air, and the remaining unpaired electrons act as charge carriers, which increases the device's conductivity. The switching characteristics of the SiO_x TF taken at −3 V are illustrated in Fig. 4a. A fast photoresponse, such as rise time (τ_r) of ~ 52 ms (time required to reach from 10 to 90% of the maximum current) and fall time (τ_f) of ~ 40 ms (90–10% of the maximum current), is depicted in Fig. 4b and c. Table 1 compares our reported work with the existing work done by other researchers.

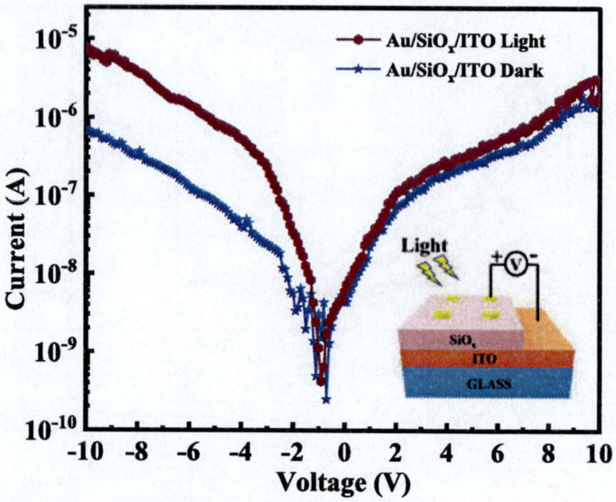

Fig. 3 I–V characteristics of the film under light and dark conditions, inset (device model)

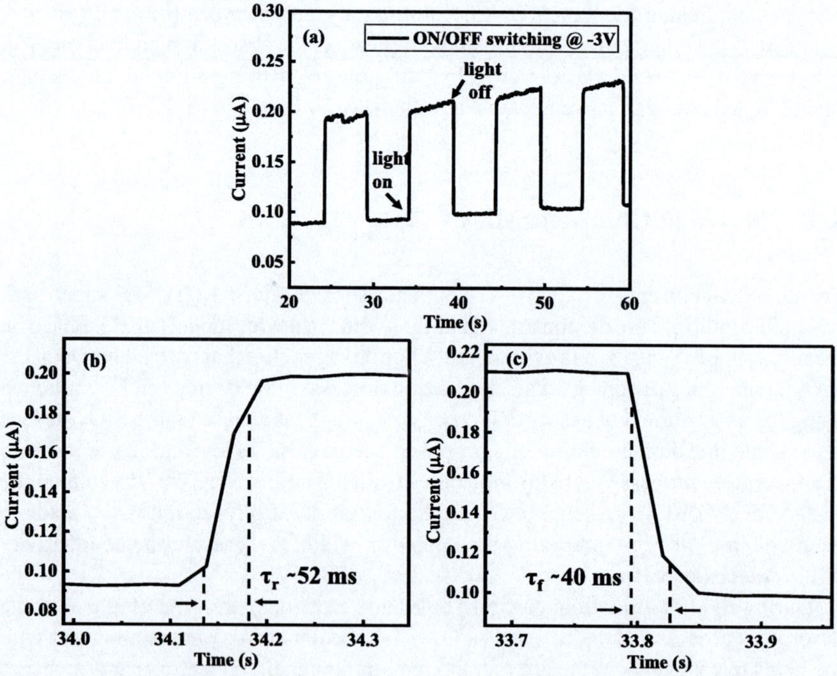

Fig. 4 **a** Response time of the fabricated device at −3 V **b** τ_r (rise time) and **c** τ_f (fall time)

Table 1 Comparison based on response time with existing literature

Device models	Rise time, τ_r (s)	Fall time, τ_f (s)
SiO_x TF (proposed work)	52 ms	40 ms
Si/SiO_x/Al-NP/SiO_x/Al-NP/Ag [12]	106 ms	108 ms
SiO_x-$In_{2-x}O_{3-y}$/Au [5]	1.82 s	1.78 s
SiO_x NW/Si [10]	0.13 s	0.13 s

4 Conclusion

We can conclude that, by using an e-beam evaporation technique, SiO_x TF has been successfully fabricated on the ITO substrate. FESEM analysis reveals uniform growth of the deposited SiO_x TF. The absorption spectrum demonstrated a broad absorption peak in the UV-Vis region. The electrical analysis shows a faster photoresponse of ~ 52 ms rise time and ~ 40 ms fall time, which validates a potential candidate for optoelectronics applications.

Acknowledgements The authors would like to appreciate the (CoE) in Advanced Material, NIT Durgapur, for FESEM measurement and NIT Nagaland for Absorption and IV measurement.

References

1. Hass G, Salzberg CD (1954) Optical properties of silicon monoxide in the wavelength region from 024 to 140 µm*. J Opt Soc Am 44:181. https://doi.org/10.1364/JOSA.44.000181
2. Park M, Park J, Kim S (2022) Compatible resistive switching mechanisms in Ni/SiOx/ITO and application to neuromorphic systems. J Alloys Compd 903. https://doi.org/10.1016/j.jallcom.2022.163870
3. Mondal A, Singh NK, Chinnamuthu P, Dhar JC, Bhattacharyya A, Choudhury S (2012) Enlarged photodetection using SiOx nanowire arrays. IEEE Photon Technol Lett 24:2020–2023. https://doi.org/10.1109/LPT.2012.2218232
4. Cheng S, Ren T, Ying P, Yu R, Zhang W, Zhang J, Li C (2012) Enhanced growth of crystalline-amorphous core-shell silicon nanowires by catalytic thermal CVD using in situ generated tin catalyst. Sci China Chem 55:2573–2579. https://doi.org/10.1007/s11426-012-4717-4
5. Singh NK, Mondal A (2015) High internal gain axial SiOx-In_2-xO_3-y/Au heterostructure nanocolumnar array based schottky detector for broad band recognition. J Nanosci Nanotechnol 15:6098–6102. https://doi.org/10.1166/jnn.2015.10289
6. Vijayalakshmi K, Monamary A (2016) Novel hydrogen sensor based on p-type Ni:TiO_2 nanorods fabricated on ITO substrate. J Mater Sci Mater Electron 27:140–145. https://doi.org/10.1007/s10854-015-3729-1
7. Chauhan HA, Rafatullah M, Ahmed Ali K, Siddiqui MR, Khan MA, Alshareef SA (2021) Metal-based nanocomposite materials for efficient photocatalytic degradation of phenanthrene from aqueous solutions. Polymers (Basel) 13:2374. https://doi.org/10.3390/polym13142374
8. Kim H, Lee G-N, Kim J (2018) Hybrid structures of ITO-nanowire-embedded ITO film for the enhanced Si photodetectors. J Nanomater 2018:1–8. https://doi.org/10.1155/2018/4178989

9. Ahmad H, Tajdidzadeh M, Thandavan TMK (2018) Heterojunction photodetector based on graphene oxide sandwiched between ITO and p-Si. J Mod Opt 65:353–360. https://doi.org/10.1080/09500340.2017.1397224
10. Devi NM, Lynrah SA, Rajkumari R, Singh NK (2021) Effect of Ag decoration on the photodetection of catalyst-free synthesized vertically oriented SiOx NW arrays. Sens Actuators A Phys 327. https://doi.org/10.1016/j.sna.2021.112744
11. Carneiro JO, Machado F, Rebouta L, Vasilevskiy MI, Lanceros-Méndez S, Teixeira V, Costa MF, Samantilleke AP (2019) Compositional, optical and electrical characteristics of SiOx thin films deposited by reactive pulsed DC magnetron sputtering. Coatings 9. https://doi.org/10.3390/coatings9080468
12. Bikesh S, Ngangbam C, Singh SS, Shougaijam B (2022) Structural and optical properties analysis of Al nanoparticle-assisted SiOx thin film for photodetector application. Bull Mater Sci 45:216. https://doi.org/10.1007/s12034-022-02802-5

Study on Structural and Optical Properties of Ta_2O_5 Nanocluster

Elangbam Rameshwar Singh, Borish Moirangthem, and Naorem Khelchand Singh

1 Introduction

Among the metal oxide, Tantalum pentoxide (Ta_2O_5) has been the most popular material in recent years due to its unique properties like electronic application and optical [1–5]. The Ta_2O_5 offers good chemical and thermal stability[6], high refractive index in the broad spectral, optical transmittance [7], and high k dielectric constant [8]. Due to these properties Ta_2O_5 is widely used in optoelectronic applications [9, 10], and memory devices [11, 12]. Deposition methods show an essential role in the physical properties of the Ta_2O_5 nanocluster [9, 13]. Fabrication techniques like electron beam (e-beam) evaporation incorporated with glancing angle deposition (GLAD) [14], atomic layer deposition [15], thermal oxidation [16], ion-assisted deposition[17], chemical vapor deposition, and radio-frequency (RF) magnetron sputtering [18] have been employed to create the Ta_2O_5 nanostructure. Among them, the GLAD technique provides a catalyst-free and user-friendly environment. Moreover, GLAD allows control over morphology and structure during deposition. In addition, there are very few reports on Ta_2O_5 nanostructure using the (GLAD) technique. In this study, we report a Ta_2O_5 nanoclusters' deposition on Fluorine-Doped Tin Oxide (FTO) substrate using the GLAD technique. We used field emission gun-scanning electron microscopy (FEG-SEM), X-ray diffraction (XRD), Energy-Dispersive X-ray spectroscopy (EDS), and UV-visible spectrum analysis to investigate the structural and optical features of a Ta_2O_5 nanocluster placed on an FTO substrate.

E. R. Singh (✉) · B. Moirangthem · N. K. Singh
Department of Electronics and Communication Engineering, NIT Nagaland, Dimapur, Nagaland 797103, India
e-mail: rameshwarela@gmail.com

B. P. Swain (ed.), *Recent Advances in Materials*, Springer Proceedings in Materials 25,
https://doi.org/10.1007/978-981-99-3844-5_12

2 Experimental

Ta_2O_5 nanocluster was fabricated on FTO substrates by an e-beam evaporation with the GLAD system (BC 300, HHV India). The proposed structure for the Ta_2O_5 nanocluster was done using Ta_2O_5 granules with a purity of ~ 99.99% as the source material. FTO substrate cleaning was performed for 3 min using isopropyl and deionized water in the ultrasonicator (KJ Group, MTI). An angle of 85° between the substrate holder and the crucible was kept. The vacuum chamber during the deposition was maintained at a pressure of 6×10^{-6} mbar. To track the thickness of the structure, a digital thickness monitor (DTM) was used.

3 Result and Discussion

3.1 *Structure Analysis*

As shown in Fig. 1, the XRD diffraction peaks show only for the FTO substrate, and there is no distinct peak for the Ta_2O_5 nanocluster, which indicates that the Ta_2O_5 is amorphous [19]. Meenal et al. also reported that amorphous Ta_2O_5 film was formed on an FTO substrate using the spin coating technique, which is used as a transporting layer for solar cell application [20].

FEG-SEM images show the successfully deposited Ta_2O_5 nanocluster over the FTO substrate using a GLAD technique, as shown in Fig. 2a. The deposition nanocluster was obtained with an average diameter of ~ 120 nm. Moreover, nanoclusters are attracting significant attention in a variety of fields due to their unique

Fig. 1 XRD peaks showing FTO crystalline and amorphous Ta_2O_5

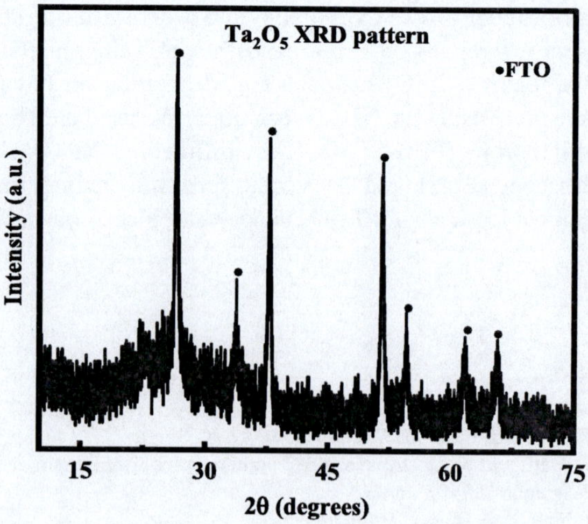

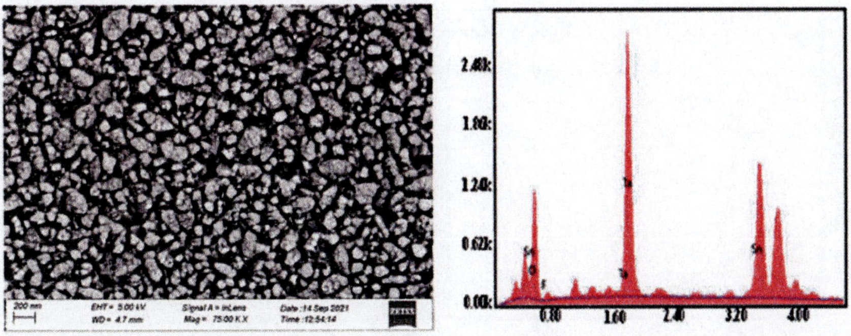

Fig. 2 **a** FE-SEM image of the Ta_2O_5 nanocluster and **b** EDS spectrum for the Ta_2O_5 nanocluster

properties, including their small size, high surface area-to-volume ratio, and optical properties [21].

EDS is a common analytical tool used to determine the elemental composition of a sample. As shown in Fig. 2b, the results of an EDS analysis on the Ta_2O_5 nanocluster showed that it is made up of only Ta, Sn, F, and O, with no other elements detected. This indicates that the sample is pure and free from impurities, which is important for understanding its properties and behavior.

3.2 *Optical Analysis*

The optical absorption spectrum was examined by UV-visible absorption measurement, giving a strong absorption in the UV region, as shown in Fig. 4a. Tauc plot was plotted from the obtained absorption using the relation [22], in which the bandgap was found to be 3.9 eV for the FTO substrate and 3.8 eV for the Ta_2O_5 nanocluster, as shown in Fig. 4b. These values are close to the reported optical bandgap [23, 24] (Table 1).

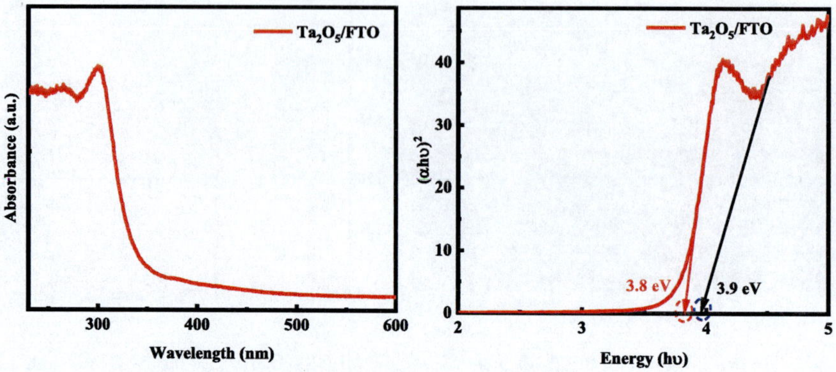

Fig. 4 **a** UV-vis absorption **b** direct band gap $(\alpha h\upsilon)^2$ versus energy (hυ)

Table 1 Comparison table with the existing structure

Material	Technique	Catalyst used
In_2O_3/TiO_2 [25]	Hydrothermal	Yes
Ta_2O_5 [26]	Hydrothermal	Yes
ZnO [27]	Pulsed laser deposition	No
This work Ta_2O_5	GLAD	No

4 Conclusion

Ta_2O_5 nanocluster was successfully deposited on an FTO substrate with the help of the GLAD, which is incorporated with the e-beam evaporator chamber. The FEG-SEM confirmed the successful growth of the nanocluster, and the amorphous nature of the Ta_2O_5 nanocluster was revealed by XRD analysis. EDS study manifests the presence of the elements Ta, Sn, F, and O without impurities during the deposition. Lastly, the absorption was found in the UV region with the obtained band gap of 3.9 eV for the FTO and 3.8 eV for the Ta_2O_5 nanocluster. From the obtained parameters, it can be concluded that the deposited Ta_2O_5 nanocluster can be helpful in developing UV light photodetector and other optoelectronics devices.

Acknowledgements The authors acknowledge NIT Durgapur for the FEG-SEM image and NIT Nagaland for facilitating UV-visible spectrum, XRD measurement, and financial support.

References

1. Thapliyal P, Kandari AS, Lingwal V, Panwar NS, Rao GM (2021) Annealing temperature-dependent structural and electrical properties of (Ta2O5)1-x–(TiO_2)x thin films, $x \leq 0.11$. Ceram Int 47:12066–12071. https://doi.org/10.1016/j.ceramint.2021.01.050

2. Zhuo M, Huang C, Zhao C, Yin J, Shen C (2022) Effects of Ta_2O_5 on the microstructure and electrical properties of ZnO linear resistance ceramics. Mater Res Express. 9:016302. https://doi.org/10.1088/2053-1591/ac4a2f
3. Wang R, Pan L, Han Q, Zhu H, Wan M, Mai Y (2021) Reactively sputtered Ta_2O_5 solid electrolyte layers in all thin film electrochromic devices. J Alloys Compd 865:158931. https://doi.org/10.1016/j.jallcom.2021.158931
4. Thapliyal P, Panwar NS, Mohan Rao G (2021) Electrical and optical properties of $(Ta_2O_5)_{1-x}$–$(TiO_2)_x$ films, $x = 0.035$, prepared by sputtering of ceramic and mosaic (Ta, Ti) metal targets. J Appl Phys 130:035304. https://doi.org/10.1063/5.0055725
5. Thapliyal P, Panwar NS, Rao GM (2021) Optical properties and current conduction in annealed (Ta_2O_5)0.94–(TiO_2)0.06 thin films. Superlatt Microstruct 158:107008. https://doi.org/10.1016/j.spmi.2021.107008
6. Shakoury R, Willey RR (2016) Optimization of Ta_2O_5 optical thin film deposited by radio frequency magnetron sputtering. Appl Opt 55:5353. https://doi.org/10.1364/AO.55.005353
7. Shang P, Xiong S, Li L, Tian D, Ai W (2013) Investigation on thermal stability of Ta_2O_5, TiO_2 and Al_2O_3 coatings for application at high temperature. Appl Surf Sci 285:713–720. https://doi.org/10.1016/j.apsusc.2013.08.115
8. Zhou J, Luo D, Li Y, Liu Z (2009) Effect of sputtering pressure and rapid thermal annealing on optical properties of Ta_2O_5 thin films. Trans Nonferrous Metals Soc China 19:359–363. https://doi.org/10.1016/S1003-6326(08)60278-2
9. Chandra Sekhar M, Nanda Kumar Reddy N, Verma VK, Uthanna S (2016) Structural, optical and electrical properties of DC reactive magnetron sputtered $(Ta_2O_5)1-x(TiO_2)x$ thin films. Ceram Int 42:18870–18878. https://doi.org/10.1016/j.ceramint.2016.09.034
10. Pai Y-H, Chou C-C, Shieu F-S (2008) Preparation and optical properties of Ta_2O_5-x thin films. Mater Chem Phys 107:524–527. https://doi.org/10.1016/j.matchemphys.2007.08.032
11. Koc K, Tepehan FZ, Tepehan GG (2005) Antireflecting coating from Ta_2O_5 and SiO_2 multilayer films. J Mater Sci 40:1363–1366. https://doi.org/10.1007/s10853-005-0566-2
12. Kao CH, Lai PL, Wang HY (2013) The comparison between Ta_2O_5 and Ti-doped Ta_2O_5 dielectrics. Surf Coat Technol 231:512–516. https://doi.org/10.1016/j.surfcoat.2012.04.045
13. Chaneliere C, Autran JL, Devine RAB, Balland B (1998) Tantalum pentoxide (Ta_2O_5) thin films for advanced dielectric applications. Mater Sci Eng R Rep 22:269–322. https://doi.org/10.1016/S0927-796X(97)00023-5
14. Wang M, Chen Y, Gao B, Lei H (2019) Electrochromic properties of nanostructured WO_3 thin films deposited by Glancing-Angle magnetron sputtering. Adv Electron Mater 5:1800713. https://doi.org/10.1002/aelm.201800713
15. Kukli K, Kemell M, Vehkamäki M, Heikkilä MJ, Mizohata K, Kalam K, Ritala M, Leskelä M, Kundrata I, Fröhlich K (2017) Atomic layer deposition and properties of mixed Ta_2O_5 and ZrO_2 films. AIP Adv 7:025001. https://doi.org/10.1063/1.4975928
16. Atanassova E, Spassov D (1998) X-ray photoelectron spectroscopy of thermal thin Ta_2O_5 films on Si. Appl Surf Sci 135:71–82. https://doi.org/10.1016/S0169-4332(98)00278-5
17. Prachachet R, Buranasiri P, Horprathum M, Eiamchai P, Limwichean S, Patthanasettakul V, Nuntawong N, Chindaudom P, Samransuksamer B, Lertvanithphol T (2017) Investigation of optical characteristics of the evaporated Ta2O5 thin films based on ellipsometry and spectroscopy. Mater Today Proc 4:6365–6371. https://doi.org/10.1016/j.matpr.2017.06.140
18. Wang S-C, Liu K-Y, Huang J-L (2011) Tantalum oxide film prepared by reactive magnetron sputtering deposition for all-solid-state electrochromic device. Thin Solid Films 520:1454–1459. https://doi.org/10.1016/j.tsf.2011.08.046
19. Deo M, Möllmann A, Haddad J, Ünlü F, Kulkarni A, Liu M, Tachibana Y, Stadler D, Bhardwaj A, Ludwig T, Kirchartz T, Mathur S (2022) Tantalum oxide as an efficient alternative electron transporting layer for perovskite solar cells. Nanomaterials 12:780. https://doi.org/10.3390/nano12050780
20. Chiu CJ, Weng WY, Hsueh TJ, Chang SJ, Huang GJ, Hsueh HT (2011) Ta_2O_5 solar-blind photodetectors. IEEE Sens J 11:2372–2373. https://doi.org/10.1109/JSEN.2011.2122332

21. Wilcoxon JP, Abrams BL (2006) Synthesis, structure and properties of metal nanoclusters. Chem Soc Rev 35:1162–1194. https://doi.org/10.1039/b517312b
22. Jia H, Xu H, Hu Y, Tang Y, Zhang L (2007) TiO_2@CdS core–shell nanorods films: fabrication and dramatically enhanced photoelectrochemical properties. Electrochem Commun 9:354–360. https://doi.org/10.1016/j.elecom.2006.10.010
23. Kumar NKB, Tejas MK, Kumar JB, Vinutha KV, Kumar SD, Mahesh HM (2016) Natural dye sensitized solar cells using anthocyanin pigment of strawberry as sensitizers quantum dot solar cell view project qunatum nano structures-an overview view project natural dye sensitized solar cells using anthocyanin pigment of strawberry as sensitizers. Imper J Interdisciplin Res (IJIR) 2
24. Shakoury R, Rezaee S, Mwema F, Luna C, Ghosh K, Jurečka S, Ţălu Ş, Arman A, Grayeli Korpi A (2020) Multifractal and optical bandgap characterization of Ta_2O_5 thin films deposited by electron gun method. Opt Quantum Electron 52. https://doi.org/10.1007/s11082-019-2173-5
25. Amoli V, Farooqui S, Rai A, Santra C, Rahman S, Sinha AK, Chowdhury B (2015) Indium oxide nanocluster doped TiO_2 catalyst for activation of molecular O_2. RSC Adv 5:67089–67092. https://doi.org/10.1039/c5ra13104a
26. Li J, Dai W, Yan J, Wu G, Li L, Guan N (2015) Hydrothermal synthesis and photocatalytic properties of tantalum pentoxide nanorods, Cuihua Xuebao/Chinese. J Catal 36:432–438. https://doi.org/10.1016/S1872-2067(14)60215-1
27. Dinh LN, Schildbach MA, Balooch M, McLean W (1999) Pulsed laser deposition of ZnO nanocluster films by Cu-vapor laser. J Appl Phys 86:1149–1152. https://doi.org/10.1063/1.370857

Non-volatile Memory Application Based on Gd_2O_3 Nanorod

Ph. Nonglen Meitei and Naorem Khelchand Singh

1 Introduction

In recent decades, rare earth oxides (REO) have received enormous interest due to their wide field of applications [1–3]. Among the REO, gadolinium oxide (Gd_2O_3) is a high-k dielectric material with a wide bandgap that can replace SiO_2 in metal oxide semiconductor (MOS) applications. Moreover, Gd_2O_3 is chemically and thermally stable with low lattice mismatch with silicon [2]. The demand for non-volatile memory (NVM) usage has grown significantly over the past few years [4]. Recently, resistive random access memory (RRAM) has drawn significant interest because of its high-density integration, low power consumption, simple design, and quick operation [5]. Most of the RRAM requires conductive filament formation to switch between a high resistance state (HRS) and a low resistance state (LRS). Forming filament requires a high-voltage stress process [6]. Therefore, eliminating the forming process is beneficial from both the design and power consumption perspectives. However, Gd_2O_3 contains enough defects for the operation of forming-free RRAM [7]. In forming-free RRAM, oxygen vacancies play an essential role in resistive switching. In addition, scaling the device's size to the nanoscale can explore untapped properties that can achieve higher density, enhance device performance, and improve resistive switching mechanisms. In this study, one (1D) nanostructure using Gd_2O_3 was deposited using the GLAD technique. The purpose structure has the potential to serve multiple functions, as the author has previously demonstrated for the application photodetector [8]. The capacitive memory and forming-free RRAM characteristics for the deposited Gd_2O_3 nanorod (NR) were analysed using Keithley 4200.

Ph. N. Meitei (✉) · N. K. Singh
Department of Electronics and Communication Engineering, National Institute of Technology Nagaland, Chumukedima, Dimapur, Nagaland 797103, India

B. P. Swain (ed.), *Recent Advances in Materials*, Springer Proceedings in Materials 25,
https://doi.org/10.1007/978-981-99-3844-5_13

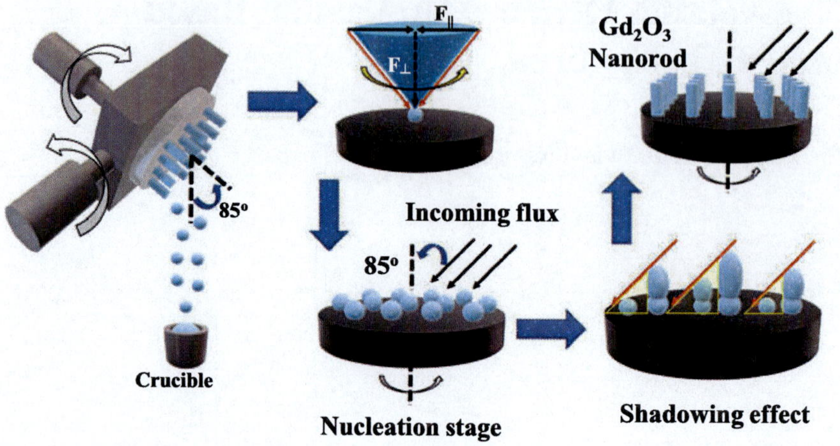

Fig. 1 Illustration of GLAD setup using the deposition

2 Experimental Details

The GLAD technique was used to deposit Gd_2O_3 (99.99% purity) NR on an n-type Si wafer. Before the deposition, the 1×1 cm^2 Si wafers were cleaned using the Radio Corporation of America (RCA) cleaning method. After cleaning, the wafer was placed on a substrate holder within the GLAD system, which was positioned at an angle of 85° relative to the incoming vapour flux from the crucible. The substrate was azimuthally rotated during the deposition process at 30 rpm with a constant deposition rate of 1.1 Å/s and a base pressure of 5.5×10^{-6} mbar. A graphic representation of a GLAD setup is illustrated in Fig. 1. Additionally, a silver contact of ~ 30 nm thickness with an area of 3.14×10^{-2} cm^2 was deposited on top of the Gd_2O_3 NR for the electrical analysis using Keithley 4200.

3 Results and Discussion

Figure 2a and b shows the current (C)—voltage (V) and conductance (G)—voltage (V) measurements for the Gd_2O_3 NR device for -10 to $+$ 10 V voltage sweep. The measurement was recorded under dark conditions for frequencies ranging from 100 kHz to 1 MHz with a 30 mV AC signal. As demonstrated in Fig. 2a, it is evident that the capacitance decreases as the frequency increases with well-defined accumulation, inversion, and depletions regions matching that of the n-type C–V curve. Similar to capacitance variation in C–V measurement with varying frequency, the G–V curve increases with the increasing frequency, as shown in Fig. 2b. This variation can be attributed to the influence of series resistance and interface trap charges affected by using the Si wafer's backside as a back contact [9]. The effect due to series resistance can be calculated from the following equation [10]:

$$R_s = \frac{G_m}{G_m^2 + (\omega C_m)^2}, \tag{1}$$

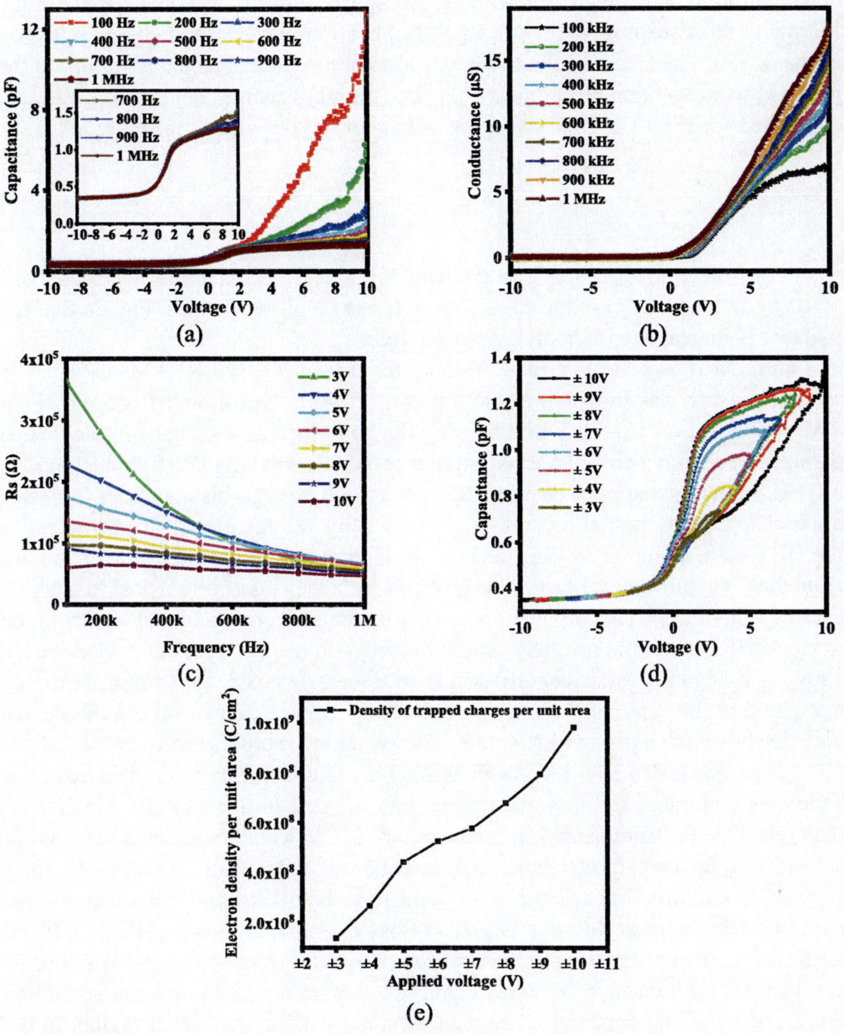

Fig. 2 Gd_2O_3 NR **a** C–V curve, inset: magnified C–V curve, **b** G–V curve, **c** series resistance for different biasing voltage, **d** C–V hysteresis, **e** trapped charge density variation with voltage

where ω = angular frequency, C_m and G_m = measured capacitance and conductance, the value of R_s from 100 kHz to 1 MHz for various biasing voltages of + 3 to + 10 V is depicted in Fig. 2c. The value of R_s is seen to decrease with the rise in frequency and vice versa, which is likely due to the interface state responding to AC signal response [4].

Further, C–V hysteresis at 1 MHz frequency was conducted (as shown in Fig. 2d) by voltage sweeping from ± 3 to ± 10 V. The devices showed a counter-clockwise behaviour loop which increases with the sweeping voltage. The enhancement in the loop with the sweeping voltage is a good indication of the strong charge injection in the region. The nature of the hysteresis loop of the C–V obtained is similar to the reported work by other researchers [11]. The trapping charges density (N) was also calculated for ± 3 to ± 10 V sweeping voltage using the following equation [4]:

$$N = \frac{C_a \times \Delta V}{qA}, \tag{2}$$

where q = electron charge, ΔV = flat band voltage, A = area of the electrode (3.14 × 10^{-2} cm^2), and C_a = oxide capacitance. It can be observed from Fig. 2e that the value of N increases with applying biased voltage.

Figure 3a shows the I–V measurement for the Gd_2O_3 NR RRAM device. The voltage biasing was provided at the top contact with the following order 0 V → 4 V → 0 V → −4 V → 0 V and keeping the bottom contact grounded. The device displayed a forming-free abnormal bipolar resistive switching (BRS) behaviour [5, 12]. The devices started with the LRS state when voltage biasing from 0 V was applied. With the gradual increment in the biasing voltage, the device switched to the HRS state, as shown in Fig. 3a. The conduction mechanism in the device arises from the slight movement of oxygen vacancies from their equilibrium position, which creates a dipole [5]. This dipole opposes the external electric field and results in the device's HRS state. Interestingly, as the biasing voltage lowered, the device started showing a negative resistance effect due to a larger electric field from the dipole compared to the external electric field, resulting in increased current in the device and slowly switching to the LRS state. The switching reproducibility of the Gd_2O_3 NR device was tested at + 1 V. As shown in Fig. 3b, the device showed an excellent switching endurance of up to 300 cycles with HRS/LRS ratio of ~ 19. Finally, the data retention measurement was conducted at + 1 V with a compliance current of 0.1 A. The device exhibits stable data retention of more than 10^3 s, as shown in Fig. 3c. In addition, the switching mechanism in both LRS and HRS regions was explored using a linear fitting in Log(I) vs Log(V) graph, as shown in Fig. 3d. In the LRS region, the linear fitting in the lower region with the value ~ 1 indicates ohmic conduction (I α V), which gradually changes to trap-controlled space charge limited conduction (SCLC) conduction denoted by slope ~ 4.6 in the higher region. In the HRS region, the conduction mechanism is mainly attributed to trap-controlled SCLC and got higher with the lower bias voltage due to the built-in electric field [12].

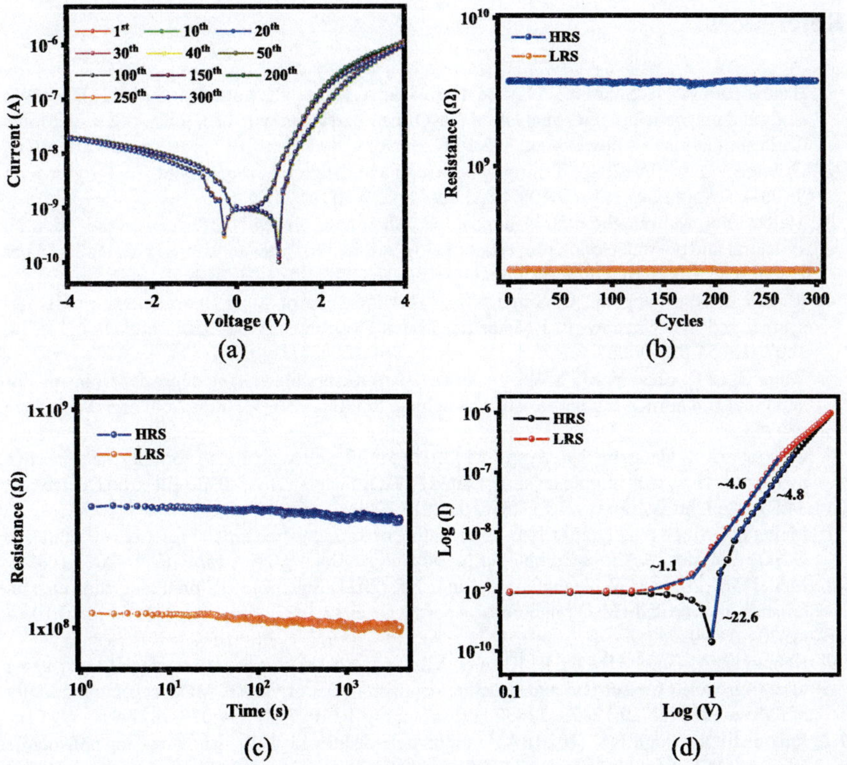

Fig. 3 Gd_2O_3 NR **a** I–V hysteresis, **b** endurance for 300 cycles, **c** retention, **d** linear curve fitting for LRS and HRS

4 Conclusion

In this study, we have successfully presented the deposition of Gd_2O_3 NR on a Si wafer using the GLAD technique for non-volatile memory application. The capacitive memory of the device was demonstrated using the C–V and G–V measurements. Also, the frequency-dependent C–V and G–V were attributed to R_s and interface states. The C–V hysteresis at 1 MHz shows a counter-clockwise behaviour loop with memory windows that increases with the increased voltage. Furthermore, it was revealed that the trapping charges density rises with the increasing biasing voltage. The Gd_2O_3 NR device exhibited an abnormal BRS with a conduction mechanism assigned to ohmic, trap-controlled space charge limited conduction. In addition, the device revealed a stable endurance with 300 cycles and good retention characteristics (10^3 s), making it suited for non-volatile memory applications.

Acknowledgements The authors would like to acknowledge NIT Nagaland for providing the facilities for electrical measurement.

References

1. Bandyopadhyay A, Sharma S, Nath M, Karmakar A, Kumari K, Sutradhar S (2021) Dielectric study and magnetic property analysis of Gd_2O_3 nanorods/nanowire in combination with Monte Carlo simulation. J Alloy Compd 882:160720. https://doi.org/10.1016/j.jallcom.2021.160720
2. Gribisch P, Fissel A (2020) Tuning of structural and dielectric properties of Gd_2O_3 grown on Si(001). J Appl Phys 128:055108. https://doi.org/10.1063/5.0007793
3. Meitei PhN, Moirangthem B, Ngangbam C, Alam MW, Singh NK (2022) Investigation on structural and photodetection properties of Gd_2O_3 thin films after annealing. J Mater Sci Mater Electron 33:10705–10714. https://doi.org/10.1007/s10854-022-08053-8
4. Rajkumari R, Ngangbam C, Singh NK (2021) Presence of capacitive memory in GLAD-synthesized WO3 nanowire. J Mater Sci Mater Electron 32:3191–3200. https://doi.org/10.1007/s10854-020-05067-y
5. Wang G, Li C, Chen Y, Xia Y, Wu D, Xu Q (2016) Reversible voltage dependent transition of abnormal and normal bipolar resistive switching. Sci Rep 6:36953. https://doi.org/10.1038/srep36953
6. Chakrabarti B, Galatage RV, Vogel EM (2013) Multilevel switching in forming-free resistive memory devices with atomic layer deposited $HfTiO_x$ nanolaminate. IEEE Electron Device Lett 34:867–869. https://doi.org/10.1109/LED.2013.2262917
7. Meitei PN, Singh NK (2023) Effect of annealing on forming-free bipolar resistive switching of Gd_2O_3 thin films. J Alloys Compound 168900. https://doi.org/10.1016/j.jallcom.2023.168900
8. Meitei PhN, Alam MW, Ngangbam C, Singh NK (2021) Enhanced UV photodetection characteristics of annealed Gd_2O_3 nanorods. Appl Nanosci 11:1437–1445. https://doi.org/10.1007/s13204-021-01787-7
9. Kahraman A, Gurer U, Lok R, Kaya S, Yilmaz E (2018) Impact of interfacial layer using ultra-thin SiO_2 on electrical and structural characteristics of Gd_2O_3 MOS capacitor. J Mater Sci Mater Electron 29:17473–17482. https://doi.org/10.1007/s10854-018-9847-9
10. Rajkumari R, Singh NK (2020) Ag nanoparticle-decorated WO_3 nanowires for nonvolatile memory. ACS Appl Nano Mater 3:12087–12094. https://doi.org/10.1021/acsanm.0c02584
11. Pandey AK, Deb P, Dhar JC (2021) Ag nanoparticles capped TiO_2 nanowires array based capacitive memory. Review. https://doi.org/10.21203/rs.3.rs-315901/v1
12. Guo B, Sun B, Hou W, Chen Y, Zhu S, Mao S, Zheng L, Lei M, Li B, Fu G (2019) A sustainable resistive switching memory device based on organic keratin extracted from hair. RSC Adv 9:12436–12440. https://doi.org/10.1039/C8RA10643F

Improved UV Detection Based on SnO_2 Nanowire Fabricated by Glancing Angle Deposition

Rosy Kimneithem Haokip, Biraj Shougaijam, and Manoj Kumar

1 Introduction

Large bandgap semiconductors namely TiO_2, SnO, In_2O_3, ZnO etc. have captivated wide attention in optoelectronic application like gas sensors [1], photodetectors [2] and solar cells [3] because of its distinctive electrical and optical properties [4] such as high on–off current ratio, swift response speed, high sensitivity as well as high thermal stability. SnO_2 is a significant n-type semiconductor having direct bandgap ~ 3.6 eV. It also exhibits high exciton binding energy ~ 130 meV and high electron mobility ~ 250 $cm^2\ V^{-1}\ s^{-1}$ [5] making it an ideal competitor in optoelectronic application [6]. 1-D structure such as nanowires, nanoarrays and nanorods have drawn large interest because of its large surface-volume ratio and surface configuration [7]. 1-D metal oxide NW increases photo-excited carriers transport efficiency, enhance photon absorption and photoconductivity of the device [8]. The vertical assembly of NW reduces noise by restricting minority carrier collection at the electrodes [9]. Shi et al. [10] reported single crystalline SnO_2 based sensor showing high external quantum efficiency and precise radiation selectiveness. Hu et al. [11] also detailed growth of high sensitivity high gain SnO_2 NW device. Many researchers also reportedly achieved growth of SnO_2 NWs by numerous deposition techniques such as RF Sputtering [12], chemical vapor deposition (CVD) [13], etc. However, not many research has been conducted on synthesis of SnO_2 NWs on p-Si substrate using Glancing angle deposition (GLAD) technique. GLAD is a straightforward, cost-efficient physical vapour deposition method utilized to synthesize 1D NWs by

R. K. Haokip (✉) · M. Kumar
Department of Electronics and Communication Engineering, National Institute of Technology Manipur, Imphal, Manipur, India
e-mail: rosy.kimneithem@gmail.com

R. K. Haokip · B. Shougaijam
Department of Electronics and Communication Engineering, Manipur Technical University, Imphal 795001, India

B. P. Swain (ed.), *Recent Advances in Materials*, Springer Proceedings in Materials 25,
https://doi.org/10.1007/978-981-99-3844-5_14

shadowing effect. In this technique, the SnO_2 deposition flux is incident to the p-Si substrate at 85°. The azimuthal rotation of substrate produces the 1-D vertically aligned structure [14]. This orderly growth morphology and surface chemistry can be adapted for improving photo detection efficiency [15].

In this work, SnO_2 thin film (TF) and NWs are successfully deposited on p-Si and glass substrate by utilizing physical vapour deposition based GLAD technique inside e-beam unit. The morphologic structural of deposited p-Si/ SnO_2 NW has been explored using field-emission scanning electron microscopy (FE-SEM) and energy-dispersive x-ray spectroscopy (EDX). The x-ray diffraction (XRD) analysis of deposited SnO_2 TF and NW was performed to investigate the crystallographic structure. The absorption spectrum was measured for SnO_2 NW and SnO_2 TF using UV–Vis Spectroscopy to examine the photocatalytic property. Lastly, the I–V and I–T characteristics of p-Si/SnO_2 TF and p-Si/SnO_2 NW were also investigated to study their electrical properties.

2 Experimental Procedure

Firstly, the *p*-type Si substrate (1 cm × 1 cm) and glass substrate (1 cm × 1 cm) is prepared using diamond cutter. Prior to deposition, p-Si is properly cleaned with acetone (1 min), methanol (1 min) and de-ionised water (1 min). Whereas glass substrate is cleaned with de-ionised water (1 min) before deposition. After cleaning, the substrates are placed on substrate holder at 24 cm distance from the evaporant source inside e-beam chamber. The SnO_2 TF was grown at constant deposition rate ~ 1.2 Å/s and base pressure ~ 2×10^{-5} mbar. For the growth of NW, the SnO_2 material is deposited at 30 rpm azimuthal rotation with 85° inclination between evaporant source SnO_2 and substrate holder [1] with the help of GLAD technique under same base pressure and growth rate.

The surface morphology of p-Si/SnO_2 NW is defined by field-emission scanning electron microscopy (FE-SEM) analysis. The chemical composition of deposited NW is analyzed using energy-dispersive x-ray spectroscopy (EDX). The x-ray diffraction (XRD) characterization of as-deposited SnO_2 TF and NW samples on glass substrate was also performed. The optical absorption examination was carried out using a UV–Vis Spectroscopy. The current (I)-voltage (V) and current (I)-time (T) switching characteristics of p-Si/SnO_2 TF and p-Si/ SnO_2 TF–NW devices under dark and UV light illumination was also investigated using Keithley 2450 source meter to study the electrical properties.

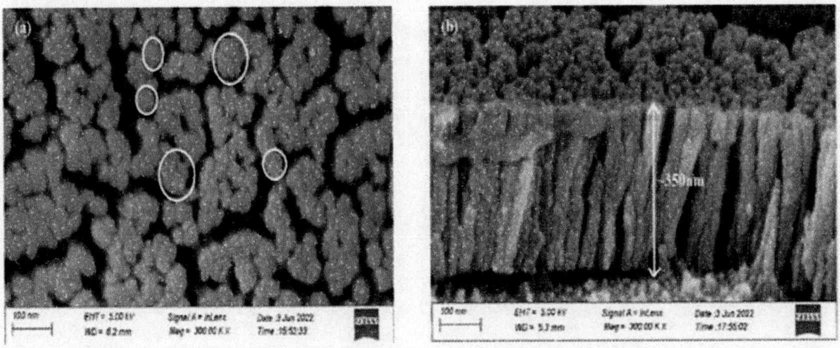

Fig. 1 **a** Top view of SnO_2 NWs **b** cross-sectional view of SnO_2 NWs

3 Results and Analysis

3.1 FE-SEM Analysis

Figure 1a and b exhibits the top view and cross-sectional view of the as-deposited 1D SnO_2 NW on Si substrate. The top view depicts the fabricated NW have an average top diameter of ~ 50 nm. Figure 1a also depicted larger diameter NW which is due to cluster formation. The cross-section view in Fig. 1b also reveal successful growth of uniform and perpendicularly positioned SnO_2 NW with average length ~ 350 nm. Some of NW were shorter length compared to adjacent NWs. This may be attributed to shadowing effect which is prevalent in GLAD technique. The perpendicular NW arrays have high surface-volume ratio, thus having the ability to trap incident light between two consecutive NW. This aspect of NW also attribute to increase absorption efficiency of the SnO_2.

3.2 EDX Analysis

The EDX analysis was performed on as-deposited p-Si/SnO_2 NW to study the chemical composition of elements present. The spectrum dispersion of EDX shows peaks corresponding to silicon (Si), oxygen (O_2) and tin (Sn) as depicted in Fig. 2a. The chemical mapping shown in inset of Fig. 2b also indicated presence of Si, Sn and O_2 in the sample as evident from the FE-SEM analysis. The highest peak revealed the presence of silicon from p-Si on which the SnO_2 NW was deposited.

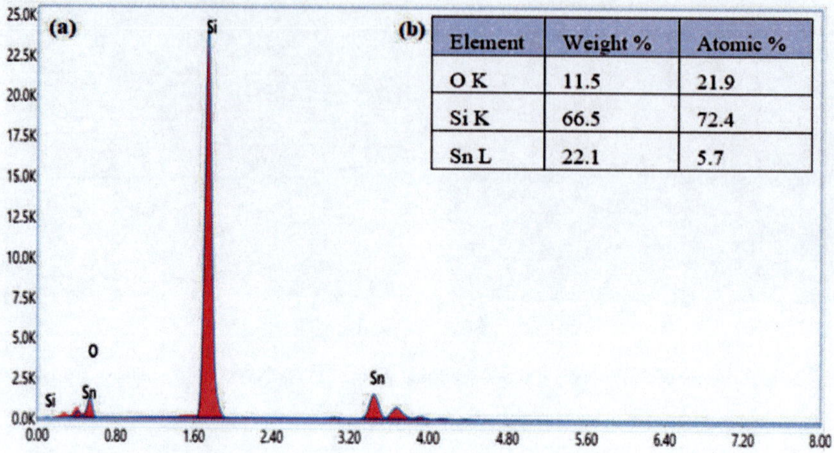

Element	Weight %	Atomic %
O K	11.5	21.9
Si K	66.5	72.4
Sn L	22.1	5.7

Fig. 2 **a** EDX spectrum of SnO_2 NWs on Si substrate **b** inset shows chemical mapping

3.3 XRD Analysis

The anatomical properties of as-deposited SnO_2 TF and NW samples deposited on glass substrate were explored using XRD analysis. As exhibited in Fig. 3, it is noted that both the samples show no significant peak. Similar XRD results have been affirmed by Tareq et al. [16], Omar et al. [17] and Chetri et al. [18] using different deposition techniques indicating that both the as- deposited SnO_2 TF and NW samples are amorphous in nature.

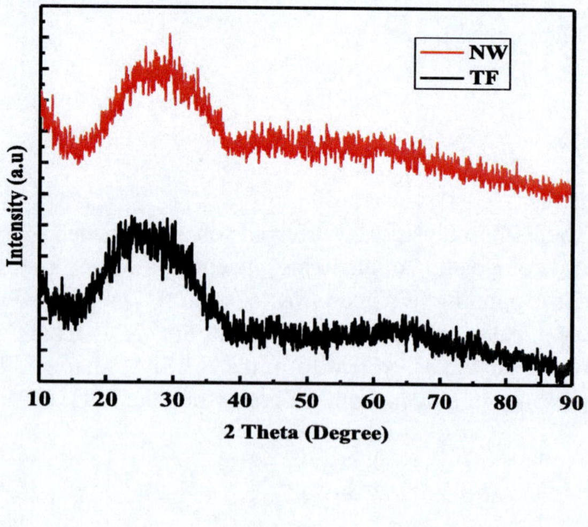

Fig. 3 XRD analysis of as-deposited SnO_2 TF and NW

3.4 Absorption Analysis

To analyse the optical properties of SnO_2 TF and NW structure, UV–Vis spectral study was performed. Figure 4a and b indicated sharp absorption edge in the UV region under 200 nm to 900 nm irradiation at room temperature. Similar result was reported by Wiktor et al. [19]. This property of SnO_2 may be useful in UV photodetector applications. Inset of Fig. 4a and b indicates the $(\alpha h\nu)^2$ versus energy (hν) curve of TF and NW attained from the absorption data. The absorption coefficient α is given by $\frac{4\pi k}{\lambda}$ where k is the absorption index and hν implies the photon energy. The optical bandgap is obtained by extrapolating linear segment of the plot to the energy axis. The bandgap thus calculated is ~ 3.23 eV for SnO_2 TF and ~ 3.7 eV for SnO_2 NW sample, as seen from the inset of Fig. 4a and b. The bandgap ~ 3.23 eV for SnO_2 TF may be ascribed to small particle size and amorphous nature of the deposited sample [19, 20], whereas band gap ~ 3.7 eV in SnO_2 NW is attributed to increase light scattering and trapping. The large surface-volume ratio of vertically positioned NW reported in the FE-SEM results is responsible large photon trapping thereby increasing the photocatalytic activity. Similar results were also reported for SnO_2 [18, 21].

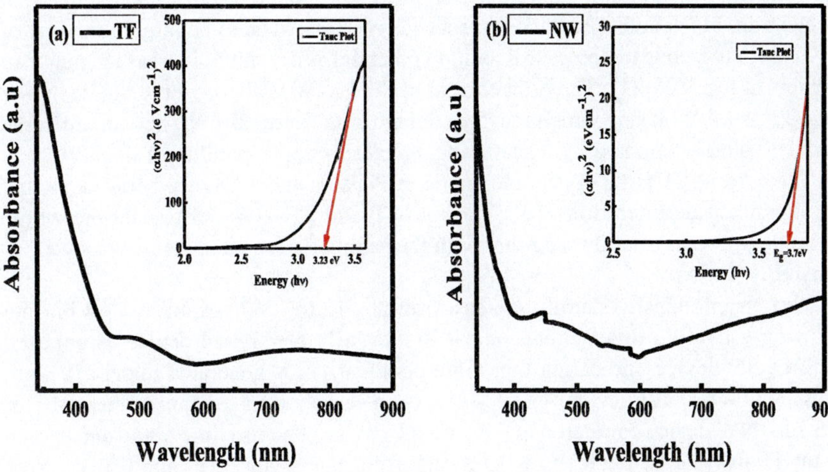

Fig. 4 **a** Absorption graph of SnO_2 TF and the inset presents the related Tauc plot [$(\alpha h\upsilon)^2$ versus $(h\upsilon)$] and **b** absorption graph of SnO_2 NW and the inset presents the related Tauc plot $(\alpha h\upsilon)^2$ versus $(h\upsilon)$)

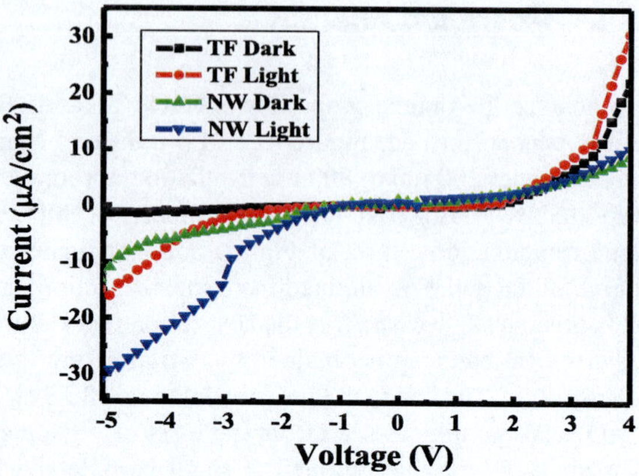

Fig. 5 I–V characteristic of SnO_2 TF and SnO_2 TF–NW

3.5 *Electrical Analysis*

Figure 5 shows is current (I)-voltage (V) characteristic of SnO_2 TF and NW under UV light and dark irradiation. The photocurrent increases rapidly for both the TF and NW under UV irradiation due to transition of valance band electrons and generation of electron–hole pairs. The dark current of NW is observed to be larger than that of TF. This may be due to decrease in built in potential as a result of increase in majority carriers of the NW [21]. The light current of SnO_2 NW is also observed to be higher than TF at −4 V biasing which may be credited to the vertically aligned nature, large surface-volume ratio and light scattering and absorption capability of the NW.

The current (I)-Time (T) characteristic in Figs. 6a and 7a showed consistent and no significant degeneration of I_{ph} for both TF and TF–NW throughout the whole switching process, manifesting that both the samples have good stability as well as reliability [21].

The current density sharply increases from ~ 12 to ~ 60 μA/cm^2 at −4 biasing showing ~ fivefold times enhancement in the SnO_2 NW based device as opposed to SnO_2 TF device, indicating that more photocurrent is generated under UV light illumination for NW device. The I_{ph}/I_{dark} ratio also showed enhanced sensitivity of ~ 6.7 for NW device compared to ~ 4.9 for TF device. The rise time t_r and decay time t_d for TF–NW SnO_2 device is ~ 0.1 s each from Fig. 7 (inset (b) and (c)) showing fast response detection compared to ~ 0.35 and ~ 0.45 s of SnO_2 TF device [Fig. 6 inset (b) and (c)].

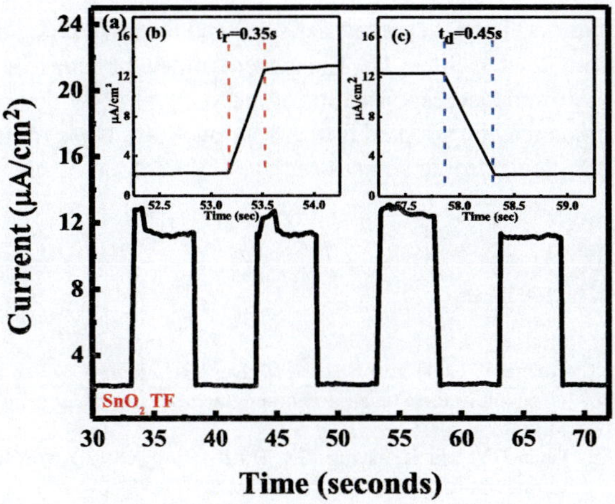

Fig. 6 **a** I–T characteristic of SnO_2 TF **b** rise time (inset) **c** decay time (inset)

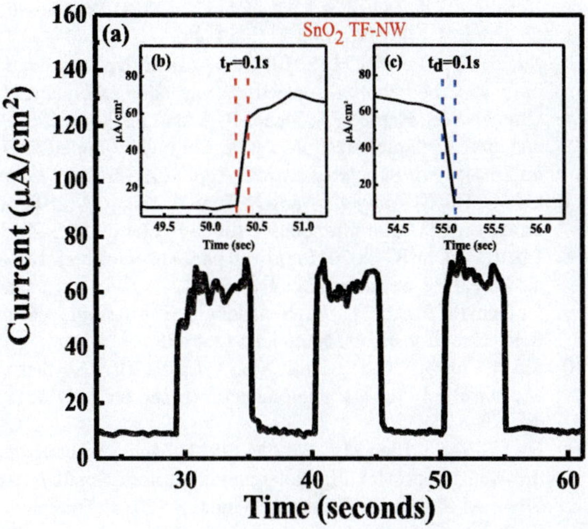

Fig. 7 **a** I–T characteristic of SnO_2 TF–NW **b** rise time (inset) **c** decay time (inset)

4 Conclusion

Tin(IV) Oxide SnO_2 thin film was successfully deposited on p-Si and glass substrate using physical vapour deposition inside an e-beam chamber. Indegineous Glancing Angle Deposition technique was successfully employed for deposition of vertically grown SnO_2 NW. FE-SEM analysis showed successful growth of the NW on glass substrate which is supported by the EDX analysis. The XRD report showed amorphous nature of the SnO_2 samples. The absorption analysis indicated sharp absorption peak for TF and SnO_2 NW in the UV region. Straight line extrapolation of Tauc Plot reported bandgap $E_g \sim 3.23$ eV (TF) and ~ 3.75 eV (NW) which is supported by many authors. The IV characteristic showed improved I_{photo} and I_{dark} for SnO_2 TF–NW as compared to SnO_2 TF. The current–time characteristic also showed improved sensitivity and fast response time of the SnO_2 TF–NW device. This GLAD synthesis technique may be adopted further to grow SnO_2 NWs for their application in fabrication of optoelectronic devices such as UV Photodetectors.

References

1. Huang Y, Li Y, Yao B, Ding Z, Deng R, Zhang L, Zhao H (2015) A facile route to realize ultraviolet emission in a nano-engineered SnO_2 based light emitting diode. J Phys D Appl Phys 48:465103
2. Chen HY, Liu H, Zhang ZM, Hu K, Fang XS (2016) Nanostructured photodetectors: from ultraviolet to terahertz. Adv Mater 28:403–433
3. Chen Z, Lai JKL, Shek CH, Chen H (2017) Synthesis and structural characterization of rutile SnO_2 nanocrystals. J Mater 18:1289–92
4. Karthik TVK, Martinez L, Agarwal V (2018) Porous silicon ZnO/SnO_2 structures for CO_2 detection. J Alloys Compd 731:853–863
5. Snaith HJ, Ducati CH (2010) SnO_2-based dye-sensitized hybrid solar cells exhibiting near unity absorbed photon-to-electron conversion efficiency. Nano Lett 10(4):1259–1265
6. Othmen WB, Hamed ZB, Sieber B, Addad A, Elhouichet H, Boukherroub R (2018) Structural and optical characterization of p-type highly Fe-doped SnO_2 thin films and tunneling transport on SnO_2: Fe/p-Si heterojunction. Appl Surf Sci 434:879–890
7. Cao C, Hu C, Wang X, Wang S, Tian Y, Zhang H (2011) UV sensor based on TiO_2 nanorod arrays on FTO thin film. Sensor Actua B-Chem 156:114–119
8. Chetri P, Dhar JC (2020) Improved photodetector performance of SnO_2 nanowire by optimized air annealing. Semicond Sci Technol
9. Graham R, Yu D (2012) High carrier mobility in single ultrathin colloidal lead selenide nanowire field effect transistors. Nano Lett 12:4360
10. Shi H, Cheng B, Cai Q, Su X, Xiao Y, Lei S (2016) Surface state controlled ultrahigh selectivity and sensitivity for UV photodetectors based on individual SnO_2 nanowires. J Mater Chem C 4:8399–8406
11. Hu LF, Yan J, Liao MY, Wu LM (2011) Fang XS Ultrahigh external quantum efficiency from thin SnO_2 nanowire ultraviolet photodetectors. Small 7:1012–1017
12. Singh A, Sharma A, Tomar M, Gupta V (2018) Tunable nanostructured columnar growth of SnO_2 for efficient detection of CO gas. Nanotechnology 29(065502):1–12
13. Lou Z, Yang X, Chen H, Liang Z (2018) Flexible ultraviolet photodetectors based on ZnO–SnO_2 heterojunction nanowire arrays. J Semicond 39(2):024002
14. Haokip RK, Shougaijam B (2022) Morphological and optical analysis of GLAD synthesize vertically oriented TiO_2 nanowires on GaN substrate for highspeed photodetector applications, micro and nanoelectronics, devices, circuits and systems, pp 307–313
15. Shougaijam B, Swain R, Ngangbam C, Lenka TR (2016) Enhanced photodetection by glancing angle deposited vertically aligned TiO_2 nanowires. IEEE Trans Nanotechnol 389–394
16. Rahman MT, Ahmed Z, Islam MJ, Kamaruzzaman MT, Khatun MA, Gafur MA, Bashar MS, Alam MM (2021) Comparative study of structural, optical and electrical properties of SnO_2 thin film growth via CBD, drop-cast and dip-coating methods. Mater Sci Appl 12(12)
17. Kassem O, Saadaoui M, Rieu M, Sao Joao S (2018) Synthesis and inkjet printing of sol–gel derived tin oxide ink for flexible gas sensing application. J Mater Sci 53(18)
18. Chetri P, Dhar JC (2019) Au/GLAD-SnO_2 nanowire array-based fast response Schottky UV detector. Appl Phys A 125(5)

19. Matysiak W, Tański T, Smok W, Polishchuk O (2020) Synthesis of hybrid amorphous/crystalline SnO_2 1D nanostructures: investigation of morphology, structure and optical properties. Nat Res
20. Hassun HK, Hussein BH, Salman EMT, Shaban AH (2020) Photoelectric properties of SnO_2: Ag/P–Si heterojunction photodetector. Energy Rep 6
21. Pan X, Zhang T, Lu Q, Wang W, Ye Z (2019)High responsivity ultraviolet detector based on novel SnO_2 nanoarrays. RSC Adv 9:37201

Biodiesel Production by Non-edible Cascabela Ovata Seeds Through Solvent Methods

M. S. Abishek, Sabindra Kachhap, and Pukhrambam Sunilkhumar Singh

1 Introduction

Renewable energy is a must in the coming future, because of endangering fossil fuel sources. Some sources of renewable energy are 1. hydroelectricity: It uses water motion to generate electricity. The most common method of power hydroelectricity is by processing and controlling the flow of water through a dam. 2. Solar: Solar energy is the light and heat of the sun that is harnessed using a range of technologies such as solar heating, solar thermal energy, and solar architecture. 3. Wind power: It is produced by air flowing through a wind turbine and hence converting the mechanical motion of the turbine to electric energy. 4. Biofuel: Any fuel that is derived from biomass, i.e., plants or algae materials or animal waste utilized hydroxide as a catalyst. The oil of Silybum Marianum is transesterification with methanol to produce biodiesel. They used catalyst as ionic, but they exchange between KF (potassium fluoride) and H_2SO_4 attapulgite by dihydroxylation of the attapulgite at 130 °C for 3 h and subsequent activation of KF at atmospheric temperature and pressure which was followed by calculation at 400 °C for 5 h. The catalyst was 12 wt.% in a 5-h reaction with an ideal temperature of 65 °C. Under these conditions, the best yield was 93.3% [1]. Even though ethanol has a high-water content of 15 wt.%, a high-fatty acid (HFA) conversion more than 90% is produced by optimizing the appropriate reaction time in both systems. It was also concluded that catalytic activity was maintained over time [2]. The biodiesel obtained meets the American Society for Testing and Materials (ASTM) D6751 standard and can be used to prevent food-use fuel competition, making it a suitable alternative to petrol-derived diesel [3]. For the homogeneous reaction, the reaction temperature is 150 and 180 °C, and the ratio of oil to methanol is 1:6 mol/mol with NaOH of 1.0 wt.% as a catalyst. For the heterogeneous reaction, the reaction temperature is 150 °C and 180 °C, and the ratio of oil to methanol is

M. S. Abishek (✉) · S. Kachhap · P. S. Singh
Department of Mechanical Engineering, NIT Manipur, Imphal 795004, India
e-mail: abishek.nitm@gmail.com

B. P. Swain (ed.), *Recent Advances in Materials*, Springer Proceedings in Materials 25,
https://doi.org/10.1007/978-981-99-3844-5_15

1:8 and 1:10, respectively. He used of catalyst amount are 5% and 4 wt.% [4]. The primary oxidation reaction takes place between 100 and 320 °C. Castor pyrolytic acid is likewise discovered to be multi-component, with a high viscosity, a wide boiling range, and low water content. If the ignition temperature was 328 °C, the burnout temperature was 513 °C, and the index of combustion characteristic was 1.991 [5]. The efficient and effective catalyst chosen for thc proccss was KOH. The maximum optimum conditions of biodiesel yield obtained at 91.76% were 0.32% of methanol and KOH 1.5% of catalyst at the temperature of 60 °C during 90 min reaction time. Along with the results predicted by Response Surface Methodology, 88.7% of biodiesel is obtained in the validation experiments fitting 96.6% of the result [6]. During the research, the temperature ranged from 23.5 to 28 °C, with a humidity of 65%. Light duty diesel (LDD) cars have higher CXHY and CO levels than the standard permitted norms, with average values of 430 and 465.4 ppm, respectively, whereas NOx has had an overall average of 99.2 ppm [7]. The results revealed that a biodiesel mixture using 20% apricot oil in diesel performed better and had lower emissions than other different percentages [8]. When compared to diesel fuel, the B20 fuel blend improves BTE by 4.7%, increases CO_2 emissions by 2.56%, and reduces SFC by 7.92%. When compared to diesel fuel, the biodiesel blend (B20) has the biggest reduction in NO_X by 14.9% and particle by 4.22%; however, smoke emission somewhat increases with an increase in fish oil in the blends [9]. Catalysts are important in the transesterification process. Because of their renewability and ease of separation, heterogeneous catalysts have gained popularity in recent years. The utilization of renewable resources to create catalysts has improved the usage of heterogeneous catalysts [10].

1.1 Selected Raw Materials

Cascabela Ovata is a flowering plant. In Manipur, we are called as Utong-Lei, and in Hindi, we are called Pile Kaner. It can be grown up to 12 ft, and its flower is 2–5 mm in size of dia. The color of this flower is yellow, and it is bloom in the summer season. This fruit is dark red–black color, and its size is 4–7 mm dia. It is mainly found in various states of India even the northeast state of Manipur also. It grows in drought tolerance to the high temperature. It is also mainly planted in the valley areas as garden plans (Figs. 1 and 2).

Fig. 1 Cascabela Ovata (Utong-Lei)

Fig. 2 Cascabela Ovata

2 Experimental and Methodology

The experiment was done to extract oil from the raw seed for transesterification process so that the biodiesel and glycerol can be separated. The oil characterization was done to find the properties of oil extracted.

Cascabela Ovata is known as "Utong-Lei" in the local name. It is found almost in all parts of Manipur as a garden plant. The fruits have to be used for oil extraction and accumulated from the local surrounding of Manipur. The collected seeds are accurately weight and noted down. The oil containing seeds extracted from its fruits by using a special mechanism. It is a soft fleshy nut. The size of oil containing seeds is measured. The removed seeds are checked for weight using Electronic Compact Scale. This is done for knowing how much oil we can extract from a known amount of seeds. Ceramic mortar and pestle are used to crush down the Cascabela Ovata seeds fibers into fine particles. It gets a better yield and also time consumed for the extraction. Then oil extraction is done using Soxhlet Apparatus using N-hexane.

2.1 *Oil Preparation*

See Figs. 3, 4, 5, 6, 7 and 8.

2.2 *Oil Characterization*

To determine free fatty acid (FFA) content in the oil. If the value of FFA is too high, then the catalyst reaction will be from soap which can prohibit the yield of methyl ester for the present study. The oil value was found 1.46 mg NaOH which is the alkaline limit and is transesterification. Transesterification is done using methanol and NaOH as a catalyst (Figs. 9, 10 and 11).

Fig. 3 Weight of the fruits

Fig. 4 Remove fibers from the fruits

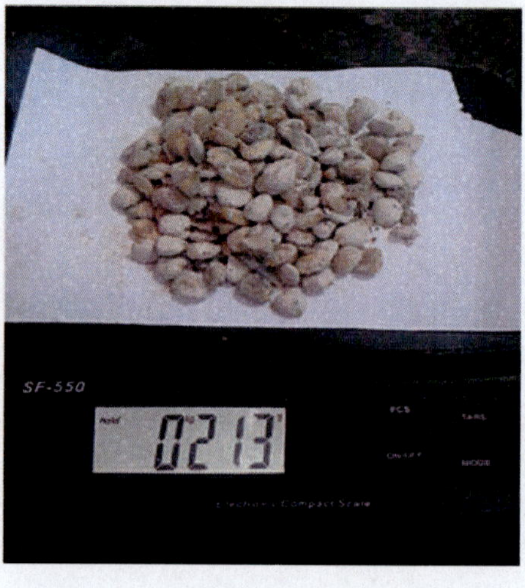

Fig. 5 Weight of the oil-content seeds

Fig. 6 Crushing seeds fibers into fine powder

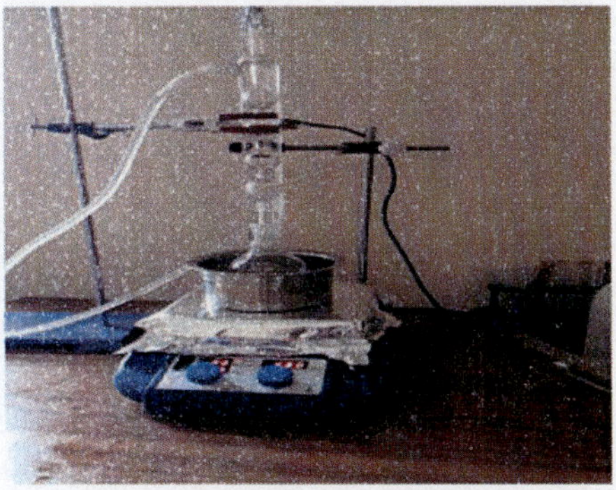

Fig. 7 Extraction of oil

Fig. 8 Removal of excess hexane using hot plate

Fig. 9 Separation of glycerol and methyl ester after transesterification

Fig. 10 Methyl ester

3 Results and Discussion

3.1 Weight Percentage of Oil

We consumed a total weight of 375 gm of oil containing seeds, and from that, we extracted 79 gm of oil with that it makes an oil yield percentage of about 68.8%. Comparing yield % between the present solvent study and previous literature on hydraulic [11] and electric [12] press methods as shown in Fig. 12:

$$\text{Oil yield} = \frac{\text{weight of oil produced}}{\text{weight of sample used}} \times 100 \tag{1}$$

Fig. 11 Glycerol

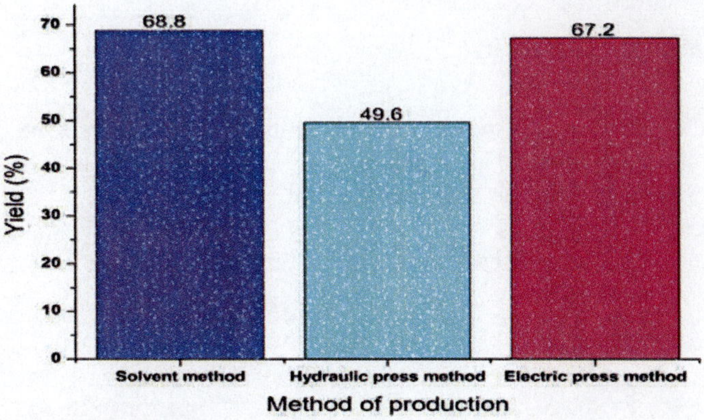

Fig. 12 Yield percentage of oil in different methods

3.2 Specific Gravity

Using the below mention formula, the specific gravity of oil is found to be 0.887 kg/ m^3 at 20 °C. Comparing specific gravity in the present solvent study and previous literature on hydraulic and electric [12] press methods as shown in Fig. 13:

$$\text{Specific Gravity} = \frac{\text{Oilfilledwt.} - \text{Emptywt.}}{\text{Distilwaterfilledwt.} - \text{Emptywt.}} \tag{2}$$

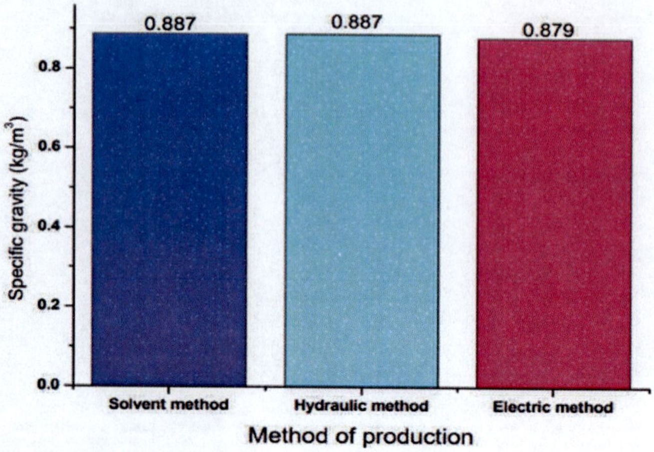

Fig. 13 Specific gravity of oil in different methods

3.3 Kinematic Viscosity

By using the below formula, the kinematic viscosity of the oil is found to be 5.78 mm^2/s at 40 °C and also compared in the present study and previous literature on hydraulic and electric press methods as shown in Fig. 14.

$$\text{Kinematic viscosity} = (\text{Time} \times \text{Tube constant})\ mm^2/s \tag{3}$$

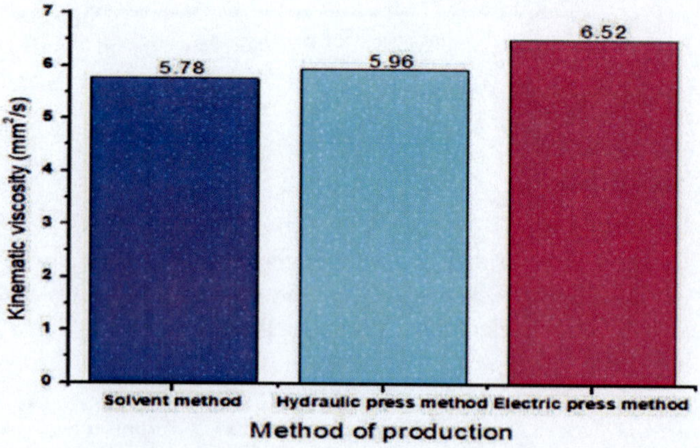

Fig. 14 Kinematic viscosity of different methods of producing oil

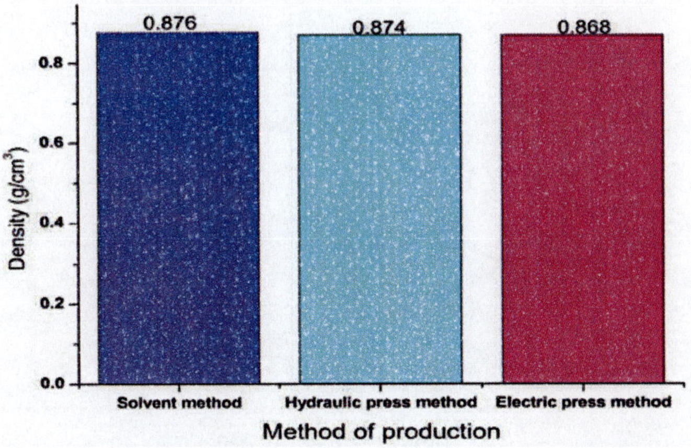

Fig. 15 Density of the oil using different methods

3.4 *Density of Oil*

Using the below-mentioned formula, we found the density of the oil as 0.876 g/cm^3. Comparing density between the present solvent study and previous literature on hydraulic [11] and electric [12] press method as shown in Fig. 15:

$$\text{Density of oil} = \frac{\text{Mass of oil}}{\text{Volume of oil}} \tag{4}$$

3.5 *Flash Point*

Cleveland Open cup flashpoint test was used to measure the flashpoint. It is the temperature at which the fuels ignite when exposed to flame. For biodiesel, the average flashpoint is 150 °C. According to the present study, the value of the flashpoint is 96 °C. Comparison of flashpoints in the present solvent study and previous literature on hydraulic [11] and electric [12] press method are shown in Figs. 16, 17 and 18.

3.6 *Free Fatty Acid*

FFA content is obtained by using the formula given by the equation.

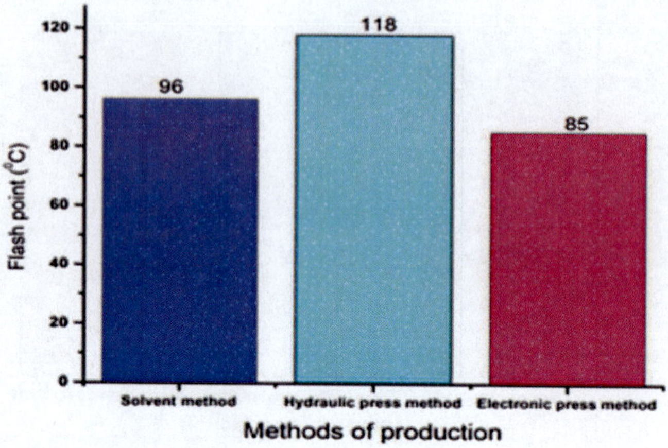

Fig. 16 Flashpoint of different methods of producing oil

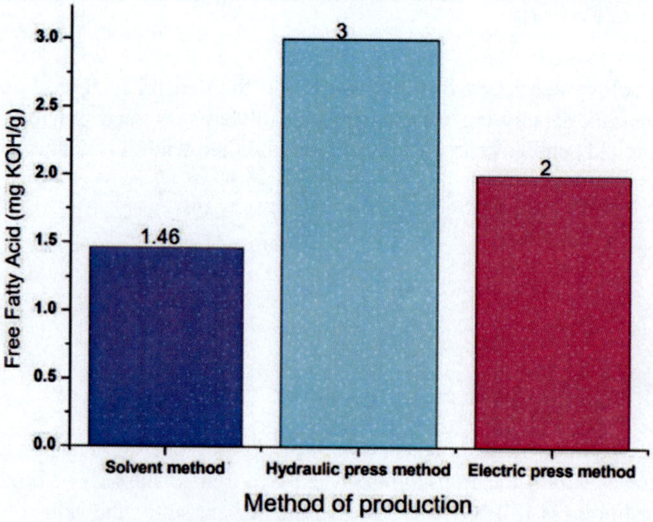

Fig. 17 FFA using different methods of producing oil

$$\text{FFA Content} = \frac{28.2 \times (\text{normality of NaOH xtitration value})}{\text{weight of oil(in gm)}} \tag{5}$$

$$\text{FFA Content} = \frac{28.2 \times (0.1 \times 5.2)}{10} \tag{6}$$

$$\text{FFA Content} = 1.46\% \tag{7}$$

3.7 Fourier Transform Infrared Spectroscopy (FTIR)

FTIR let us know which group the testing sample belongs to. Here is this study by analyzing the graph. We can check for ester groups whether they are strongly stretched or not. In this study, from Figs. 18 and 19, the oil characteristic peaks are found in the range of 3000.95–3008.05 cm^{-1} due to O–H stretching vibration and at 2913.16–2999.26 cm^{-1} due to alkenes C–H stretching vibration. Peaks of C=O stretching vibration of triglyceride ester appear at 1738.65–1743.72 cm^{-1} for the atomic compound of C–H bending at 1451.18–1452.34 cm^{-1}. Also, peaks at 1152.26–1149.15 cm^{-1} are due to C–O–C stretching vibration of esters and that of 729.38–728.71 cm^{-1} due to methylene rocking vibrations are also observed.

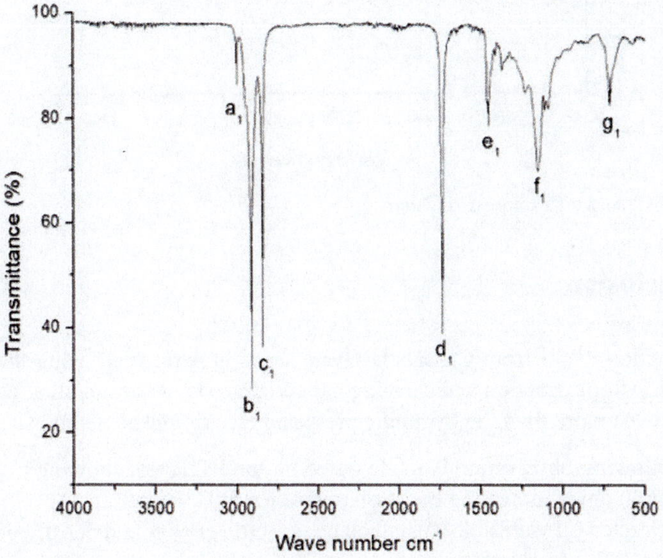

Fig. 18 FTIR test of CO obtained through the solvent

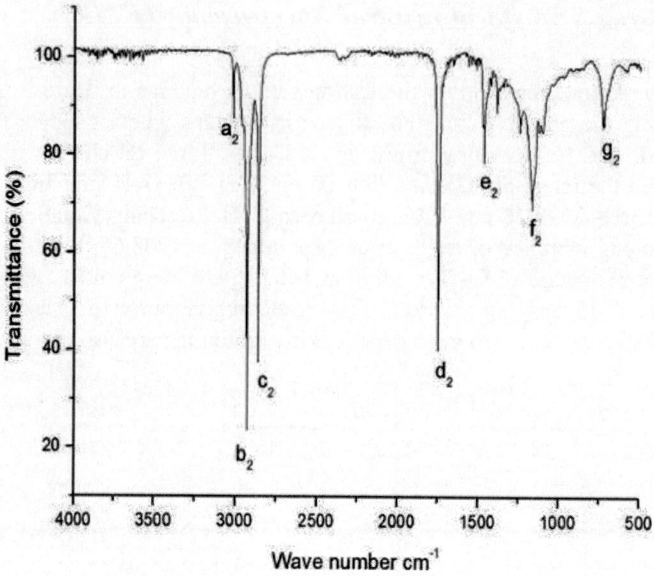

Fig. 19 FTIR tests after transesterification

4 Conclusion

The extraction of oil from Cascabela Ovate seeds is performed using the Soxhlet apparatus. After extraction and carrying out certain tests, it is found that the solvent method yields more than the hydraulic press and electric press methods.

- The following observation is made based on the FTIR test showing a stretching at 1743.65 cm^{-1} indicating the strong presence of C=0.
- The extracted oil yields 68.8%, which has specific gravity and density of 0.88 kg/m^3 and 0.876 g/cm^3, respectively.
- The amount of fatty acid (FFA) is 1.46% along with kinematic viscosity of 5.78 cSt.
- The extracted oil has a fire point of 96 °C which shows that feedstock used has the potential to be a biofuel.
- As we know, biofuel is a much cleaner and eco-friendly fuel as compared with fossil fuel. Also, by using biofuel, we can reduce environmental pollution, thus giving us a cleaner and safer environment.

4.1 Scope for Further Research

For future works, Cascabela Ovate can be extracted using the expeller press method, and it can find the characteristic of Gas Chromatography Mass Spectrometer (GCMS), Scanning Electron Microscopy (SEM) and Engine Performance of the oil.

Acknowledgements I am thankful for Department of Chemistry, NIT Manipur for providing necessary chemicals and instruments for testing of fuel.

References

1. Takase M, Nii A, Pappoe M, Afrifa EA (2018) High performance heterogeneous catalyst for biodiesel production from non-edible oil. Reinf Plast 25(00):24–30. https://doi.org/10.1016/j.ref.2018.03.002
2. Siddique N, Suzue M, Kato M, Hiromori K, Shibasaki-kitakawa N (2020) Process optimization for continuous production of sustainable biodiesel from completely non-edible biomass, lignocellulosic hydrous ethanol and waste fatty acids. Fuel 289:119884. https://doi.org/10.1016/j.fuel.2020.119884
3. Kafuku G, Kee M, Kansedo J, Teong K, Mbarawa M (2010) Bioresource technology croton megalocarpus oil: a feasible non-edible oil source for biodiesel production. Bioresour Technol 101(18):7000–7004. https://doi.org/10.1016/j.biortech.2010.03.144
4. Kansedo J, Lee KT, Bhatia S (2009) Cerbera odollam (sea mango) oil as a promising non-edible feedstock for biodiesel production. Fuel 88(6):1148–1150. https://doi.org/10.1016/j.fuel.2008.12.004
5. Chen G, Li Y, Lan C, Lin H, Chao Y (2017) Micro-explosion and burning characteristics of a single droplet of pyrolytic oil from castor seeds. Appl Therm Eng 114:1053–1063. https://doi.org/10.1016/j.applthermaleng.2016.12.052
6. Muthukumaran C, Praniesh R, Navamani P, Swathi R, Sharmila G, Manoj N (2017) Process optimization and kinetic modeling of biodiesel production using non-edible Madhuca indica oil. Fuel 195:217–225. https://doi.org/10.1016/j.fuel.2017.01.060
7. Singh TS et al (2022) Exhaust emission characteristics study of light and heavy-duty diesel vehicles in India. Case Stud Therm Eng 29:101709. https://doi.org/10.1016/j.csite.2021.101709
8. Karishma SM, Dasore A, Rajak U, Verma TN, Rao KP, Omprakash B (2021) Experimental examination of CI engine fueled with various blends of diesel-apricot oil at different engine operating conditions. Mater Today Proc 49:307–310. https://doi.org/10.1016/j.matpr.2021.02.105
9. Sharma DK, Verma TN (2020) Characteristics of fish oil biodiesel with the impact of diesel fuel addition on a ci engine. J Comput Appl Res Mech Eng 10(1):245–256. https://doi.org/10.22061/jcarme.2019.4737.1571
10. Dwivedi G, Jain S, Shukla AK, Verma P, Verma TN, Saini G (2022) Impact analysis of biodiesel production parameters for different catalyst. Environ Dev Sustain 0123456789. https://doi.org/10.1007/s10668-021-02073-w
11. Wali I, Naeem A, Farooq M, Ud I, Ali Z (2021) Reusable Na-SiO^2 @ CeO_2 catalyst for efficient biodiesel production from non-edible wild olive oil as a new and potential feedstock. Energy Convers Manage 231:113854. https://doi.org/10.1016/j.enconman.2021.113854
12. Roschat W, Siritanon T, Yoosuk B, Sudyoadsuk T, Promarak V (2017) Rubber seed oil as potential non-edible feedstock for biodiesel production using heterogeneous catalyst in Thailand. Renew Energy 101:937–944. https://doi.org/10.1016/j.renene.2016.09.057

RSM Optimization of Biodiesel from Waste Cooking Oil Using Snail Shell Derived Heterogeneous Catalyst

Wangkhem Robinson Singh and Huirem Neeranjan Singh

1 Introduction

The need for biofuels has emerged as a result of the imminent threat of the depletion of fossil fuel, and as per statistics [1], the proven reserve of oil, natural gas, and coal will last for 54, 49, and 139 years respectively. When compared to fossil fuels, biodiesel mostly consists of fatty acid methyl esters (FAME) and provides several environmental advantages, such as low CO_2 and CO emissions, low sulfur content, and biodegradable nature [2]. Additionally, biodiesel can be combined with diesel and utilized in engines without any modifications or with minimal changes [3]. The presence of a catalyst in the transesterification of biodiesel accelerates the reaction time [4]. Even though researchers claim that acid transesterification produces a better yield, it is constrained by the higher catalyst cost. Commercially, sodium hydroxide and potassium hydroxide are frequently used as a catalyst for transesterification as it is affordable and widely accessible. But the fundamental problem with homogeneous catalysts is that it leads to saponification when used with high free fatty acid (FFA) feedstock such as WCO and also produce toxic water waste while washing [5]. Although acids can esterify FFA, the process is less desirable due to its sluggish rates and high cost. Many researchers have already identified potential sources such as chicken egg shells, scallop shells, animal bones, etc. as a source for base heterogeneous catalysts [6]. Previously, researchers claimed that heterogeneous catalysts produced higher ester conversion yields than homogeneous catalysts [7].

When snail shell is utilized as a heterogeneous catalyst during the synthesis of biodiesel, it is stated to have a conversion rate of 92.5% when used cooking oil is used as the feedstock [8]. As a result, it is crucial in developing a low-cost, environmentally friendly catalyst that can overcome the shortcomings of base homogeneous and acid

W. Robinson Singh (✉) · H. Neeranjan Singh
Department of Mechanical Engineering, National Institute of Technology Manipur, Langol, Imphal, Manipur 795004, India
e-mail: wangkhem@gmail.com

B. P. Swain (ed.), *Recent Advances in Materials*, Springer Proceedings in Materials 25,
https://doi.org/10.1007/978-981-99-3844-5_16

Fig. 1 **a** Shell of ningkhabi tharoi (*Bellamya crassa*), **b** lai tharoi (*Thiara* tuberculate) and **c** labuk tharoi (*Angulyagra oxytropis*)

catalysts. Therefore, given the above-mentioned facts, three different varieties of snail shells namely labuk tharoi (*Angulyagra oxytropis*), ningkhabi tharoi (*Bellamya crassa*), and lai tharoi (*Thiara tuberculate*) were selected in this study to produce calcium oxide heterogeneous catalyst, and the best catalyst was chosen out of the three based on crystal size and elemental composition.

2 Experimental Methodology

2.1 Materials

Five liters (5 L) of waste cooking oil and two kilograms (2 kg) of Waste snail shells known locally as labuk tharoi (*Angulyagra oxytropis*), ningkhabi tharoi (*Bellamya crassa*), and lai tharoi (*Thiara tuberculate*) were collected from hotels situated in Imphal, Manipur, as shown in Fig. 1. Chemicals utilized throughout the experiment were procured locally from Imphal, India and of standard grade.

2.2 Catalyst Preparation

To remove organic materials from the snail shells, the shells were washed several times with distilled water and then sundried for 7 days. The snail shells were grounded separately using ceramic mortar and pestle and sieved through a mesh size of 200–250 μm. The powdered snail shells were calcinated separately in a muffle furnace at a temperature range of 800–1000 °C for a time interval of 3–5 h. A total of 27 catalyst samples, were obtained. Figure 2 shows nine calcinated samples for one type of snail shell.

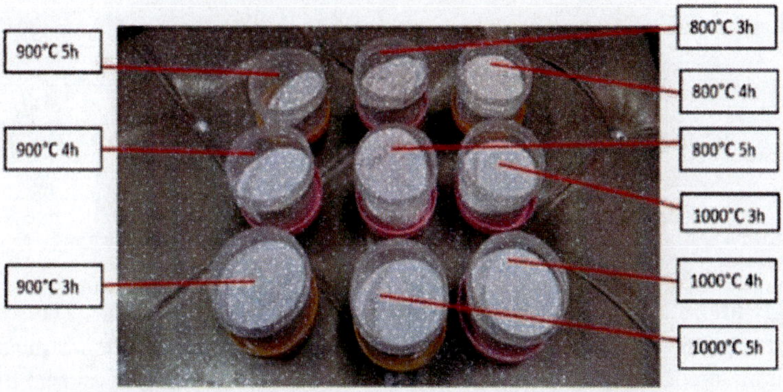

Fig. 2 Snail shell calcinated at a temperature of 800, 900, and 1000 °C and time of 3, 4, and 5 h

2.3 Characterization of the Catalysts

X-ray diffractometer (XRD), Fourier transform infrared spectroscopy (FTIR), and Energy dispersive X-ray (EDX) were performed for raw and calcined snail shells to analyze the crystal structure, chemical makeup, and elemental composition of the catalyst. The XRD diffractions patterns of the CaO samples were recorded using Bruker AXS; Model: D8-Advance powder XRD equipped with 1 $\lambda = 1.5406$ Å of Cu-Kα radiation over the Bragg angle 2θ range of 20°–80° and a step size of 0.04° at room temperature. FTIR was performed for uncalcined and calcined snail shells using PerkinElmer; Model: Spectrum Two FTIR spectrometer to investigate the functional group present in the samples in the spectrum range of 2200–400 cm^{-1}.EDX analysis was carried out for both uncalcined and calcined snail shells to determine the elemental composition.

2.4 Transesterification Reaction and FAME Analysis

The Waste Cooking oil (WCO) was filtered using filter paper and is transferred to a glass container and heated to about 90 °C in an electric oven for about 15 min to remove moisture content from the oil. The FFA content of the WCO was determined by titration of the WCO with 0.1 M NaOH with methanol as solvent and phenolphthalein as indicator. The value of the FFA content was found to be 0.47%, hence there was no requirement for esterification. The transesterification reaction was carried out using WCO and the synthesized catalyst under different reaction parameters considered for the present study. Alcohol-to-oil molar ratio, catalyst loading, and reaction time have been fixed at, 12:1, 5 wt.%, and 1.5 h respectively for all the experimental runs. Transesterification of the WCO was carried out by mixing the prescribed amount of catalyst with 30.39 g of methanol preheated at 40 °C and

Table 1 Parameter levels employed for calcination

S. no.	Parameters	Symbol	Levels		
			−1	0	+1
1	Calcination temperature (°C)	A	800	900	1000
2	Calcination time (h)	B	3	4	5
3	Reaction temperature (°C)	C	55	65	75

mixing with 80 g of WCO in a 500 ml conical flask (Make: Borosil). The solution was then heated for transesterification as per the prescribed reaction temperature using a magnetic hot plate stirrer (Make: Tarsons Digital Spinot) at an rpm of 800. The FAME yield of the WCO has been calculated using Eq. (1).

$$\text{Biodiesel yield} = \frac{\text{weight of biodiesel}}{\text{weight of WCO}} \times 100\% \tag{1}$$

2.5 *Design of Experiment Using CCD-RSM*

The response surface methodology (RSM) is used in the current study to examine how different parameters influence the outcome and how the parameters interact with each other. Three independent variables were considered for this study namely calcination temperature (800–1000 °C), calcination time (3–5 h), and reaction temperature (55–75 °C) as shown in Table 1. Based on the 3 independent parameters with 3 levels each, the total number of experiments was found to be twenty as shown in Table 2.

3 Results and Discussion

3.1 *XRD of Calcinated Snail Shells*

XRD patterns of the snail shells of Lai tharoi, Ningkhabi tharoi, and labuk tharoi calcined at 900 °C for 4 h were recorded and analyzed. From Fig. 3. Intensity peaks of Lai tharoi, Ningkhabi tharoi and Labuk tharoi calcined at 900 °C for 4 h were observed at 2 theta values of 32.60, 37.83, 54.43, 64.86, 67.65 and 32.63, 37.78, 54.28, 64.56, 67.7, and 32.64, 37.79, 54.29, 64.57, 67.76 respectively, which correspond to the (111), (200), (220), (311) and (222) planes as reported by other researchers as well [9]. The high-intensity peaks of ningkhabi tharoi and labuk tharoi show good crystalline formation in the synthesized catalyst.

Table 2 Design of experiments using RSM (CCD)

Run order	Calc. temp. °C	Calc. time h	Reaction temp. °C	Actual value	Predicted value	Residual
1	1000	3	55	94.52	94.51	0.0111
2	900	4	65	94.85	94.83	0.0242
3	1000	5	55	94.23	94.22	0.0071
4	800	3	75	94.41	94.4	0.0111
5	800	3	55	94.22	94.21	0.0071
6	900	4	65	94.81	94.83	-0.0158
7	800	5	75	94.3	94.29	0.0071
8	1000	5	75	94.96	94.95	0.0111
9	1000	4	65	94.53	94.58	-0.0493
10	900	4	65	94.88	94.83	0.0542
11	900	3	65	94.84	94.89	-0.0493
12	1000	3	75	95.18	95.16	0.0201
13	900	4	65	94.9	94.83	0.0742
14	900	4	75	95.17	95.22	-0.0493
15	900	4	65	94.87	94.83	0.0442
16	900	5	65	94.67	94.69	-0.0233
17	800	4	65	94.08	94.1	-0.0233
18	900	4	65	94.79	94.83	-0.0358
19	900	4	55	94.74	94.76	-0.0233
20	800	5	55	93.53	93.53	-0.0019

3.2 *FTIR of Calcinated Snail Shells*

Lai tharoi, Labuk Tharoi, and Ningkhabi shells were calcined at 900 °C for 4 h, and FTIR spectra were taken to identify the material's absorption bands. Figure 4 displays the FTIR spectra for calcined snail shells, and the major absorption bands are found at 891 cm^{-}1, 865 cm^{-1}, 863 cm^{-1}, 1449 cm^{-1}, 1443 cm^{-1}, and 1442 cm^{-1} respectively [10]. The $CO_3{}^{2-}$ lost during the calcination process of the snail shells, caused a shift in the absorption bands when exposed to high energy due to a reduction in the decreased mass.

3.3 *EDX of the Calcinated Snail Shells*

EDX analysis of the Labuk Tharoi shells calcined at 900 °C for 4 h was carried out to find the major elements present in the calcined snail shells. The major elements Ca (95.22 wt.%), O (2.93 wt.%), C (1.58 wt.%) and Mg (0.27 wt.%) were found in the samples.

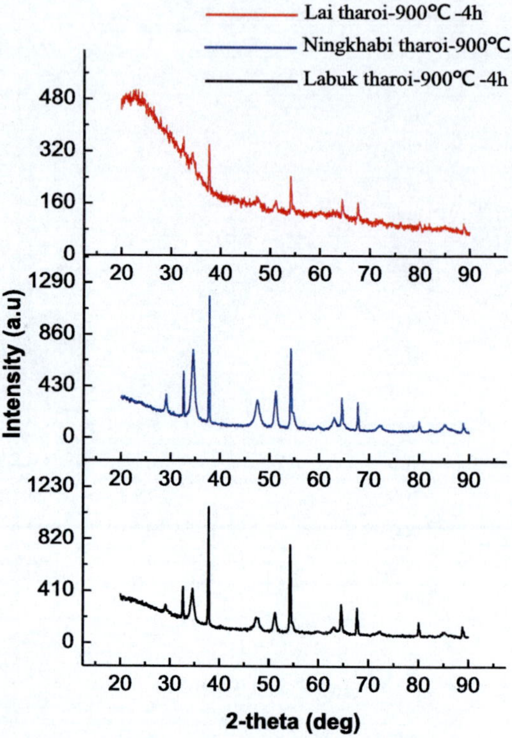

Fig. 3 XRD of lai tharoi, labuk tharoi, and ningkhabi tharoi calcinated at 900 °C for 4 h

3.4 *Selection of the Best Catalyst*

Labuk tharoi was selected as the most suitable catalyst since it showed the highest XRD intensity peak with a crystal grain size of 39.23 nm when compared to lai tharoi (121.68 nm) and ningkhabi tharoi (40.89 nm). EDX analysis also shows a high CaO composition of 95.22% for the selected catalyst.

3.5 *Design of Experiment and Statistical Analysis By RSM*

A quadratic model with a high *F*-value of 109.45 and a low *p*-value of 0.0001 was selected as the best model. The correlation between the response i.e., yield and the independent variables in biodiesel synthesis is shown in Eq. (2) where A represents calcination temperature, B calcination time, and C reaction temperature.

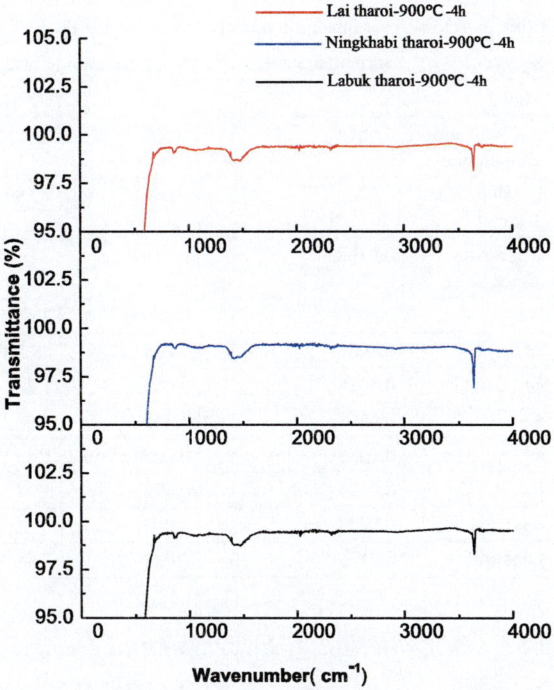

Fig. 4 FTIR of lai tharoi, labuk tharoi, and ningkhabi tharoi calcinated at 900 °C for 4 h

$$\begin{aligned}\text{Biodiesel Yield} = {} & 65.1265 + 0.0830919 * \text{A} + 0.292739 * \text{B} \\ & - 0.304416 * \text{C} - 0.0002625 * \text{AB} \\ & + 0.00011625 * \text{AC} + 0.001875 * \text{BC} \\ & - 0.000048 * \text{A}^2 - 0.0345455 * \text{B}^2 + 0.00165455 * \text{C}^2 \end{aligned} \quad (2)$$

To determine the significance of the independent parameters, the *p*-value and *F*-value of ANOVA were examined as shown in Table 3. Thus, for the synthesis of biodiesel from WCO using waste snail shell catalyst, calcination temperature ($F = 253.55$), calcination time ($F = 42.99$), and reaction temperature ($F = 232.69$) were found with a significance of 24.94%, 4.22%, and 22.89% respectively. The minor difference of 0.2 between the predicted R^2 of 0.9676 and adjusted R^2 of 0.9809 indicates that all the paraments taken are significant in the biodiesel synthesis. The high value of predicted R^2 of 0.9676 indicates that the predicted response is in line with the experimental yield [11].

Table 3 ANOVA for quadratic model

Source	Sum of squares	*df*	Mean square	*F*-value	*p*-value	Remarks
Model	2.2	9	0.244	109.45	< 0.0001	Significant
A-Calcination temperature	0.5664	1	0.566	253.55	< 0.0001	
B-calcination Time	0.096	1	0.096	42.99	< 0.0001	
C-reaction temperature	0.5198	1	0.519	232.69	< 0.0001	
AB	0.0055	1	0.005	2.47	0.1473	
AC	0.1081	1	0.108	48.39	< 0.0001	
BC	0.0028	1	0.003	1.26	0.2881	
A^2	0.6457	1	0.645	289	< 0.0001	
B^2	0.0033	1	0.003	1.47	0.2534	
C^2	0.0753	1	0.075	33.7	0.0002	
Residual	0.0223	10	0.002			
Lack of fit	0.0133	5	0.002	1.48	0.3382	Not significant

3.6 *Confirmation of Experimental Design*

The optimum solution of calcination temperature 931.8 °C, calcination time 3.5 h, and reaction temperature 70.8 °C was obtained from the RSM (CCD) and a confirmation test was carried out with these observed values. Three consecutive experiments were conducted, and the biodiesel yield obtained was 95.91 ± 0.15% which is superior to the predicted yield of 95.24% as predicted by RSM.

3.7 *Effect of Parameters on Biodiesel Production*

To study the effect of the independent parameters on biodiesel yield, the 3-D surface graphs of the CCD were used as shown in Fig. 5. From Fig. 5a it is observed that with an increase in reaction temperature from 55 to 75 °C, the biodiesel yield increases [12]. This is because, with an increase in reaction temperature, the kinetic energy between the triglycerides of the Waste Cooking Oil, catalyst, and the methanol increases [13]. The catalyst temperature also plays a very significant role in improving the yield of biodiesel. From Fig. 5c it can be seen that with an increase in calcination temperature from 800 to 900 °C, the biodiesel yield increases from 93.53 to 95.22 [14]. Further increasing the calcination temperature reduces the biodiesel yield which may be due to the formation of ash at higher temperatures in the catalyst sample. From Fig. 5b it is observed that the calcination time is not so significant in improving the yield of biodiesel.

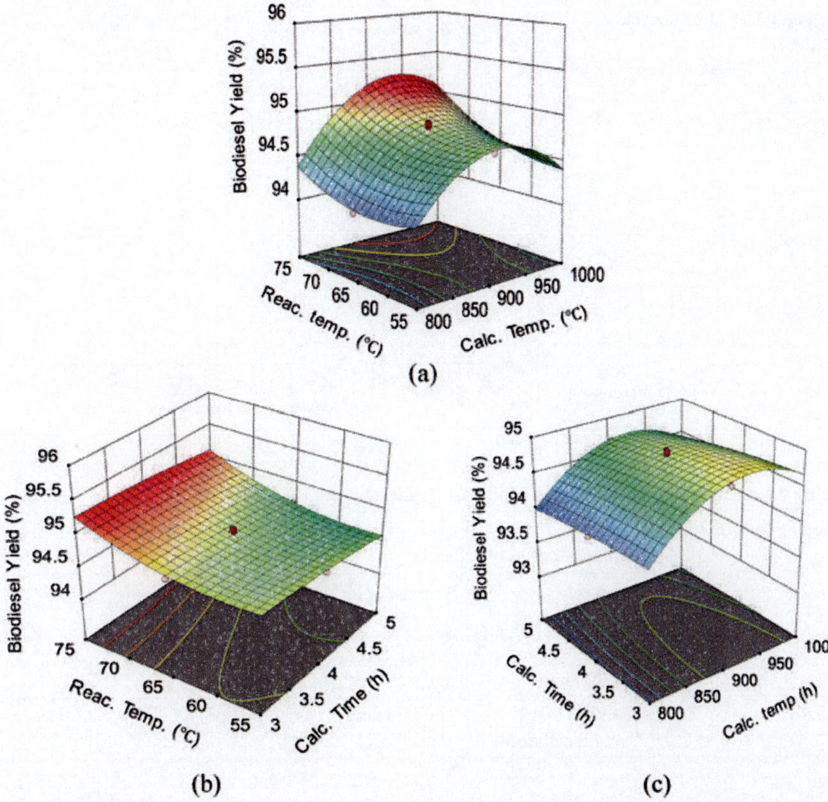

Fig. 5 Surface plot of RSM **a** calcination and reaction temperature versus biodiesel yield **b** calcination time and reaction temperature versus biodiesel yield **c** calcination temperature and calcination time vs biodiesel yield

3.8 Properties of Biodiesel

3.8.1 FTIR of Biodiesel

FTIR of the extracted FAME is shown in Fig. 6. It is observed that the stretching at 3010.312 cm^{-1} with 93.99% transmittance shows C-H bond presence, 2923.035 cm^{-1}, and 2853.706 cm^{-1} and 723.093 cm^{-1}shows CH2 presence. C=O ester presence is detected at 1461.654 cm^{-1}. (CO)–O–CH3 is detected at 1435.553 cm^{-1} and C–O ester is detected at 1169.710 cm^{-1}.

Fig. 6 FTIR of biodiesel (FAME)

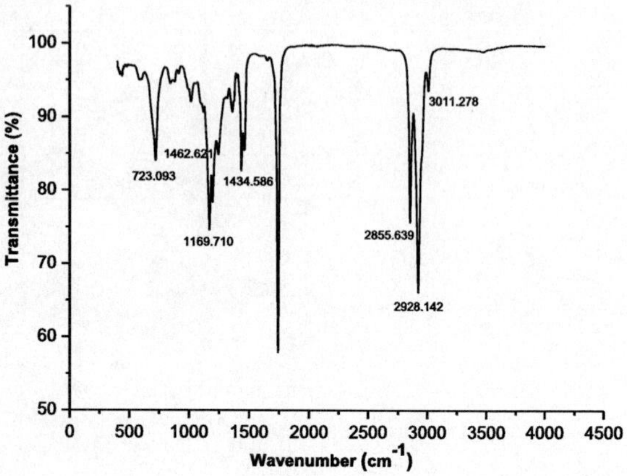

Table 4 Methyl ester and fatty acid profile of biodiesel

Retention time (min)	Identified compounds	Corresponding acids	Chemical formulae	Values
42.22	Methyl docosenoate	Behenic acid	$C_{23}H_{46}O_2$	45.67
35.55	Methyl oplopanone	Pentadecadienoic acid	$C_{15}H_{26}O_2$	24.33
35.43	Methyl 9–12 octadecadienoic	Linoleic acid	$C_{19}H_{34}O_2$	16.89

3.8.2 GCMS of Biodiesel

A gas chromatography-mass spectrometer was used to find out the composition of FAME. Table 4 shows the composition of the esters and the corresponding fatty acids present in the sample. The result shows the presence of behenic acid, pentadecadienoic acid, and linoleic acid, as the main composition.

4 Conclusion

An eco-friendly, cost-effective, and reusable heterogeneous catalyst was developed by calcination of three different types of waste snail shells known locally as labuk tharoi (*Angulyagra oxytropis),* ningkhabi tharoi (*Bellamya crassa*) and lai tharoi (*Thiara tuberculate*) found at Imphal, Manipur. Labuk tharoi was selected as the most suitable catalyst as it contains the highest percentage of CaO (98.15%) as per EDX analysis. The catalyst was found to be highly crystalline as per XRD analysis. WCO was employed as feedstock as it is readily and cheaply available. A high biodiesel yield of 95.91 ± 0.15% was achieved under optimized reaction conditions of calcination temperature of 931.8 °C, calcination time of 3.5 h, and reaction temperature of 70.8 °C which is more than the RSM predicted yield value of 95.24%. Most of the researchers fail to study the effect of independent parameters such as calcination temperature and calcination time while optimizing biodiesel yield using RSM. In this study, it was observed that calcination temperature (24.94%) plays a significant contribution in optimizing biodiesel yield. The characterization of the biodiesel and the synthesized catalyst shows that high-grade biodiesel has been obtained by deploying WCO as a feedstock and waste snail shell as a source for heterogeneous catalysts.

References

1. Statistical Review of World Energy globally consistent data on world energy markets. and authoritative publications in the field of energy (2021)
2. Parida S, Singh M, Pradhan S (2022) Biomass wastes: a potential catalyst source for biodiesel production. https://doi.org/10.1016/j.biteb.2022.101081
3. Shrivastava P, Verma TN, Pugazhendhi A (2019) An experimental evaluation of engine performance and emisssion characteristics of CI engine operated with Roselle and Karanja biodiesel. Fuel 254:115652. https://doi.org/10.1016/j.fuel.2019.115652
4. Adepoju TF, Victor E, Ekop EI, Emberru RE, Balogun TA, Adeniyi AD (2022) Residual wood ash powder: A predecessor for the synthesis of $CaO–K_2O–SiO_2$ base catalyst employed for the production of biodiesel from Asimina triloba oil seed. Case Stud Chem Environ Eng 6:100252. https://doi.org/10.1016/J.CSCEE.2022.100252
5. Brahma S, Basumatary B, Basumatary SF, Das B, Brahma S, Rokhum SL, Basumatary S (2023) Biodiesel production from quinary oil mixture using highly efficient Musa chinensis based heterogeneous catalyst. Fuel 336:127150. https://doi.org/10.1016/j.fuel.2022.127150
6. Abdullah SHYS, Hanapi NHM, Azid A, Umar R, Juahir H, Khatoon H, Endut A (2017) A review of biomass-derived heterogeneous catalyst for a sustainable biodiesel production. Renew Sustain Energy Rev 70:1040–1051. https://doi.org/10.1016/j.rser.2016.12.008
7. Singh TS, Verma TN (2019) An assessment study of using Turel Kongreng (river mussels) as a source of heterogeneous catalyst for biofuel production. Biocatal Agric Biotechnol 20:101185. https://doi.org/10.1016/j.bcab.2019.101185
8. Kaewdaeng S, Sintuya P, Nirunsin R (2017) Biodiesel production using calcium oxide from river snail shell ash as catalyst. Energy Proc 138:937–942. https://doi.org/10.1016/j.egypro.2017.10.057
9. Ashine F, Kiflie Z, Prabhu SV, Tizazu BZ, Varadharajan V, Rajasimman M, Joo S-W, Vasseghian Y, Jayakumar M (2023) Biodiesel production from Argemone mexicana oil using chicken eggshell derived CaO catalyst. Fuel 332:126166. https://doi.org/10.1016/j.fuel.2022.126166
10. Pandit PR, Fulekar MH (2017) Egg shell waste as heterogeneous nanocatalyst for biodiesel production: optimized by response surface methodology. J Environ Manage 198:319–329. https://doi.org/10.1016/j.jenvman.2017.04.100
11. Shi X, Karachi A, Hosseini M, Yazd MS, Kamyab H, Ebrahimi M, Parsaee Z (2020) Ultrasound wave assisted removal of ceftriaxone sodium in aqueous media with novel nano composite g-C_3N4/MWCNT/Bi2WO6 based on CCD-RSM model. Ultrason Sonochem 68:104460. https://doi.org/10.1016/j.ultsonch.2019.01.018
12. Nongbe MC, Ekou T, Ekou L, Yao KB, Le Grognec E, Felpin FX (2017) Biodiesel production from palm oil using sulfonated graphene catalyst. Renew Energy 106:135–141. https://doi.org/10.1016/j.renene.2017.01.024

13. Abukhadra MR, Sayed MA (2018) K+ trapped kaolinite (Kaol/K+) as low cost and eco-friendly basic heterogeneous catalyst in the transesterification of commercial waste cooking oil into biodiesel. Energy Convers Manag 177:468–476. https://doi.org/10.1016/j.enconman.2018.09.083
14. Singh TS, Verma TN (2019) Taguchi design approach for extraction of methyl ester from waste cooking oil using synthesized CaO as heterogeneous catalyst: response surface methodology optimization. Energy Convers Manag 182:383–397. https://doi.org/10.1016/j.enconman.2018.12.077

A Comprehensive Review of Aerogels for Oil Spill Cleanup and Heat Storage Applications

Karuna Kumari and Vikash Kumar Singh Chauhan

1 Introduction

According to IUPAC, aerogels are gels composed of microporous solids where the dispersed phase is a gas rather than a liquid. Aerogels are low-density solids and composed of a complex, cross-linked structure of chains of constituent molecules, with a huge number of air-filled pores occupying most of the volume where the liquid in a pore has been replaced by gas without the structure collapsing [1]. Aerogels are usually prepared by supercritical drying or freeze-drying, which allows the liquid to dry slowly without collapsing the gel matrix due to capillary action. Due to their complex internal cross-linked structure and the large number of air-filled pores occupying the bulk of the volume, these materials have extremely unique properties compared to all other materials, and therefore a significant number of such materials have been developed with peculiar applications for them. Aerogels can be transparent, opaque, and colourless, as well as different colours, such as red, black, depending on the precursors and additives used to make them [2].

More and more applications are being developed as the scientific and engineering community is becoming aware of aerogels' unusual and exceptional physical properties, such as their large internal surface area and open-cell structure, making them ideal materials for catalysts in the chemical industry. The high porosity and surface area of aerogels have led to their use as filters and desiccants. Aerogels, also have many commercial applications such as catalysts, thermal insulation, impact-resistant windows, particle detectors, supercapacitors, automotive thermal insulation, capacitive deionization electrodes, carbon dioxide capture, oil spill cleanup [3]. This article will first discuss using aerogels as oil sorbent for oil spill cleanup. In the latter part,

K. Kumari
Usha Martin University, Ranchi, Jharkhand 835103, India

V. K. S. Chauhan (✉)
SANDIP University, Madhubani, Bihar 847235, India
e-mail: vikashkumarsinghchauhan@gmail.com

B. P. Swain (ed.), *Recent Advances in Materials*, Springer Proceedings in Materials 25,
https://doi.org/10.1007/978-981-99-3844-5_17

we will discuss the benefits of using phase change materials reinforced in aerogel composites for making thermal storage systems.

2 Aerogels for Oil Spill Cleaning

Crude oil plays important role in sustaining everyone's day-to-day life and it is also important to note that crude oil is difficult to find everywhere around the globe. The oil is prominently transported via shipping lanes and thus causing the problem of oil spills in seawater [4, 5]. Oil spills cause severe loss of financial assets, cause loss of biodiversity and disturb the ecological balance. There have been several methods of removing crude from seawater. However, owing to the effective recovery of oils, aerogels have gained the attention of industries because of their numerous structured oleophilic-hydrophobic interconnected pores and low density [6, 7]. Chitosan is a naturally occurring polymer that is found in several biopolymers and is an excellent source for making aerogel due to its compatibility, biodegradability, renewability, and hydrogel-binding nature [8–13]. The chitosan has fragile scaffolds [14] and thus requires chemical treatment for increasing its mechanical strength and durability so that it can be used for multiple cycles of oil–water separation [15, 16]. Graphene oxide (GO) can be used due to its property of dispersion in organic solutions which oxidizes the surficial functional groups and make them able to react with several other polymers [17–19]. This platform is more adheres to oil in water for oil–water separation [20, 21].

Reynolds and team [14] reported one of the earliest possibilities of cleaning seawater oil spills using aerogels. They prepared the aerogels from $CF_3(CH_2)$ and CF_3-aerogel groups using $(CH_3O)_4$ Si. The product was hydrophobic in nature. The temperature of hydrophobic to hydrophilic transition was reported at 375 °C. Bigoli and team [22] synthesized an aerogel from cotton cellulose via low surface moieties then dissolving and cross-linking the matrix in an organic medium. The freeze-drying method was also used in the process. They observed that the swelling tendency of fibres played an important role with the advantage of capillary pressure working in favour of improved oil retention. The aerogel prepared in this manner exhibited high oil absorptivity, good mechanical strength, superior oil-water selectivity, and outstanding recyclability. Khoshnevis and team [23] synthesized carbon nanotubes-based aerogel with a three-dimensional porous network of high thermal stability (650 °C) required for proper oil recovery. This aerogel was found to be highly permeable and porous with the ability to absorb 107 kilos of oil per kilo of aerogel mass. Wang and team [24] prepared polyimide (PI) aerogel having a three-dimensional structure of interconnected pores. Owing to its high hydrophobicity by heredity, it was found to be a very good oil–water separator in harsh conditions. Apart from oil spill cleaning, their aerogel also performed excellently in civil and industrial wastewater treatment.

Wang and Liu [25] synthesized aerogel for oil spill cleaning using raw cotton fibres. The authors reported that the oil absorption capacity was of 20 g per gram

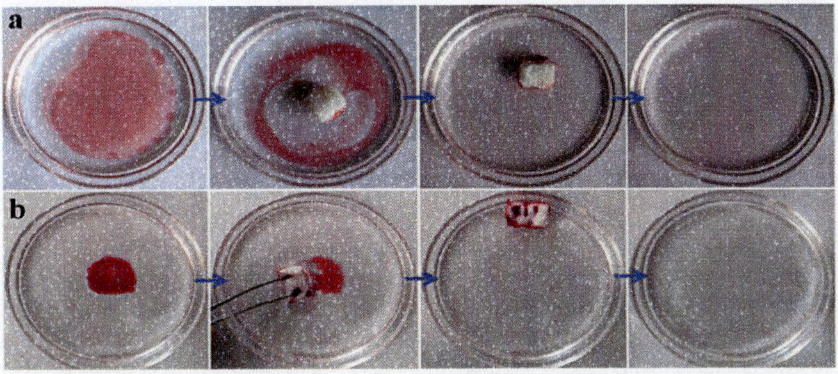

Fig. 1 Removal of **a** red-coloured kerosene on the water surface and **b** red-coloured chloroform underwater using superhydrophobic cellulose aerogel [25]

of the aerogel material and even after reusability of 80 times, it still maintained its initial absorption capacity. Figure 1 depicts the cleaning of red-coloured Kerosine (a) and chloroform (b) from water using the aerogel.

Yi and team [26] synthesized mechanically superior aerogel through the directional freezing and siltation method. A special spring-like structure composed of porous channels provides fast elastic recovery of the material. The oil absorptivity was 63 times the original mass and the recyclability/separation can be achieved through simple squeezing. Parmar and team [27] synthesized a CNT aerogel by decomposing CH_4 in a vapour depositor. The CNT was reinforced in mero-hydrophobic carboxymethyl cellulose and silica matrix making the structure highly flexible and superhydrophobic. Whereas Lang and team2019 prepared polypropylene-based aerogel from a Xylene base. This aerogel is then cooled to induce thermal phase separation of the molecular network of polypropylene and then by supercritical CO_2 drying. Both of the aerogels showed superhydrophobicity and excellent recyclability with inexpensive operation.

Thai and team [28] synthesized aerogel from recycled rubber of polymer structure and ultra-low density and high porosity. This aerogel was coated with methoxytrimethylsilane (MTMS) to convert the surface hydrophobic. This aerogel has excellent thermal insulation, thermal stability, and better rigidity than Styrofoam. The rubber aerogel had good elastic properties and can perform as a noise reducer as well with an oil absorptivity (19 g/g). In another study, Gonzalez and team[29] used extracts from Posidonia oceanica and prepared aerogel based on adsorbent pads for oils with oil absorption capacity of 43 times of aerogel mass through the freeze-drying method. This product has great usability not only for oil spill cleanup but also for food packaging applications. Tran and team [30] used rice straw as the raw material for making aerogels with PVA and cationic starch and then freeze-drying. They reported, when coated with MTMS, the aerogel turns highly hydrophobic. They mentioned the nature of the aerogel as thermally and acoustically insulating. This aerogel was able to absorb 13 g of oil per gram of aerogel. Kumar and team [2] developed biodegradable sugarcane-based aerogel for a wide range of applications ranging from thermal to noise insulators in buildings and for oil spill cleanup. This aerogel also had more than 90% porosity and can absorb 25–50 times more oil than the original aerogel mass.

3 Aerogels for Thermal Storage

Renewable energy harvesting in today's era brings with it the major problems of meeting energy supply–demand side fluctuations, low energy consumption and high environmental impact. Energy storage mainly occurs in mechanical systems solid or fluid (compression, expansion, rotation, or elevation), and chemical storage systems [31] where two or more chemicals absorb or release energy, e.g. battery, or fuel cell. Magnetic storage systems such as supercapacitors [32, 33] or thermal energy storage by increasing/decreasing temperature or by phase changing [34].

The two main types of thermal energy storage systems are sensible heat storage systems and latent heat storage systems. The former is widely used in the solar power sector (concentration), while the latter uses phase change materials and thermo-chemical reactions. Thermal storage systems have higher and more robust storage capacities, while thermo-chemical-storage systems use intermediaries and the lifetime of the system depends on the operation cycle [35].

Sensible heat storage systems raise the temperature of the medium by absorbing heat energy and storing the heat until later use. The efficiency of the system mainly depends on the specific heat of the material and the working temperature range. Latent heat storage systems make use of the comparatively large latent heat required to change the phase of matter. Latent heat storage systems make extensive use of the solid–solid and solid–liquid schemes. Gasification of liquids results in large volume changes, which are prone to unwanted energy leakage. Solid-state PCMs vary from amorphous, crystalline and in their lattice structure [36]. However, solid–liquid phase change materials show excellent thermal efficiency with no significant change in temperature, low volume change, and very large latent heat for storage of heat energy, thus justifying its wide application.

The heat energy denoted as "Q" in the thermal energy storage systems is illustrated in Fig. 2 and can be modelled mathematically by

$$Q = m\left(\int_{T_1}^{T_m} C_1 dT + \int_{T_m}^{T_2} C_2 dT + \Delta H_a + \Delta H_b\right)$$

where m is the quantity of material in moles, T_1 and T_2 are lower and higher temperatures, respectively, T_m is the melting point, C_1 and C_2 are the specific thermal capacities of the solid and liquid phase, respectively, ΔH_a is the energy required for phase change, and ΔH_b is the energy required for the chemical phase transition [37].

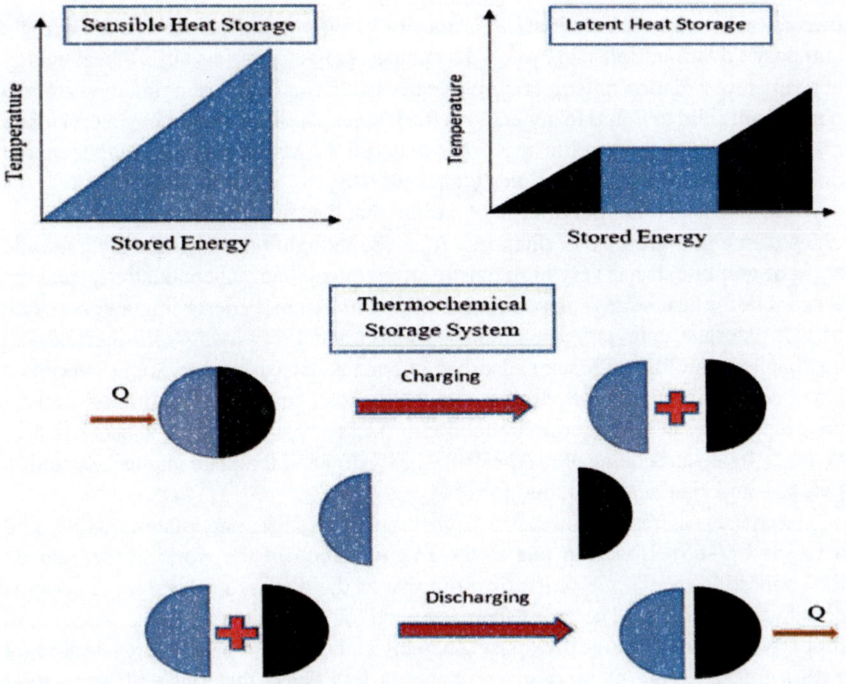

Fig. 2 Energy storage in form of SHS, LHS, and TCS [3]

Some common applications of phase change materials include microelectronics [13], solar-chilling plants[38], solar photovoltaics [39], solar thermal packages [40], thermal insulation in extreme weather [41, 42], and concrete panels [43], temperature stabilizers for packaged foods [44], and on battlefields it is used for thermal camouflage [36].

PCMs with good thermal conductivity have low latent heat of fusion and vice versa, raising serious challenges in rapidly charging heat into and out of PCMs. A possible solution is to reinforce the PCM with a good conductor as a filler or to incorporate the PCM into it. However, this creates an additional problem of reducing the fraction of PCM in the overall structure, and on the other hand, the higher the filler fraction, the lower the phase change we obtain out of the structure. The work, therefore, requires less reactive and more porous materials with higher thermal conductivity to be used as fillers/encapsulations.

Graphene aerogel is 99% porous, that is, 99% of the volume occupied by the aerogel bulk is void and can be filled with PCM, so the invention of graphene aerogels has provided a way to solve the above problems. On the other hand, it has good thermal conductivity for charging and thermal storage systems. When used with PCM, it forms a phase change composite (PCC) with negligible change in the desired properties. Therefore, the graphene aerogel-reinforced PCM helps in lowering the thermal resistance, phase separation, density variation, and stability of the aerogel composite over multiple use cycles. The phase change materials absorb heat energy at a very low variation of temperature (Nearly isothermal) and changing the system's phase from solid to liquid in general. As the latent heat of fusion has in general much higher values than the specific heat of the material, PCM offers a much higher energy density for storing around 10 times higher [45–48].

Graphene is another polymorph of carbon that has much better mechanical properties than either graphite or diamond. It can be thought of as a single-layer atomic sheet of graphite. It has very high tensile strength and thermal conductivity, making it suitable for heat storage applications. The main issues in designing a good batch of PCM storage units is to maintain thermal conductivity along with thermal and mechanical stability. PCMs are adsorbed or encapsulated, or fillers are incorporated into the matrix for the above purpose. Graphene sheets based on monolithic structures are prepared by a hydrothermal reduction process [49, 50]. Similar sheets can also be made by electrochemical synthesis [51, 52], directed template vapour deposition [53, 54], and chemical reduction [55, 56].

Graphene spheres are prepared from graphene oxide, nanoplates, foams, and aerogels [57–65]. However, due to the fragmentation of the working medium by the supporting media, the properties improve as the density of the storage systems continues to decrease. Therefore, there is a need for a material that increases a certain property for a small change in density. Zhejiang University prepared graphene aerogel with a freeze-drying method and produced a less dense material with very good strength, making it the most suitable material for this purpose. However, graphene aerogels were first prepared via microwave irradiation/sol–gel technique [56]. Hu et al. [66] prepared lightly compressed graphene aerogels (ULGA) by functionalized lyophilization-microwave treatment, which reduced the ability of atomic graphene planes to stick together in aerogel assembly process. This ultra-lightweight graphene aerogel was found to be highly compressible and extensible.

Although graphene aerogels have lower thermal conductivity than graphene foam, their shape is more stable. The bonding of graphene aerogels is because of weak Van der wall forces, π–π stacking and hydrogen bonds, which support the properties of PCMs and also provide a route for heat conduction. Since phase change composites do not require high phase change enthalpy or have very low leakage, graphene aerogels with low filler loading are the most promising substrates. To prepare graphene aerogel and PCM composite, graphene aerogel is first oxidized and then self-assembled to form rGO, which easily adsorbs phase change materials and is suitable for phase change composites [1] (Fig. 3).

Graphene aerogels are usually grown by rGO and it readily adsorbs phase change materials into the matrix. Thus, the obtained phase change composites have good heat storage density, shape stability, and thermal conductivity. The phase change enthalpy of PCM graphene composite depends on the phase change enthalpy of PCM. Generally, PCM-based composites are prepared by the hydrothermal reduction method. One such method was developed by Zhong and team [49] combined graphene oxide sheets with octadecanoic acid to form a graphene aerogel in a hydrothermal reaction. The stack of graphene sheets formed a network of microscopic structures. This

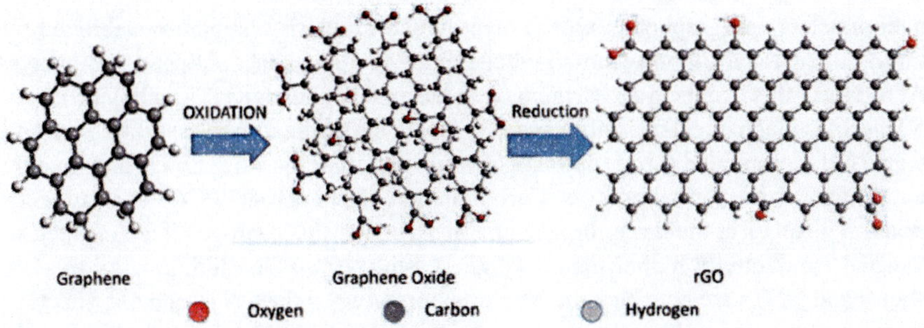

Fig. 3 Formation of rGO [3]

compound results in a 14 times higher thermal conductivity (2.64 W/mK). However, a slight decrease in heat storage capacity of less than 5% was observed because of the addition of graphene filler.

Li et al. [67] modified the hydrothermal reduction method when they produced a microporous monolith graphene from graphene oxide emulsion consisting of hexane droplets. In this synthesis process, a reticulated and precise three-dimensional network was formed around hexane by rGO grown from graphene oxide. The ability of graphene to assemble in a shell shaped structure was used by Ye and team [50]. A modified hydrothermal process yields hollow shells of graphene from graphene oxide sheets. At the same time the paraffin is covered in these micrometre scale spheres as shown in Fig. 4.

This composite resulted in a better encapsulation ratio for PCM (97% by weight), phase transition enthalpy ($\Delta H_m = 202$ J/g and $\Delta H_f = 213$ J/g) and 100 retentions of PCM after 20 cycles. It also resulted in superior conductivity, both electrical and thermal and shape stability. Yang and team [68] synthesized a hybrid graphene aerogel by impregnating polyethylene glycol in a mixture of graphene

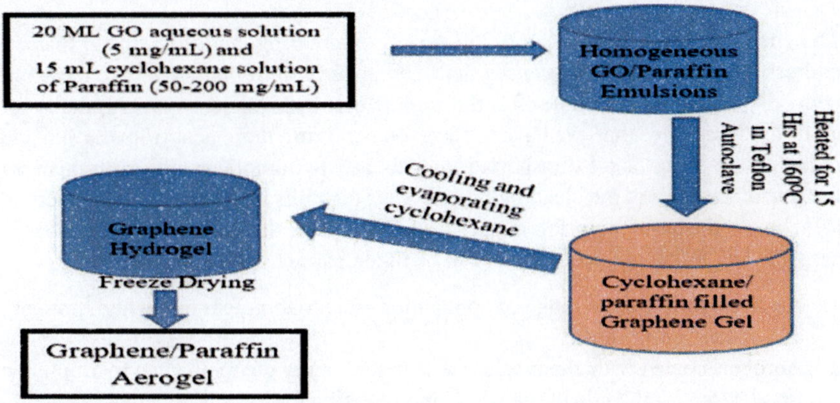

Fig. 4 Modified hydrothermal process [50]

nanoplatelets and graphene oxide. Polyethylene glycol is a structure stabilized by three-dimensional support from graphene oxide to allow heat conduction pathways. At full load, this results in a 31% increase in thermal conductivity. This alloy results in 50 thermos mechanical reliability and 100 fusion crystallization reusability. Tian and team [69] investigated the properties of carbon nanotube-based aerogels prepared by a modified hydrothermal process. Carbon nanotubes in aerogel PCC act as an additional reinforcer in the three-dimensional network of PCC aerogel. This resulted in thermal conductivity higher than 2.1 W/mK with 98% PCM addition. As the GA sheets and CNTs are attached, the heat transfer between the CNTs and GA sheets is limited, and due to leakage of more than 13% at about 323 K, this composite is only suitable for low-temperature applications. Its phase change enthalpy is 223 kJ/kg.

While the performance of CNTs as a reinforcer material in PCC is surprising, uncontrolled thermal conductivity is a major challenge. Tang et al. [70] studied different oxygenated groups by preparing GAs from EG to GO by modified Hummer's method [71] and ultrasonic treatment, freeze-drying and analysed its effect on physical and chemical properties and compared it with PEG/GA. The synthesized PCM shows better shape stability and capillary force through pores and synergistic interaction between work and support material. Although it shows a slightly lower phase change enthalpy, it has better light for thermal energy conversion, making it a potential candidate for storage systems. Various efforts have been made to improve the mechanical properties of graphene aerogel phase change composites, at different temperatures and strain rates, which provide critical information on their behaviour. When the weight of paraffin in the GO aerogel mixture is less than 5%, the liquid paraffin shows no leakage, which shows the stability of the mixture [72]. Furthermore, this aerogel composite showed approximately 3–7 times higher strength and less sensitivity to stress.

4 Conclusion

The energy demands of humanity are steadily increasing, and it is crucial that we address the potential consequences that may arise if we do not prioritize the well-being of our environment. The oil and sun contribute to the major supply of energy for all of us. To achieve that, we have to focus on capturing more heat from the sun and reducing the harm of oil transportation. This article comprehensively summarizes the role of aerogels in the cleaning of malicious oil spills in the oceans and a method to enhance the heat-capturing capacity of PCM with the help of graphene-based aerogel. The following conclusions can be made based on the above article:

1. Aerogels have a wide range of applications due to their extremely high porosity and active surface area.
2. Aerogels coated with chemicals can be made highly oleophilic and hydrophobic for absorbing oil spills in sea water and reducing their environmental impact.

3. The oil recovery methods of aerogels range from thermally heating to mechanical squeezing. The usefulness of the aerogel for oil spill cleanup is highly dependent on the oil absorptivity after a repetitive number of oil recovery cycles and the cost of production.
4. Graphene aerogel-based PCM composited increases the thermal storage and release capacity many folds than the pure PCM-based storage system due to enhancement in thermal conductivity with a small reduction in phase change enthalpy.
5. Graphene aerogel and foam-based paraffin show a significant increase in thermal conductivity.
6. The CGA-phase change compound has surprisingly higher heat conductivity and very poor structural stability at high temperatures, but its use has increased in lower-temperature thermal energy storage systems.

Acknowledgements The Authors of the Article thank SANDIP University, Madhubani for providing the motivational and financial support to conduct this review work at their premises. I also thank Dr. Sandip Jha for his consistent support in favour of research-related activities inside the University campus and nearby areas for the betterment of the civilian society.

References

1. McNaught AD, Wilkinson A (2006) Aerogel. In: IUPAC compendium of chemical terminology. IUPAC, Research Triangle Park
2. Kumar G, Dora DTK, Jadav D, Naudiyal A, Singh A, Roy T (2021) Utilization and regeneration of waste sugarcane bagasse as a novel robust aerogel as an effective thermal, acoustic insulator, and oil adsorbent. J Clean Prod 298:126744
3. Kashyap S, Kabra S, Kandasubramanian B (2020) Graphene aerogel-based phase changing composites for thermal energy storage systems. J Mater Sci 55(10):4127–4156
4. Yuan D, Zhang T, Guo Q, Qiu F, Yang D, Ou Z (2018) Recyclable biomass carbon@ SiO_2@ MnO_2 aerogel with hierarchical structures for fast and selective oil-water separation. Chem Eng J 351:622–630
5. Yoon H, Na SH, Choi JY, Latthe SS, Swihart MT, Al-Deyab SS, Yoon SS (2014) Gravity-driven hybrid membrane for oleophobic–superhydrophilic oil–water separation and water purification by graphene. Langmuir 30(39):11761–11769
6. Lavoine N, Bergström L (2017) Nanocellulose-based foams and aerogels: processing, properties, and applications. J Mater Chem A 5(31):16105–16117
7. Wu M, Shi Y, Chang J, Li R, Ong C, Wang P (2018) Sunlight induced rapid oil absorption and passive room-temperature release: an effective solution toward heavy oil spill cleanup. Adv Mater Interf 5(14):1800412
8. Wang J, Gong C, Wen S, Liu H, Qin C, Xiong C, Dong L (2019) A facile approach of fabricating proton exchange membranes by incorporating polydopamine-functionalized carbon nanotubes into chitosan. Int J Hydrogen Energy 44(13):6909–6918
9. Pipattanawarothai A, Suksai C, Srisook K, Trakulsujaritchok T (2017) Non-cytotoxic hybrid bioscaffolds of chitosan-silica: sol-gel synthesis, characterization and proposed application. Carbohyd Polym 178:190–199
10. de Luna MS, Ascione C, Santillo C, Verdolotti L, Lavorgna M, Buonocore GG, Ambrosio L et al (2019) Optimization of dye adsorption capacity and mechanical strength of chitosan aerogels through crosslinking strategy and graphene oxide addition. Carbohyd Polym 211:195–203

11. Ramasamy DL, Wojtuś A, Repo E, Kalliola S, Srivastava V, Sillanpää M (2017) Ligand immobilized novel hybrid adsorbents for rare earth elements (REE) removal from waste water: assessing the feasibility of using APTES functionalized silica in the hybridization process with chitosan. Chem Eng J 330:1370–1379
12. Baldino L, Cardea S, Reverchon E (2018) Nanostructured chitosan–gelatin hybrid aerogels produced by supercritical gel drying. Polym Eng Sci 58(9):1494–1499
13. Sun W, Du A, Gao G, Shen J, Wu G (2017) Graphene-templated carbon aerogels combining with ultra-high electrical conductivity and ultra-low thermal conductivity. Microporous Mesoporous Mater 253:71–79
14. Reynolds JG, Coronado PR, Hrubesh LW (2011) Hydrophobic aerogels for oil-spill cleanup? Intrinsic absorbing properties. Energy Sour 23(9):831–843
15. Li Z, Shao L, Hu W, Zheng T, Lu L, Cao Y, Chen Y (2018) Excellent reusable chitosan/cellulose aerogel as an oil and organic solvent absorbent. Carbohyd Polym 191:183–190
16. Li Z, Yang L, Cao H, Chang Y, Tang K, Cao Z, Li M (2017) Carbon materials derived from chitosan/cellulose cryogel-supported zeolite imidazole frameworks for potential supercapacitor application. Carbohyd Polym 175:223–230
17. Chen L, Li Y, Du Q, Wang Z, Xia Y, Yedinak E, Ci L (2017) High performance agar/graphene oxide composite aerogel for methylene blue removal. Carbohyd Polym 155:345–353
18. Mi X, Huang G, Xie W, Wang W, Liu Y, Gao J (2012) Preparation of graphene oxide aerogel and its adsorption for Cu2+ ions. Carbon 50(13):4856–4864
19. Zhai T, Zheng Q, Cai Z, Turng LS, Xia H, Gong S (2015) Poly (vinyl alcohol)/cellulose nanofibril hybrid aerogels with an aligned microtubular porous structure and their composites with polydimethylsiloxane. ACS Appl Mater Interf 7(13):7436–7444
20. Ren F, Li Z, Tan WZ, Liu XH, Sun ZF, Ren PG, Yan DX (2018) Facile preparation of 3D regenerated cellulose/graphene oxide composite aerogel with high-efficiency adsorption towards methylene blue. J Colloid Interf Sci 532:58–67
21. Mi HY, Jing X, Huang HX, Peng XF, Turng LS (2018) Superhydrophobic graphene/cellulose/silica aerogel with hierarchical structure as superabsorbers for high efficiency selective oil absorption and recovery. Ind Eng Chem Res 57(5):1745–1755
22. Bidgoli H, Khodadadi AA, Mortazavi Y (2019) A hydrophobic/oleophilic chitosan-based sorbent: toward an effective oil spill remediation technology. J Environ Chem Eng 7. https://doi.org/10.1016/j.jece.2019.103340
23. Khoshnevis H, Mint SM, Yedinak E, Tran TQ, Zadhoush A, Youssefi M, Duong HM (2018) Super high-rate fabrication of high-purity carbon nanotube aerogels from floating catalyst method for oil spill cleaning. Chem Phys Lett 693:146–151
24. Wang J, Zhang W, Zhang C (2019) Versatile fabrication of anisotropic and superhydrophobic aerogels for highly selective oil absorption. Carbon 155:16–24
25. Wang J, Liu S (2019) Remodeling of raw cotton fiber into flexible, squeezing-resistant macroporous cellulose aerogel with high oil retention capability for oil/water separation. Sep Purif Technol 221:303–310
26. Yi L, Xia Y, Tan Z, Fang X, Zhao L, Wu H, Guo S (2020) Design of tubelike aerogels with macropores from bamboo fungus for fast oil/water separation. J Clean Prod 264:121558
27. Parmar KR, Dora DTK, Pant KK, Roy S (2019) An ultra-light flexible aerogel based on methane derived CNTs as a reinforcing agent in silica-CMC matrix for efficient oil adsorption. J Hazard Mater 375:206e215. https://doi.org/10.1016/j.jhazmat.2019.04.017
28. Thai QB, Nguyen ST, Ho DK, Tran TD, Huynh DM, Do NHN, Luu TP, Le PK, Le DK, Phan-Thien N, Duong HM (2019) Cellulose-based aerogels from sugarcane bagasse for oil spill-cleaning and heat insulation applications. Carbohyd Polym 228:115365. https://doi.org/10.1016/j.carbpol.2019.115365
29. Benito-González I, López-Rubio A, Gómez-Mascaraque LG, Martínez-Sanz M (2020) PLA coating improves the performance of renewable adsorbent pads based on cellulosic aerogels from aquatic waste biomass. Chem Eng J 390:124607
30. Tran DT, Nguyen ST, Do ND, Thai NNT, Thai QB, Huynh HKP, Nguyen VTT, Phan AN (2020) Green aerogels from rice straw for thermal, acoustic insulation and oil spill cleaning

applications. Mater Chem Phys 253:123363. https://doi.org/10.1016/j.matchemphys.2020.123363
31. Dincer I, Rosen MA (2010) Energy storage systems. Therm Energy Storage Syst Appl 51–82
32. Cambero C, Sowlati T (2016) Incorporating social benefits in multi-objective optimization of forest-based bioenergy and biofuel supply chains. Appl Energy 178:721–735
33. Mahlia TMI, Saktisahdan TJ, Jannifar A, Hasan MH, Matseelar HSC (2015) A review of available methods and development on energy storage; technology update. Renew Sustain Energy Rev 33:532–545
34. Kalaiselvam S, Parameshwaran R (2019) Thermal energy storage technologies for sustainability: systems design, assessment and applications. Elsevier
35. Li PW, Chan C (2017) Thermal energy storage analyses and designs. Academic Press
36. Qu Y, Li Q, Cai L, Pan M, Ghosh P, Du K, Qiu M (2018) Thermal camouflage based on the phase-changing material GST. Light: Science & Applications, 7(1), 1–10 (2018).
37. Jordan S (2014) E-assessment for learning? Exploring the potential of computer-marked assessment and computer-generated feedback, from short-answer questions to assessment analytics. Open University (United Kingdom)
38. Plytaria MT, Bellos E, Tzivanidis C, Antonopoulos KA (2019) Numerical simulation of a solar cooling system with and without phase change materials in radiant walls of a building. Energy Convers Manag 188:40–53
39. Zhang Q, Uchaker E, Candelaria SL, Cao G (2013) Nanomaterials for energy conversion and storage. Chem Soc Rev 42:3127–3171
40. Ye Q, Tao P, Chang C et al (2019) Form-stable solar thermal heat packs prepared by impregnating phase-changing materials within carbon-coated copper foams. ACS Appl Mater Interf 11:3417–3427
41. Qiu L, Niu RP, Tan Z (2012) Experimental research of PCMs-TH29 using on building energy storage. Adv Mater Res 569:202–206
42. Sharma A, Tyagi VV, Chen CR, Buddhi D (2009) Review on thermal energy storage with phase change materials and applications. Renew Sustain Energy Rev 13:318–345
43. Lee D, Lee S-G, Kim S (2017) Composite phase-change material mold for cost-effective production of free-form concrete panels. J Constr Eng Manag 143:04017012
44. Bal LM, Satya S, Naik SN (2010) Solar dryer with thermal energy storage systems for drying agricultural food products: a review. Renew Sustain Energy Rev 14:2298–2314
45. Pielichowska K, Pielichowski K (2014) Phase change materials for thermal energy storage. Prog Mater Sci 65:67–123
46. Farid MM, Khudhair AM, Razack SAK, Al-Hallaj S (2004) A review on phase change energy storage: materials and applications. Energy Convers Manag 45:1597–1615
47. Tyagi VV, Buddhi D (2007) PCM thermal storage in buildings: a state of art. Renew Sustain Energy Rev 11:1146–1166
48. Prajapati DG, Kandasubramanian B (2019) Biodegradable polymeric solid framework-based organic phase-change materials for thermal energy storage. Ind Eng Chem Res
49. Zhong Y, Zhou M, Huang F, Lin T, Wan D (2013) Effect of graphene aerogel on thermal behavior of phase change materials for thermal management. Sol Energy Mater Sol Cells 113:195–200
50. Ye S, Zhang Q, Hu D, Feng J (2015) Core–shell-like structured graphene aerogel encapsulating paraffin: shape-stable phase change material for thermal energy storage. J Mater Chem A 3(7):4018–4025
51. Sheng K, Sun Y, Li C, Yuan W, Shi G (2012) Ultrahigh-rate supercapacitors based on eletrochemically reduced graphene oxide for ac line-filtering. Sci Rep 2(1):247
52. Chen K, Chen L, Chen Y, Bai H, Li L (2012) Three-dimensional porous graphene-based composite materials: electrochemical synthesis and application. J Mater Chem 22(39):20968–20976
53. Yang J, Qi GQ, Bao RY, Yi K, Li M, Peng L, Yang W et al (2018) Hybridizing graphene aerogel into three-dimensional graphene foam for high-performance composite phase change materials. Energy Storage Mater 13:88–95

54. Ji H, Sellan DP, Pettes MT, Kong X, Ji J, Shi L, Ruoff RS (2014) Enhanced thermal conductivity of phase change materials with ultrathin-graphite foams for thermal energy storage. Energy Environ Sci 7(3):1185–1192
55. Pham HD, Pham VH, Cuong TV, Nguyen-Phan TD, Chung JS, Shin EW, Kim S (2011) Synthesis of the chemically converted graphene xerogel with superior electrical conductivity. Chem Commun 47(34):9672–9674
56. Worsley MA, Pauzauskie PJ, Olson TY, Biener J, Satcher JH Jr, Baumann TF (2010) Synthesis of graphene aerogel with high electrical conductivity. J Am Chem Soc 132(40):14067–14069
57. Jain V, Kandasubramanian B (2020) Functionalized graphene materials for hydrogen storage. J Mater Sci 55:1865–1903
58. Bhalara PD, Balasubramanian K, Banerjee BS (2015) Spider–web textured electrospun composite of graphene for sorption of Hg (II) ions. Mater Focus 4(2):154–163
59. Thakur K, Kandasubramanian B (2019) Graphene and graphene oxide-based composites for removal of organic pollutants: a review. J Chem Eng Data 64:833–867
60. Joshi A, Bajaj A, Singh R, Alegaonkar PS, Balasubramanian K, Datar S (2013) Graphene nanoribbon–PVA composite as EMI shielding material in the X band. Nanotechnology 24(45):455705
61. Yadav R, Subhash A, Chemmenchery N, Kandasubramanian B (2018) Graphene and graphene oxide for fuel cell technology. Ind Eng Chem Res 57(29):9333–9350
62. Sahoo BN, Kandasubramanian B (2014) Photoluminescent carbon soot particles derived from controlled combustion of camphor for superhydrophobic applications. RSC Adv 4(22):11331–11342
63. Palaniappan N, Cole IS, Kuznetsov AE, Balasubramanian K, Thomas KJ (2019) Experimental and computational studies of a graphene oxide barrier layer covalently functionalized with amino acids on Mg AZ13 alloy in salt medium. RSC Adv 9(56):32441–32447
64. Palaniappan N, Cole IS, Caballero-Briones F, Balasubaramanian K, Lal C (2018) Praseodymium-decorated graphene oxide as a corrosion inhibitor in acidic media for the magnesium AZ31 alloy. RSC Adv 8(60):34275–34286
65. Kant K, Shukla A, Sharma A, Biwole PH (2017) Heat transfer study of phase change materials with graphene nano particle for thermal energy storage. Sol Energy 146:453–463
66. Hu H, Zhao Z, Wan W, Gogotsi Y, Qiu J (2013) Ultralight and highly compressible graphene aerogels. Adv Mater 25(15):2219–2223
67. Li Y, Chen J, Huang L, Li C, Hong JD, Shi G (2014) Highly compressible macroporous graphene monoliths via an improved hydrothermal process. Adv Mater 26(28):4789–4793
68. Yang J, Qi GQ, Liu Y, Bao RY, Liu ZY, Yang W, Yang MB (2016) Hybrid graphene aerogels/phase change material composites: thermal conductivity, shape-stabilization and light-to-thermal energy storage. Carbon 100:693–702
69. Tian B, Yang W, He F, Xie C, Zhang K, Fan J, Wu J (2017) Paraffin/carbon aerogel phase change materials with high enthalpy and thermal conductivity. Fullerenes, Nanotubes, Carbon Nanostruct 25(9):512–518
70. Tang LS, Yang J, Bao RY, Liu ZY, Xie BH, Yang MB, Yang W (2017) Polyethylene glycol/graphene oxide aerogel shape-stabilized phase change materials for photo-to-thermal energy conversion and storage via tuning the oxidation degree of graphene oxide. Energy Convers Manage 146:253–264
71. Marcano DC, Kosynkin DV, Berlin JM, Sinitskii A, Sun Z, Slesarev A, Tour JM et al (2010) Improved synthesis of graphene oxide. ACS Nano 4(8):4806–4814
72. Xu Y, Fleischer AS, Feng G (2017) Reinforcement and shape stabilization of phase-change material via graphene oxide aerogel. Carbon 114:334–346

Repurposing Waste Foundry Sand as a Sustainable Building Material with Improved Thermal Performance

A. Rahman, D. Mazumder, R. Haque, G. Sutradhar, and S. Haidar

1 Introduction

Sand has been utilized widely as a construction material from the dawn of human civilization. Sand can be found in fired earthen bricks, mud mortars, and lime mortars. Nowadays, concrete is mostly used as building medium, and sand is an essential component of concrete. Metals were employed to make tools and utensils after the stone era. Foundry operations are generally necessary for the manufacturing of metal goods. Sand has been utilized for mould manufacturing since the beginning of the human being's use of metals in everyday life. Sand is used in foundries for two key applications: mould building and core making [1]. Landfilling has also become unfeasible due to high national average tipping prices for foundry wastes. After China and the USA, India's foundry sector is the world's third largest casting maker. With roughly 6000 units of foundry with total capacity for production is 15 million metric tonnes per year. Yearly production for 2020–21 is reported to be nearly 11.31 million metric tonnes. The installed production capacity and actual production may be larger than estimated because foundry sector is mostly disorganized (approximately 85%) and does not report publicly. The waste generated by these foundries (WFS) is roughly 1,710,000 tonnes (1.71 MT) each year [2, 3, 11].

Because of the massive amount of WFS, depositing it in landfills has a significant impact on the ecosystem and the environment. As a result, reuse of leftover foundry sand (WFS) in construction industry other than land filling benefits society in both economic and environmental terms. The building sector is the greatest consumer of natural sand; hence, research on repurposing leftover WFS as a substitute of natural sand in construction industry is critical. In fact, the natural supplies of sand are being

A. Rahman · D. Mazumder · R. Haque · S. Haidar (✉)
Department of Mechanical Engineering, Aliah University, Kolkata, India
e-mail: sh.mech@aliah.ac.in

G. Sutradhar
National Institute of Technology, Imphal, Manipur, India

B. P. Swain (ed.), *Recent Advances in Materials*, Springer Proceedings in Materials 25,
https://doi.org/10.1007/978-981-99-3844-5_18

depleted as a result of over-consumption. Every step should be taken to stop the over exploitation of natural sand. On of such endeavour is the use of waste foundry sand as a complete or partial replacement for natural sand that not only results the conservation of natural resources but also safely disposes the hazardous industrial waste sand to a useful application that protects the ecology and environment. Furthermore, the usage of old foundry sand, as proposed by numerous researchers in a range of construction-related sectors, is supported with suitable experimental findings. Because concrete is so important in the construction sector, a detailed research of the concrete properties with WFS is also included for a clear knowledge of the use of WFS in concrete manufacturing.

2 Experimental Details

2.1 Characterization of Constituents

2.1.1 Cement

For sample preparation, commercially available PPC with fly ash in accordance with IS 1489 Part 1:15 [4] was used. The physical and chemical properties of used cement listed under Table 1 with required stipulation made under IS code.

2.1.2 Waste Foundry Sand (WFS)

WFS was obtained from an iron casting foundry in Howrah cluster, West Bengal. The chemical and physical properties of WFS are depicted in Tables 2 and 3, respectively.

Sieve analysis of WFS was done in accordance with IS 383:1970 [5] and the grading carve, shown in Fig. 1, comes under the zone III as specified by the code.

2.1.3 Sand

Sand was gathered from a quarry on the Mayurakshi River bed in the Birbhum region of West Bengal, India. The sand used in this investigation was graded according to BIS: 383-1970 [5] grading zone II as shown in Fig. 2, and Table 3 shows the specific gravity and fineness modulus value of sand.

Table 1 PPC cement properties as per test certificate

Properties	Requirement of IS 1489	Result obtained
Fineness (m^2/Kg)	Minimum 300	345
Starting setting time (min)	Minimum30	160
Final setting time (min)	Maximum 600	260
Drying Shrinkage (%)	Maximum 0.15	0.03
Soundness by Autoclave (%)	Maximum0.8	0.04
3 days CS (MPa)	Minimum 16	22
7 days CS (MPa)	Minimum 22	31
28 days CS (MPa)	Minimum 33	51
Soundness by Le chatelier (mm)	Maximum 10	0.5
% of pozzolana	Minimum 15, Maximum 35	25.21
Sulfuric anhydride (%) by mass	Maximum 5	2.48
Insoluble residue (%) by mass	Maximum 37.6	23.21
Chloride content (%) by mass	Maximum 0.1	0.019
Magnesia (%) by mass	Maximum 6	0.84
Sulfuric anhydride (%) by mass	Maximum 5	2.86
Total loss on ignition (%) by mass	Maximum 2.85	5.00

Table 2 Chemical characteristics of WFS

Characteristics	Result obtained
SiO_2 (%) by mass	89.25
CaO (%) by mass	1.63
Al_2O_3 (%) by mass	0.51
Fe_2O_3(%) by mass	2.19
Na_2O (%) by mass	0.39
Loss on ignition (%) by mass	4.35

Table 3 Physical characteristics of aggregates

Characteristics	Stone aggregate	Sand	Waste foundry sand
Fineness modulus	2.789	2.379	1.857
Specific gravity	2.949	2.27	2.25

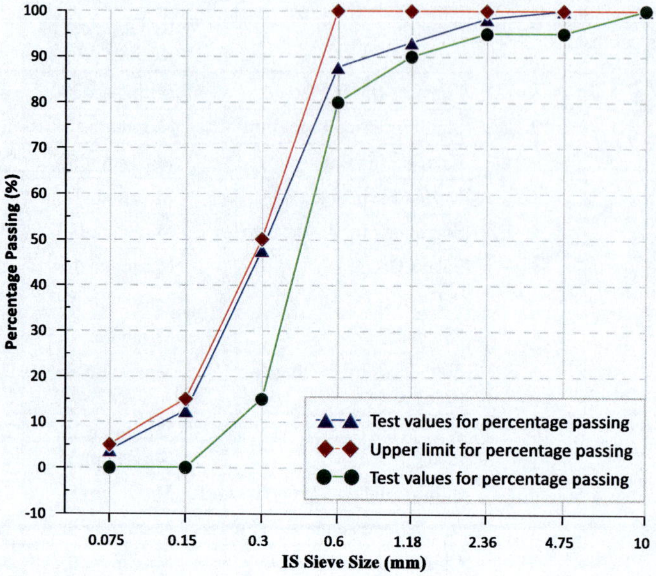

Fig. 1 WFS grading curve

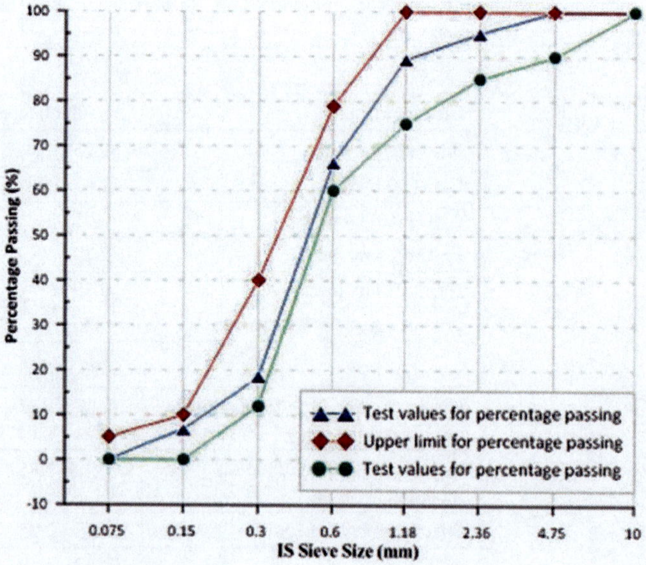

Fig. 2 Sand grading curve

2.1.4 Stone Aggregate

The crushed stone aggregate having 10 mm down size utilized in this study was obtained from a quarry in West Bengal's Birbhum area. Table 3 shows the specific gravity and fineness modulus of coarse aggregate.

2.1.5 Casting of Samples

Five sets of M20 grade concrete mix were prepared as specified in IS 456-2000 [6] with WFS 0%, 5%, 10%, 20%, 30%, and 40%, respectively, in lieu of sand. The concrete materials were blended in a blender. Weighed amounts of cement, sand, and WFS were dry mixed until a uniform mixture was obtained with no cement, sand, or WFS lump. The coarse aggregate was then weighed and blended in a dry state until a homogeneous mix was achieved. The total quantity of water added is equal to half of cement used and it was poured in two stages. Before casting the specimens, all of the moulds were lubricated. Each concrete mixture's 3nos of 15 cm * 15 cm * 15 cm cube was casted to carry out compressive strength test. A 50 cm * 50 cm * 10 cm roof sample casted of each concrete mix reinforced with 8 mm TMT bar for determination overall heat transfer (U value) value.

2.1.6 Compressive Strength

Concrete samples were tested for compressive strength at 28 days as per IS 516 [7] standard. The compressive strength of concrete using WFS as replacement of natural sand is shown in Fig. 3. Concrete compressive strength increases with increasing WFS percentage till 40%; however, further increasing WFS percentage decreases compressive strength. WFS has a lower particle size than natural sand, so it fills the microscopic pores inside concrete, causing pores to be minimized. Furthermore, the chemical characteristics of WFS are comparable to those of cement. These two factors help to boost compressive strength. A higher amount of WFS percentage impedes cement hydration, which may be the cause of decrease in compressive strength.

3 Thermal Performance

3.1 Guarded Hot Box

The guarded hot box method is used to find out the total thermal transmission coefficient (U value) of a building enclosure in a specified thermal environment. One similar facility has been established at Aliah University's Building Energy Efficiency Laboratory, Department of Mechanical Engineering. The testing facility was

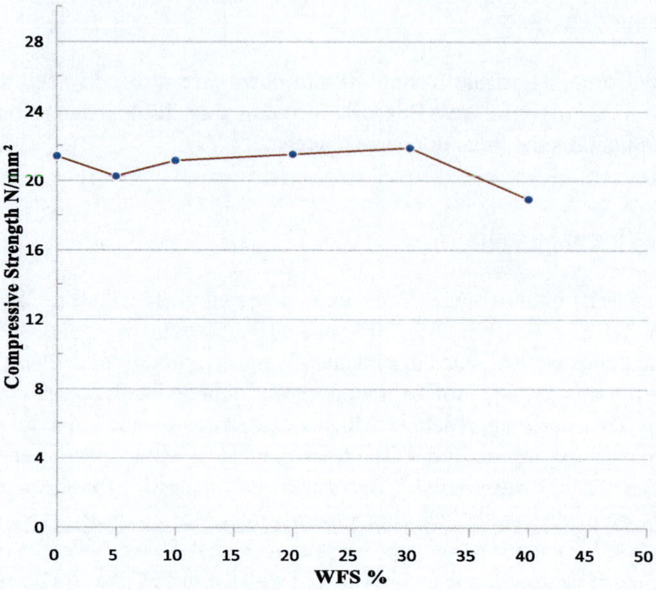

Fig. 3 Change in compressive strength with WFS%

built in compliance with BS-EN-ISO 8990:1996 [8] and BS-EN-ISO 874: Part 3: Sect. 3.1: 1987 [9]. Figure 4 shows an actual representation of the same. It is made up of 4 major parts, namely metering box (MB), guard box (GB), cold box (CB) and a surround panel which holds sample between the metering -box and the cold-box. This setup is capable to determine U value of any structural envelop with a range of 0.12–20 W/m^2 K for testing at temperatures ranging from −20 to 50 °C. The setup's walls are built with 25 cm extruded polystyrene reinforced with plywood. The sample can be inserted in sample holding split of the surround panel, which is located in the midst of MB and the CB. This cut-out is exactly in the middle of the surround panel's two edges. A comprehensive overview of the facilities was already published [10]. The temperature inside the MB and CB is kept at consistent levels. Temperatures in the guard-box which depict the environmental temperatures are maintained as prescribed by BS-EN-ISO 8990:1996, so that the cumulative metering box 'wall loss' is less than 10% of the total heat input from fans and heaters located inside the metering box. Because, as the fan motors are installed inside the metering box, it is assumed that under steady-state conditions, the electrical power into the fans is totally turned into heat. The metering box, guard box, and cold box all together assembled and tied against the surround panel using a tie bar assembly to create an airtight chamber, as shown in Fig. 4. Under steady-state conditions, all heat added to the metering box now via the sample and surround panels flows into the cold box.

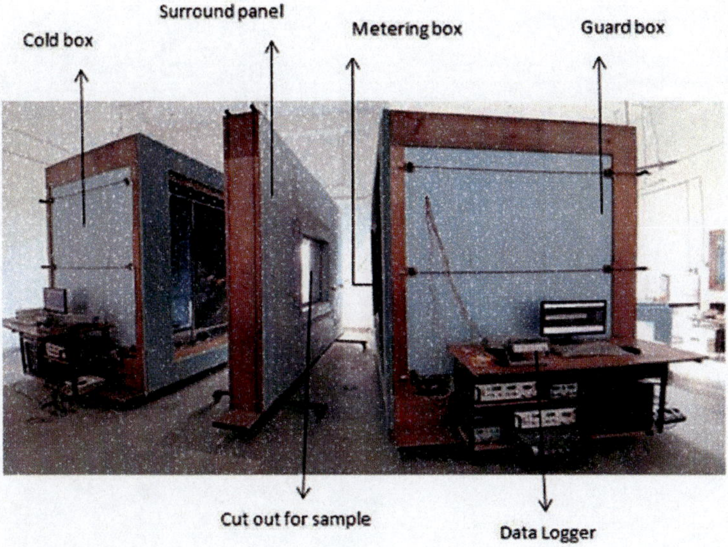

Fig. 4 Guarded hot box setup

3.2 Calibration of Guarded Hot Box

Calibration of guarded hot box apparatus before actual sample test is a prerequisite as it determines flanking loss and heat transfer to ambient from metering box. Detailed calibration process has been described by D. Choudhury [10].

3.3 Sample and Testing Condition

Roof samples are being cured for 28 days with water. Samples are loaded into sample holder made with FRP materials. Edges of the samples are sealed with silicon sealant to prevent air pass. Samples are then mounted into surround panel with the help of hydraulic trolley lifter (Figs. 5 and 6).

The tests are being carried out in accordance with BS EN ISO 8990:1996. The total time taken for each test is around 10 $\approx$ 20 h. Testing temperature of cold box and hot box which resembles outer ambient temperature and room temperature respectively was set to 15–30, 15–35, 20–35, 20–40 °C. Each roof sample was tested with same set of temperature difference. Concrete roof sample with more than 30% replacement of sand with WFS was excluded from test as further increase in WFS % compromised compressive strength beyond specified limit of IS standard.

Fig. 5 Sample in FRP frame

Fig. 6 Sample mounted in guarded hot box

3.4 Result and Discussions

This research was carried out to investigate the viability of replacing natural sand with WFS as a sustainable building material, as well as the influence of WFS on overall thermal transmittance (U value).It was found that U value decreases with the increase percentage of replacing sand with WFS (shown in Fig. 7).

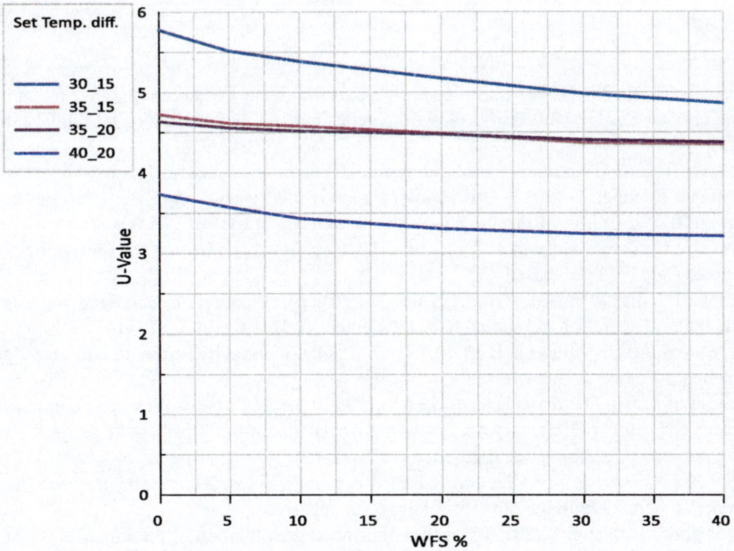

Fig. 7 Overall thermal transmittance with the % of WFS

4 Conclusion

Following can be concluded from this work

1. WFS, which is an industrial waste, can be effectively utilize as a sustainable building material. Up to 30% replacement of WFS in lieu of natural sand is safe in M20 grade concrete.
2. Effective use of WFS in building material cuts down the river sand extraction thus saves the environment.
3. Use of WFS in building is economical in two ways; first, it reduces the sand uses in building construction and reduces the thermal transmittance, and it will cut down the cooling energy requirements.
4. The environmental risks associated with industrial waste disposal can be mitigated by making sensible use of WFS in building construction.
5. Rapid mining of river bed sand has a negative impact on the environment that can be avoided.

Acknowledgements The authors acknowledge the contribution towards this research to M/S Met Alloy Corporation, Howrah, West Bengal.

References

1. Javed S, Lovell CW (1994) Use of foundry sand in highway construction. Department of Civil Engineering Purdue University, West Lafayette, Indiana. https://doi.org/10.5703/1288284316152
2. Oudhia SP (2015) An overview of indian foundry industry. Metalworld, pp 3–4
3. Siddique R, Singh G (2011) Utilization of waste foundry sand (WFS) in concrete manufacturing. Resour Conserv Recycl 55:885–892. 1016/j.resconrec.2011.05.001
4. Bureau of Indian Standards B (n.d.) IS 1489-1 (1991): specification for Portland pozzolana cement, Part 1: Flyash based
5. Bureau of Indian Standards B (n.d.) IS 383-1 (1970): specification for coarse and fine aggregates from natural sources for concrete [CED 2: cement and concrete]. In IS, vol 383
6. Bureau of Indian Standards B (n.d.) IS 456 (2000): plain and reinforced concrete—code of practice
7. Bureau of Indian Standards B (n.d.) IS 516 (1959): method of tests for strength of concrete
8. BS EN ISO 8990 (1996) Thermal insulation. Determination of steady-state thermal transmission properties. Calibrated and guarded hot box
9. BS EN ISO 874: Part 3: Section 3.1 (1987) Methods for determining thermal insulating properties with definitions of thermal insulating terms
10. Chowdhury D, Neogi S (2019) Thermal performance evaluation of traditional walls and roof used in tropical climate using guarded hot box. Constr Build Mater 218:73–89. https://doi.org/10.1016/j.conbuildmat.2019.05.032
11. http://foundryinfo-india.org/profile_of_indian.aspx

Fuzzy Logic-Based Model for Predicting Material Removal Rate of Machined Cupola Slag-Reinforced Aluminum Metal Matrix Composite

Soumyabrata Chakravarty, Partha Haldar, Titas Nandi, and Goutam Sutradhar

1 Introduction

The modern era of innovation needs economic new-age materials which have high strength-to-weight ratios and can accommodate tailored properties. Composites can be a great alternative for that fact, although the cost of composite fabrication is very high. The fabrication cost of composites can be reduced by using low-cost fabrication processes like stir casting and using economic reinforcement derived from industrial wastes such as cupola slag. Cupola slag is an industrial waste that can be reused as reinforcement in metal matrix as it contains hard ceramics like Al_2O_3, SiO_2, CaO, etc. [1]. Aluminum and its alloys are great to be used as matrix materials in composites due to their lightweight and higher conductivity and low melting point. The development of economic composites requires successful fabrication which should be followed by quantitative and qualitative studies of finishing operations such as machining. Machinability can be defined as the ease of machining. One of the most important parameters of machinability is the material removal rate (MRR). Turning is one of the most used machining processes. The MRR analysis in turning requires extensive experimentation and analysis. The experiment should be designed statistically to reduce materials and time loss following techniques such as Taguchi L9 orthogonal array. Taguchi L9 orthogonal array provides the most accurate analysis

S. Chakravarty (✉) · T. Nandi
Mechanical Engineering Department, Jadavpur University, Kolkata 700032, India
e-mail: soumyachakravarty.edu@gmail.com

P. Haldar
Mechanical Engineering Department, Government College of Engineering and Ceramic Technology, Kolkata 700010, India

G. Sutradhar
National Institute of Technology, Imphal 795004, India

B. P. Swain (ed.), *Recent Advances in Materials*, Springer Proceedings in Materials 25,
https://doi.org/10.1007/978-981-99-3844-5_19

with a minimal number of experiments. Further analysis of MRR requires mathematical modeling. Regression analysis is a simple and well-established mathematical model for MRR prediction. New soft computing-based prediction techniques like fuzzy prediction are believed to provide more accurate predictions. Fuzzy logic-based prediction is based on the principle of human intuition. Fuzzy logic takes various true values between 0 to 1 mimicking human behavior. The crisp input values are converted to linguistic variables which are then passed through pre-defined sets of fuzzy IF–THEN rules to predict the linguistic values of output. These outputs are converted to crisp prediction values.

Attempts have been made to analyze machinability in dry turning of aluminum metal matrix composites (AMCs) in terms of MRR in the present academia. Kesarwani et al. [2] have investigated MRR in hybrid eggshell, boron carbide-reinforced AMCs and concluded that MRR has been improved in presence of a second phase when compared with monolithic alloy. MRR in dry turning of discontinuous SiC_p-reinforced Al 7075 matrix has been analyzed by the gray Taguchi approach in the works of Das et al [3]. The MRR was observed to have an increasing trend with increasing cutting speed, feed, weight percentage of reinforcements. Kumar et al. [4] have investigated stir-casted Al/SiC/Mo-reinforced AMCs MRR in dry turning. The results indicate that maximum MRR has been achieved while turning at a high depth of cut, feed and cutting speed. The fuzzy logic-based prediction of MRR in turning has been discussed in the works of Saradhi et al. [5]. The results show valid fuzzy logic prediction with a percentage error of 7.49%. Sharma et al. [6] predicted cutting force in hard turning operations using fuzzy logic and compared it with regression prediction. The authors have concluded that fuzzy prediction is more accurate. The literature survey indicates that MRR analysis in turning of AMCs is adequate, but fabrication and machinability studies on cupola slag-reinforced AMCs are found to be scarce.

In this work, cupola slag-reinforced LM11 matrix composites have been fabricated using the stir casting route. Machinability of cast composites has been analyzed in terms of MRR in dry turning. Spindle speed, feed rate and weight percentage of cupola slag have been chosen as process parameters for turning. The effect of process parameters on MRR has been studied in detail. MRR prediction using regression and fuzzy model has been established and an elaborate comparative study between the models has been presented.

2 Materials and Methods

2.1 Materials

Al-4.5-Cu or LM 11 has been chosen as the matrix of the composites due to its lightweight and high strength moreover LM11 has shown great response to solution heat treatment. 99% pure LM 11 has been procured from M/S Kolkata Die Casting,

Table 1 Chemical composition of LM 11

Elements	Cu	Mg	Si	Fe	Mn	Ni	Zn	Pb	Sn	Ti	Al
Wt.%	4.5	0.1	0.25	0.25	0.1	0.1	0.1	0.05	0.05	0.3	Bal

Table 2 Chemical composition of Cupola slag

Elements	SiO_2	Fe_2O_3	Al_2O_3	CaO	MnO	MgO	TiO_2	K_2O	SiO_2	Fe_2O_3	Oxides
Wt.%	53.1	16.1	11.1	10.7	3.33	1.94	1.22	1.05	53.1	16.1	Bal

Liluah, Howrah. The chemical composition of LM 11 as observed using XRF has been shown in Table 1. Composites can be fabricated by using low-cost reinforcement developed from industrial waste cupola slag. Bolder like cupola slag chunks has been fetched from M/s Binoy Udyog Pvt. Ltd., Andul, Howrah. The slag has been ball milled and sieved to an average particle size of 100 μm. The chemical composition of developed cupola slag reinforcement particles has been analyzed using XRF, and the results have been reported in Table 2.

2.2 *Fabrication of Cast Composites*

The stir casting has been performed in vacuum-assisted bottom pouring type stir casting setup supplied by SWYAMEQUIP, Chennai. The weighted amount of LM 11 ingots has been melted in a graphite crucible at 750 °C and at the same time 3% weight percent of developed cupola slag particles (average particle size 100 μm) has been pre-heated to 300 °C in a powder pre-heating furnace. After the complete melting of the LM 11 ingots, mechanical stirrer has been pulled down into the molten metal, and a stirring speed of 500 rpm has been maintained to initiate a vortex in the molten metal. The pre-heated cupola slag particles have been poured into the vortex while stirring. The stirring continues for another 8–10 min with a stirring speed of 600 rpm to ensure homogeneous particle distribution. The spilt-type steel mold has been positioned parallel to the bottom pouring valve after pre-heating to 400 °C to achieve defect less casting. Vacuum pressure has been set to 10^{-2} mbar using a vacuum pump. The bottom pouring valve has been opened after the proper mixing of particles. The molten metal has been poured to mold due to gravity and left for solidification for 24 h. Upon solidification, 3wt.% cupola slag-reinforced LM 11 matrix composite has been fabricated. Similar methods have been followed for the fabrication of 5 and 7 wt.% cupola slag-reinforced cast composites.

Table 3 Input parameters along with their levels [10]

Parameters	Unit	Low (L)	Medium (M)	High (H)
Spindle speed (N)	rpm	495	620	800
Feed rate (f)	mm/rev	0.083	0.109	0.125
Weight percentage (w)	Wt.%	3	5	7

2.3 Taguchi Experimental Design

The machinability studies in terms of material removal rate have been performed in dry turning. The process parameters for turning have been chosen as spindle speed (rpm), feed rate (mm/rev) and weight percentage (wt.%) as these are the most effective parameters in turning of cast composites as per established academia [7]. MRR has been chosen as a response parameter as it is one of the most important aspects of machinability. The depth of cut has been deliberately kept constant as 1 mm, in the investigation as the effect of depth of cut on MRR has been found to be nominal in the literature [8, 9].

The experiments have been designed by using 3 factors 3 levels Taguchi L9 orthogonal array. The experimental design and analysis have been performed on Minitab 18 software. The levels of process parameters have been selected as per intuition from academia and pilot experiments and are shown in Table 3 [7–9].

2.4 Fuzzy Prediction Modeling

In this work, experimental investigation has been used to formulate a fuzzy prediction model in MATLAB R2022b. A Mamdani max–min fuzzy interface system has been designed using three inputs, and one output as the Mamdani system is one of the simplest and most accurate processes to predict response [11]. The input variables and output response have been fuzzified by converting the crisp value to linguistic variables of low (L), medium (M) and high (H) as shown in Table 3. The membership functions have been defined as per intuition from the experimental results and literature survey [11]. Triangular membership function has been chosen for all the variables to simplify the prediction model which is given as,

$$\mu_{\text{Triangle}}(x; L, M, H) = \begin{cases} 0, & x \le L \\ \frac{x-L}{M-L}, & L \le x \le M \\ \frac{H-x}{H-M}, & M \le x \le H \\ 0, & H \le x \end{cases}$$
$$= \max\left(\min\left(\frac{x-L}{M-L}, \frac{H-x}{H-M}\right), 0\right) \quad (1)$$

Table 4 Fuzzy IF–THEN rules

Rule no.	IF (AND operation between the parameters)			THEN
	Spindle speed	Feed rate	Weight percentage	MRR
1	L	L	L	L
2	L	M	M	M
3	L	H	H	H
4	M	L	M	L
5	M	M	H	M
6	M	H	L	M
7	H	L	H	M
8	H	M	L	H
9	H	H	M	H

The fuzzy rules in terms of linguistic variables have been shown in Table 4. In Table 4, parameters under IF are the inputs, and the parameter under THEN column is the output or conclusion.

2.5 Methodology

Setup. The dry tunning experiments have been carried out in conventional engine lathe in Blue Earth Machine Shop, Jadavpur University, Kolkata. CNMG 120408 coated carbide tool with tool holder DCLNR2020K12 has been used as cutting tool. The cutting length has been taken as 50 mm. The weight of samples before and after machining has been measured using high precision industrial weighing machine Mettler Toledo—BBA236-4A3N.

Procedure. The MRR has been measured using weight loss method, the weight of cast composite has been measured before and after machining and the machining time has been noted. Then the MRR has been calculated by,

$$\text{MRR} = \frac{W_i - W_f}{t}\ \text{g/min}, \tag{2}$$

where, W_i is the initial weight, W_f is the final weight and t is machining time.

The measured MRR has been analyzed using Taguchi main effect plots. A regression model has been developed to predict the MRR for same set of data points. Fuzzy prediction model has been applied to predict the same. The comparison between experimental, regression and fuzzy prediction has been investigated. The flow chart of the methodology has been shown in Fig. 1.

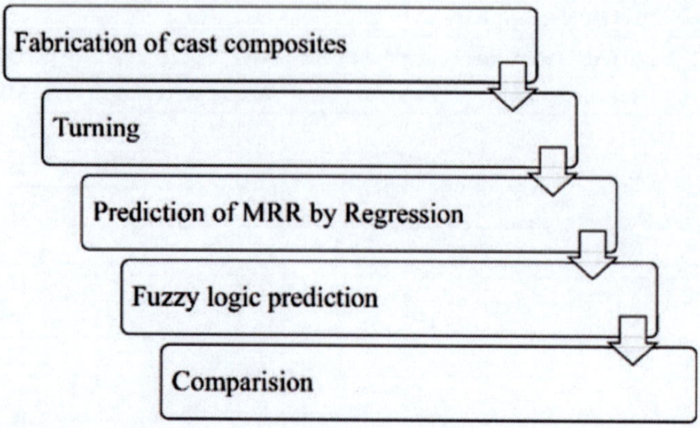

Fig. 1 Flow chart of experimental investigation

3 Results and Discussions

3.1 Taguchi Results

The results of the experimental investigations have been reported in Table 5. It is evident from main effect plot shown in Fig. 2 that with increasing spindle speed, MRR decreases slightly and then increases at highest speed. This is due to the fact that with increasing spindle speed, increases the tool workpiece temperature which initiates strain hardening which reduces the MRR [12]. Further increase in spindle speed increases the interface temperature to an extent, where MRR increases due to thermal softening of the workpiece [13]. The effect of spindle speed on MRR found in agreement with established academia [14]. The MRR observed to be increase with increasing feed rate. This is attributed to the fact that with increasing feed rate, the contact area between cutting tool and work piece increases which results in higher MRR. The similar trend of increasing MRR with feed rate increase has been reported in the works of Yin et al. [15] and Sharma et al. [6]. The MRR has been investigated to increase with an increasing weight percentage of cupola slag. This is due to the fact that an increase in weight percentage increases the presence of cupola slag particles in the matrix. These particles increase porosities and dislocations in the matrix which act as crack generation sites and eases the machining process. This leads to an increase in MRR with an increasing weight percentage.

Table 5 Experimental results

Exp no.	Spindle speed	Feed rate	Weight percent	MRR (g/min)		
				Experimental	Regression	Fuzzy
1	495	0.083	3	7.50	7.91	7.79
2	495	0.109	5	13.85	11.86	11.7
3	495	0.125	7	15.00	14.46	14.5
4	620	0.083	5	9.23	9.51	9.08
5	620	0.109	7	11.54	13.46	10.7
6	620	0.125	3	13.33	14.71	13.4
7	800	0.083	7	12.00	11.63	11.7
8	800	0.109	3	15.00	14.22	14.9
9	800	0.125	5	17.14	16.82	16.8

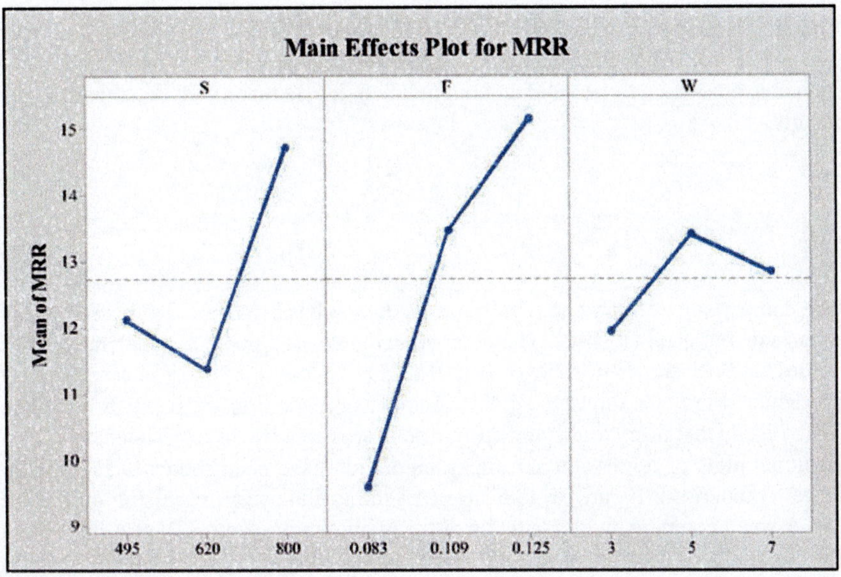

Fig. 2 Main effect plots of MRR

3.2 Regression Results

The results of regression analysis have been shown in Table 5. The linear regression has been adopted for finding the regression equation. The regression equation has been given as,

$$\text{MRR} = -8.50 + 0.00923S + 134.5F + 0.225W, \quad (3)$$

where S, F and W are spindle speed, feed rate and weight percentage, respectively. The value predicted from the regression equation has been presented in Table 5.

3.3 Fuzzy Logic Prediction

The prediction using fuzzy logic has been reported in Table 5. The predictions have been observed using rule viewer in MATLAB and reported in Table 5. Table 5 indicates prediction to be valid as the predicted data is maps with the experimental results.

3.4 Comparison

The comparison of fuzzy and regression prediction has been analyzed based on root mean squared error (RMSE). The comparison table along with residual and RMSE values has been shown in Table 6. The RMSE value found to be less in case of fuzzy prediction. Figure 3a shows the prediction plot for regression and fuzzy. It is evident from Fig. 3a that fuzzy logic prediction results are nearer to the experimental results. Residual plots of regression and fuzzy prediction have been shown in Fig. 3b for better comparison. Figure 3b indicates that the residuals for regression model are larger when compared with residuals of fuzzy prediction. Figure 3b also shows that the residuals for both fuzzy prediction and regressions are random which indicates that the models are free from inherent error thus can be treated as valid. The similar results in comparison of fuzzy logic prediction and regression prediction have been found in the works of Sharma et al. [6] and Amir et al. [11]. Comparison of predictions along with experimental values has yield error of 1.10% and 0.81% for regression and fuzzy-based model, respectively, as per Table 6. The fuzzy-based model yields more accurate result due to the fact that fuzzy logic is soft computing-based method in which model learns the trends of output by intuition from the experimentation thus yields more accurate results [6].

Table 6 Comparison between experimental, regression and fuzzy prediction in terms of residual and RMSE

Exp no	MRR			Residuals		Residual squared	
	Exp	Regression	Fuzzy	Regression	Fuzzy	Regression	Fuzzy
1	7.50	7.91	7.79	−0.41	−0.29	0.17	0.08
2	13.85	11.86	11.7	1.99	2.15	3.95	4.61
3	15.00	14.46	14.5	0.54	0.50	0.29	0.25
4	9.23	9.51	9.08	−0.28	0.15	0.08	0.02
5	11.54	13.46	10.7	−1.92	0.84	3.70	0.70
6	13.33	14.71	13.4	−1.38	−0.07	1.90	0.00
7	12.00	11.63	11.7	0.37	0.30	0.14	0.09
8	15.00	14.22	14.9	0.78	0.10	0.61	0.01
9	17.14	16.82	16.8	0.32	0.34	0.10	0.12
RMSE (%)						1.10	0.81

4 Conclusions

The objective of this experimental work is to successfully fabricate cupola slag-reinforced LM11 metal matrix composite along with machineability studies in terms of MRR. The regression and fuzzy-based prediction model have been developed and validated. The experimental investigation can be concluded as:

- Successful fabrication of 3, 5, 7 wt.% cupola slag-reinforced LM11 composites has been achieved using stir casting route.
- The MRR has been observed to have increasing trend with increasing spindle speed, feed and weight percentage.
- The regression and fuzzy-based prediction model have been compared with experimental results and has an RMSE error of 1.10% and 0.81%, respectively.
- The fuzzy prediction model observed to have better prediction when compared with regression model.

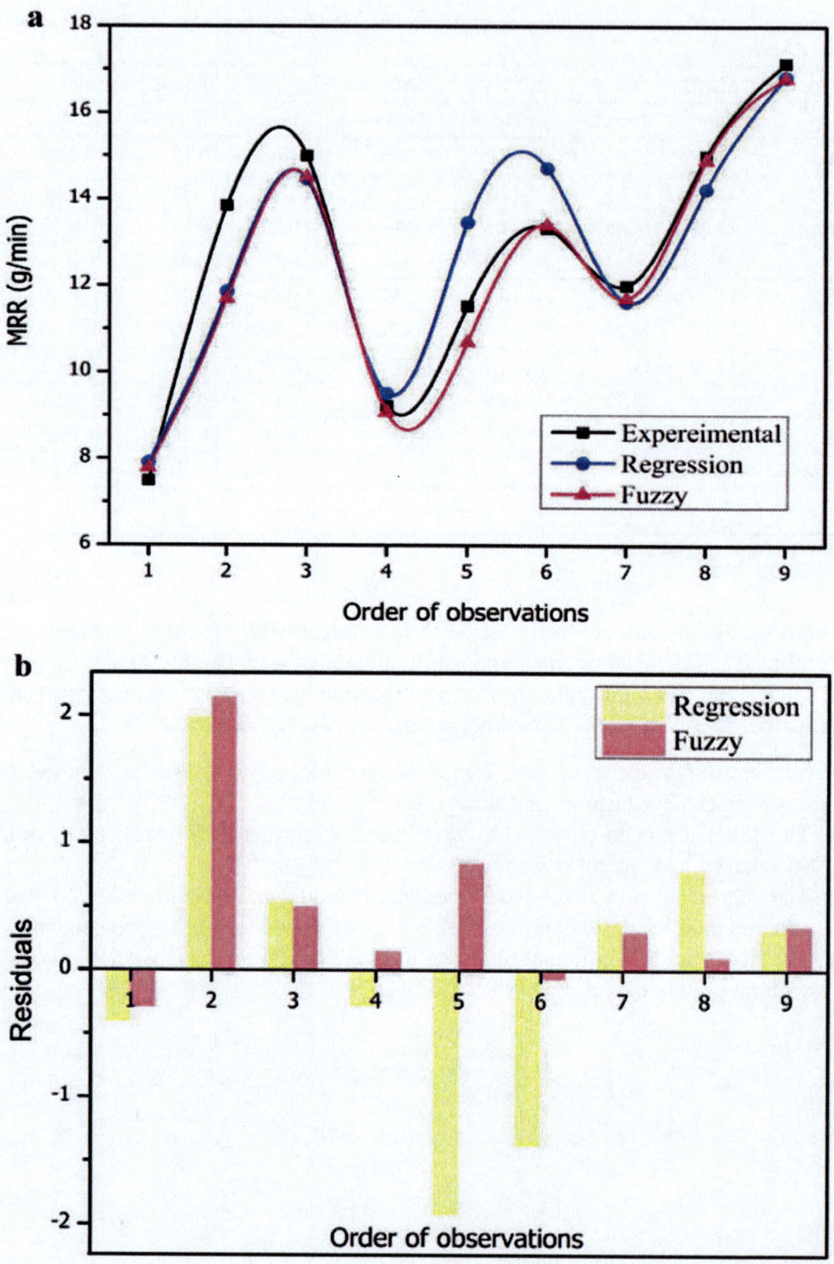

Fig. 3 **a** Prediction plot for regression and fuzzy logic, **b** residual plot for regression and fuzzy

Acknowledgements Department of Science & Technology and Biotechnology, Government of West Bengal (Memo No.: 114 (Sanc.)/STBT—11012(16)/16/2021—ST SEC dated 28.04.2022) has been acknowledged by authors for funding this work.

References

1. Chakravarty S, Haldar P, Nandi T, Sutradhar G (2021) Cupola slag reutilization for sustainable waste management: review and economic analysis. Int J Environ Sci Technol, pp 1–16
2. Kesarwani S, Niranjan MS, Singh V (2020) To study the effect of different reinforcements on various parameters in aluminium matrix composite during CNC turning. Compos Commun 22:100504
3. Das D, Saha S, Nayak BB, Nanda BK (2018) Surface quality and material removal rate in turning aluminium matrix composite–a grey relational approach. Mater Today Proc 5(11):23568–23575
4. Kumar J, Singh D, Kalsi NS, Sharma S, Mia M, Singh J, Rao KV (2021) Investigation on the mechanical, tribological, morphological and machinability behavior of stir-casted Al/SiC/Mo reinforced MMCs. J Market Res 12:930–946
5. Saradhi VP, Shashank V, Anbarasu G, Bharat A, Jagadesh T (2018) Prediction of surface roughness and material removal rate in laser assisted turning of aluminium oxide using fuzzy logic. Mater Today Proc 5(9):20343–20350
6. Sharma V, Kumar P, Misra JP (2020) Cutting force predictive modelling of hard turning operation using fuzzy logic. Mater Today Proc 26:740–744
7. Subramanian AVM, Nachimuthu MDG, Cinnasamy V (2017) Assessment of cutting force and surface roughness in LM6/SiCp using response surface methodology. J Appl Res Technol 15(3):283–296
8. Manna A, Bhattacharayya B (2005) Influence of machining parameters on the machinability of particulate reinforced Al/SiC–MMC. Int J Adv Manuf Technol 25:850–856
9. Agrawal SS, Yadava V (2013) Modeling and prediction of material removal rate and surface roughness in surface-electrical discharge diamond grinding process of metal matrix composites. Mater Manuf Processes 28(4):381–389
10. Chakravarty S, Haldar P, Nandi T, Sutradhar G (2023) Fabrication and machinability studies on cupola slag reinforced Aluminium metal matrix composites using Taguchi method. Mater Today Proc
11. Aamir M, Tu S, Tolouei-Rad M, Giasin K, Vafadar A (2020) Optimization and modeling of process parameters in multi-hole simultaneous drilling using Taguchi method and fuzzy logic approach. Materials 13(3):680
12. Giraud E, Rossi F, Germain G, Outeiro JC (2013) Constitutive modelling of AZ31B-O magnesium alloy for cryogenic machining. Proc CIRP 8:522–527
13. Pawade RS, Joshi SS, Brahmankar PK, Rahman M (2007) An investigation of cutting forces and surface damage in high-speed turning of Inconel 718. J Mater Process Technol 192:139–146
14. Hu L, Cai W, Shu L, Xu K, Zheng H, Jia S (2021) Energy optimisation for end face turning with variable material removal rate considering the spindle speed changes. Int J Precis Eng Manuf Green Technol 8(2):625–638
15. Yin G, Shen J, Wu Z, Wu X, Jiang F (202) Experimental investigation on the machinability of PCBN chamfered tool in dry turning of gray cast iron. Processes 10(8):1547

Comparative Performance and Emission Characteristics of Diesel-Ethanol Fuel Blends on a CRDI

Pradeep Kumara, B. Akhil, Riyaz Bashaa, Venu Gopal, B. Anil, Upendra Rajak, K. Thirupathi Reddy, Tikendra Nath Verma, and Manoj Arya

1 Introduction

Bioenergy is an unconventional and sustainable fuel for compression ignition engines that may be produced from a variety of edible and non-edible vegetable oils derived from first-, second-, and third-generation renewable fuels [1] Due to the consistent expansion in the use of vehicles in industry, populace development, and economic growth, energy depletion has increased quickly [2]. Due to the quantity of oxygen they contain, alcohols frequently offer superior properties [3]. Sardemir et al. [4] investigated the influence of maize oil methyl-ester and diesel mixtures on engine presentation and discharge features. At fuel injection pressures of 210 and 230 bar, three unique fuel samples (10, 20, and 50%) were evaluated. They found that when the injection pressure rose, the breadth of the droplets reduced. Also observed at 230 bar were reductions in CO by 66.67% and HC by 52.3%, but increases in NO_x by 22.45%. Rajak et al. [5] studied the things of the Aegle methyl-ester and diesel mixes on engine characteristics. According to the findings, there is a 3.0%, 4.5%, and 8.5% increase in brake-specific fuel consumption, CO_2 releases, and NO_x releases, respectively. Moreover, there is a rapid reduction of 21.4%, 20.83%, and 24.41% in the discharges of particulate matter, smoke, and soot, correspondingly. Using the EGR technique, Sindhu et al. calculated the fuel injection timing and discovered that higher fuel injection timing led to higher NO emissions but decreased soot discharge [6]. Karagoz et al. [7] examined the engine operation and ignition properties of a

P. Kumara · B. Akhil · R. Bashaa · V. Gopal · B. Anil · U. Rajak (✉) · K. T. Reddy
Department of Mechanical Engineering, RGM College of Engineering and Technology, Nandyala 518501, India
e-mail: upendrarajak86@gmail.com

T. N. Verma (✉) · M. Arya
Department of Mechanical Engineering, Maulana Azad National Institute of Technology Bhopal-, 462003, Bhopal, India
e-mail: verma.tikks@gmail.com

B. P. Swain (ed.), *Recent Advances in Materials*, Springer Proceedings in Materials 25,
https://doi.org/10.1007/978-981-99-3844-5_20

Table 1 Fuel sample

Properties	Diesel	Ethanol	E5D95	E10D90	E15D85
Density (kg/m^3)	830	800	828.5	827.2	825.5
Viscosity (mm^2/s)	3.4	1.1	3.2	3.05	2.88
Cetane number	48	6–8	46.1	44	42
Heat value (MJ/kg)	45.5	26–27	44.5	43.8	42.93

single/cylinder, direct/injected, diesel engine using tyre pyrolytic oil-diesel mixes at various engine loads (three, six, nine, and twelve Nm) at a speed of two thousand revolutions per minute. The highest thermal efficiencies for TPO10D90 were found to be 28.15%. In this work, studies of engine performance were conducted on a single-cylinder, four-stroke, naturally aspirated, compression ignition diesel engine fuelled with ethanol–diesel blends of varying volumetric fractions, under varying engine loads and constant engine speed. The findings of combustion investigations conducted on ethanol diesel blends were presented and compared with diesel fuel.

2 A Fuel Sample and Method

2.1 *Fuel Properties*

In this investigation, base diesel fuel (95, 90, and 85%) was combined with ethanol (5, 10, and 15%) as E5D95, E10D90, and E15D85. Table 1 provided a brief overview of the characteristics of ethanol and diesel. The numerical employed three ethanol blends with different ratios to base diesel.

2.2 *Experimental Set-up*

The Kirloskar, single cylinder, four-stroke, water-cooled, 110 mm stroke, 87.5 mm bore, 661 cc was used in the experiments for this study. Table 2 contains the test engine's technical specs. In Fig. 1, the experimental set-up for this study is depicted schematically. At a 200 bar injection pressure, four different loads 25%, 50%, 75%, and 100% were tested. The gas spindle was maintained at this speed throughout all experiments when the engine speed was changed to 1500 rpm. The experimental set-up instrument's uncertainty as stated in Table 3.

Table 2 Test engine condition

Constraint	Value
Bore/stroke	(87.5mm*110mm)
Compression ratio	18
Swept volume	661.45 cc
Power	3.5 kW @1500 rpm

Fig. 1 Engine set-up

Table 3 Uncertainty of instruments

Constraint	Value (%)
Smoke	±1.0
Temperature	±0.5
Pressure	±0.5
Load meter	±0.2
Encoder	±0.2

3 Results

3.1 Validation of Model

This study used first law thermodynamic models to examine the combustion of the Diesel-RK engine [8–10]. Numerical data and trial data from the engine test were compared. The error was estimated to be up to 8.3% for the peak value of the engine

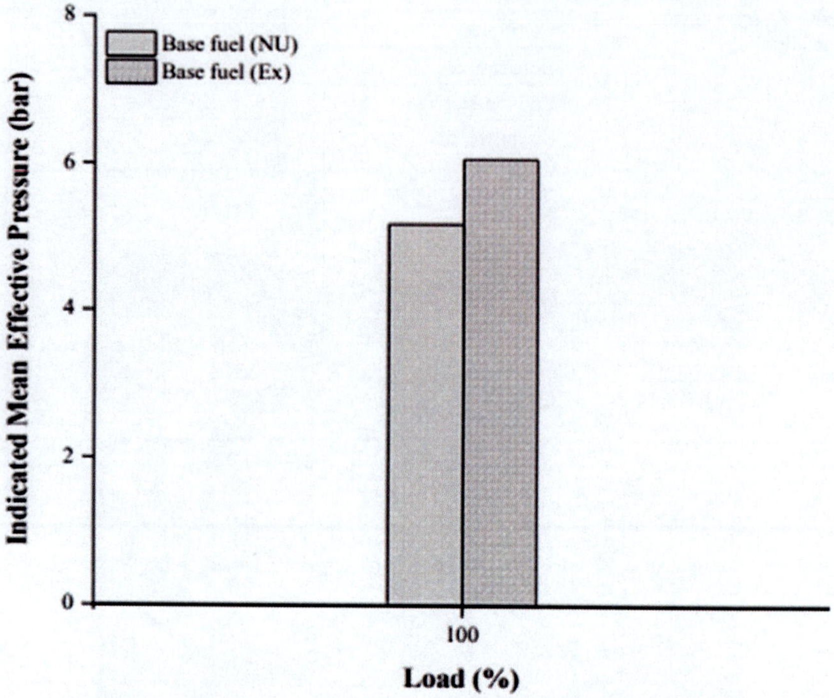

Fig. 2 IMEP with load (100%)

IMEP at the same operating state, as seen in Fig. 2's variance between trial and numerical results at full load (the operating conditions are the same as in Table 2).

3.2 *Indicated Mean Effective Pressure (IMEP)*

IMEP change in relation to load is seen in Fig. 3 along with the blend ratio of ethanol (5, 10, and 15%) to base fuel (95, 90, and 85%) at constant injection pressure and speed. The pressure operating on the piston during its whole stroke, which would result in the same amount of work output, can be compared with the given IMEP [11]. The value of 5% ethanol has the highest IMEP when compared with others blend ratio. For base fuel (BF100), 5%, 10%, and 15% of ethanol with full load conditions, the IMEP (bar) was determined to be 5.2, 5.5, 5.1, and 5.0, respectively.

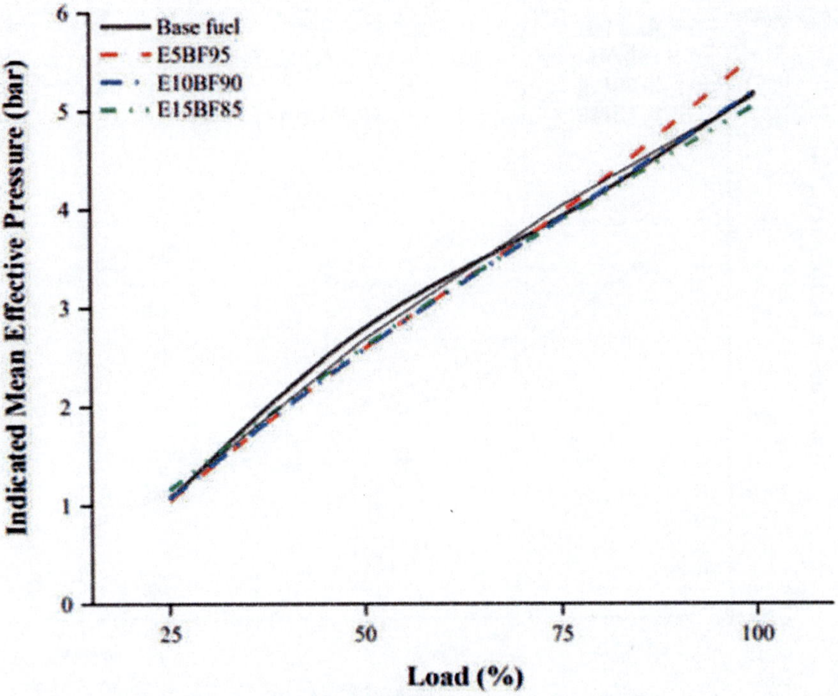

Fig. 3 Indicated mean effective pressure with loads

3.3 Engine Volumetric Efficiency (EVE)

EVE change in relation to load is seen in Fig. 4 along with the blend ratio of ethanol (5, 10, and 15%) to base fuel (95, 90, and 85%) at constant injection pressure and speed. The value of 10% ethanol has the lowest EVE when compared with others blend ratio. For base fuel (BF100), 5%, 10%, and 15% of ethanol with full load conditions, the EVE (%) was determined to be 94.87, 94.9, 94.7, and 95.1, respectively. The data demonstrates that a greater proportion of blends has a greater volumetric efficiency than base gasoline owing to the addition of extra air to the cylinder. It was determined that the E15D85 mixture had a higher volumetric efficiency at full load [5].

3.4 Cylinder Temperature (CT)

CT change in relation to load is seen in Fig. 5 along with the blend ratio of ethanol (5, 10, and 15%) to base fuel (95, 90, and 85%) at constant injection pressure and speed. The value of 5% ethanol has the highest CT when compared with others blend ratio. The in-cylinder temperature, oxygen concentration, thermal conductivity, water

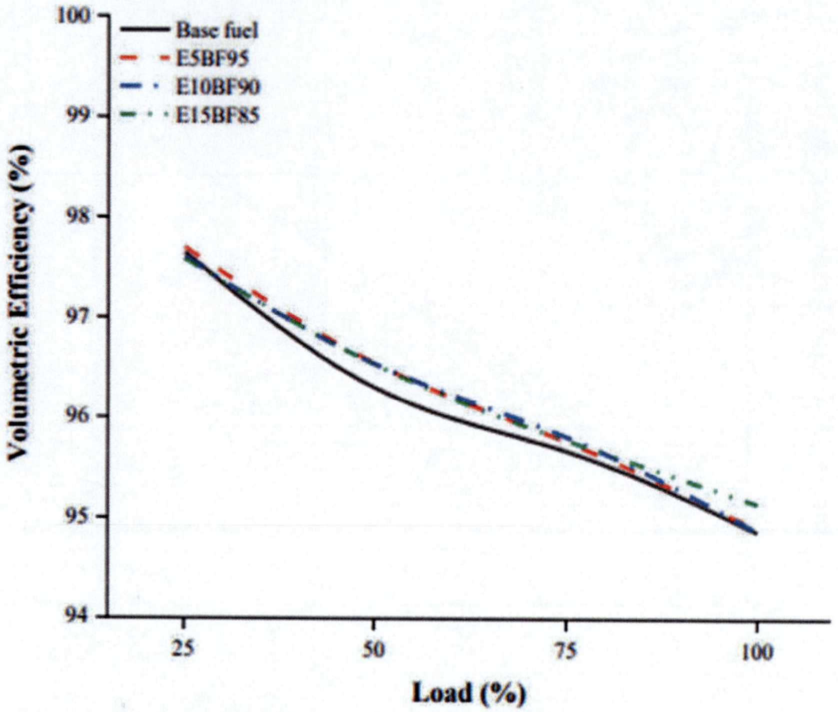

Fig. 4 Volumetric efficiency with loads

content, and the amount of time it takes for the reaction to occur in the test fuels are all closely correlated with the generation of nitrogen oxide emissions [12, 13].

3.5 *Factor of Absolute Light Absorption (FALA)*

FALA change in relation to load is seen in Fig. 6 along with the blend ratio of ethanol (5%, 10%, and 15%) to base fuel (95%, 90%, and 85%) at constant injection pressure and speed. The value of 15% ethanol has the highest FALA when compared with others blend ratio. This may be CT, oxygen concentration [14], water content, and the time required for the reaction to proceed in the test fuels are all strongly correlated with FALA production.

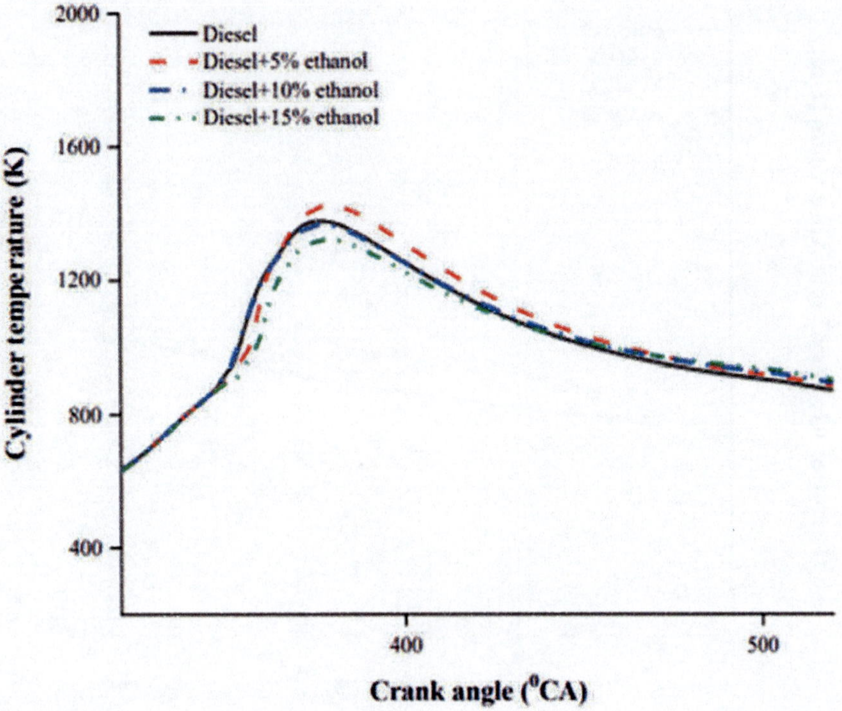

Fig. 5 Cylinder temperature with crank angle

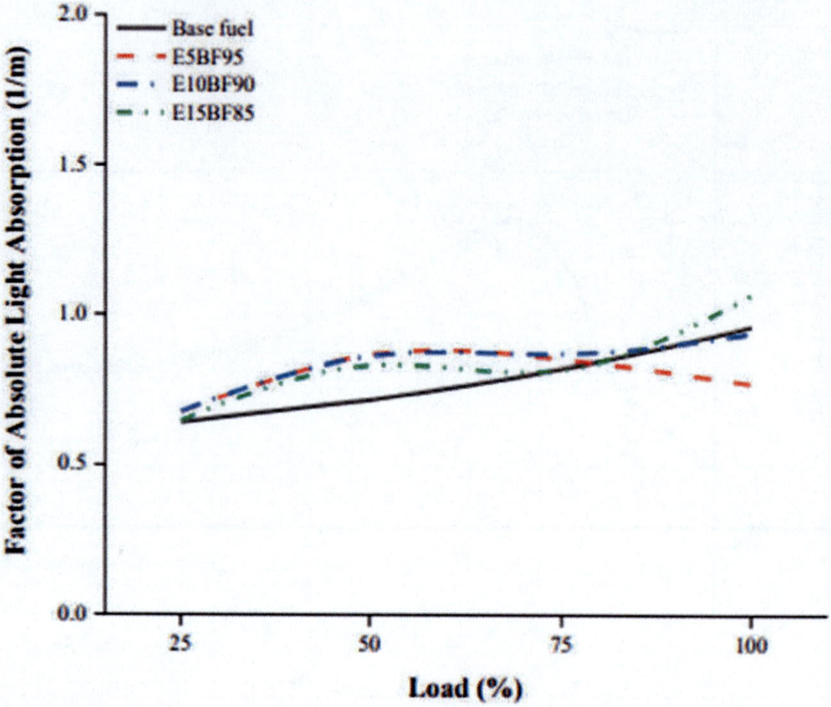

Fig. 6 FALA with load

4 Conclusion

This research evaluates the effects of 5%, 10%, and 15% base gasoline and ethanol mixtures on presentation, ignition, and discharges in a diesel engine with natural aspiration. The list below includes the main numerical findings.

- At 100% load, 5% ethanol increased IMEP by 7.2%. Reduced viscosity of ethanol improves its combustion rate.
- The volumetric efficiency at 100% load is observed to increase as the ethanol enrichment percentage (15%) rises.
- The amount of factor of absolute light absorption (FALA) emission dramatically decreased with the addition of 5% ethanol. In comparison with base fuel, the FALA emission is significantly lower at 5% by 19.7%.

References

1. Rajak U, Nashine P, Verma T (2020) Comparative assessment of the emission characteristics of first, second and third generation biodiesels as fuel in a diesel engine. J Therm Eng 211–225. https://doi.org/10.18186/thermal.818036
2. Verma TN, Shrivastava P, Rajak U, Dwivedi G, Jain S, Zare A et al (2021) A comprehensive review of the influence of physicochemical properties of biodiesel on combustion characteristics, engine performance and emissions. J Traffic Transp Eng (English Ed). https://doi.org/10.1016/j.jtte.2021.04.006
3. Ağbulut Ü, Sarıdemir S, Karagöz M (2020) Experimental investigation of fusel oil (isoamyl alcohol) and diesel blends in a CI engine. Fuel 267:117042. https://doi.org/10.1016/j.fuel.2020.117042
4. Sarıdemir S, Etem Gürel A, Ağbulut Ü, Bakan F (2020) Investigating the role of fuel injection pressure change on performance characteristics of a DI-CI engine fuelled with methyl ester. Fuel 271:117634. https://doi.org/10.1016/j.fuel.2020.117634
5. Rajak U, Nashine P, Verma TN (2019) Performance analysis and exhaust emissions of aegle methyl ester operated compression ignition engine. Therm Sci Eng Prog 12:100354. https://doi.org/10.1016/j.tsep.2019.05.004
6. Sarıdemir S, Ağbulut Ü (2022) Combustion, performance, vibration and noise characteristics of cottonseed methyl ester–diesel blends fuelled engine. Biofuels 13:201–210. https://doi.org/10.1080/17597269.2019.1667658
7. Karagoz M, Uysal C, Ağbulut Ü, Sarıdemir S (2020) Energy, exergy, economic and sustainability assessments of a compression ignition diesel engine fueled with tire pyrolytic oil–diesel blends. J Clean Prod 264:121724. https://doi.org/10.1016/j.jclepro.2020.121724
8. Kesharvani S, Verma TN, Dwivedi G (2023) Computational analysis of chlorella protothecoides biofuels on engine combustion, performance and emission. Sustain Energy Technol Assess 55:102972. https://doi.org/10.1016/j.seta.2022.102972
9. Singh A, Sinha S, Choudhary AK, Panchal H, Elkelawy M, Sadasivuni KK (2020) Optimization of performance and emission characteristics of CI engine fueled with Jatropha biodiesel produced using a heterogeneous catalyst (CaO). Fuel 280:118611. https://doi.org/10.1016/j.fuel.2020.118611
10. Khan K, Kumar G, Sharma AK, Kumar PS, Mandal C, Chintala V (2018) Performance and emission characteristics of a diesel engine using complementary blending of castor and karanja biodiesel. Biofuels 9:53–60. https://doi.org/10.1080/17597269.2016.1256552
11. Gentle R, Edwards P, Bolton B (2001) Introduction: the basis of engineering. In: Gentle R, Edwards P, Bolton BBT-MES (eds) IIE core textbook series, Butterworth-Heinemann, Oxford, p 1–6. https://doi.org/10.1016/B978-075065213-1/50001-5
12. Rajak U, Nashine P, Verma TN (2019) Effect of fuel injection pressure in a diesel engine using microalgae-diesel emulsion. Int J Eng Adv Technol 8:263–271
13. Martos FJ, Doustdar O, Zeraati-Rezaei S, Herreros JM, Tsolakis A (2023) Impact of alcohol–diesel fuel blends on soot primary particle size in a compression ignition engine. Fuel 333:126346. https://doi.org/10.1016/j.fuel.2022.126346
14. Kesharvani S, Dwivedi G. Algae as a feedstock for biodiesel production in Indian perspective. Mater Today Proc 2021. https://doi.org/10.1016/J.MATPR.2021.04.295.

Swan Eggshell Derived CaO as a Heterogeneous Catalyst for Biodiesel Production from Castor Oil

Mohd Rakimuddin Khan and Huirem Neeranjan Singh

1 Introduction

The need for alternative energy sources has arisen as a result of the global's accelerating energy utilization and the quickly dwindling supply of fossil fuels. The production of biodiesel from vegetable oils and other sources, which mostly contain triglycerides and free fatty acids, is one of the most promising techniques for sustainable fuels [1]. Green fuel biodiesel is a promising and renewable option to fossil fuels that are increasingly widely acknowledged. Due to its renewable, non-toxic, and clean burning qualities, biodiesel shows to be a better option for dual crises of fossil fuel depletion and environmental damage [2]. As a result, it is considered the greatest petrodiesel supplement. Additionally, biodiesel is eco-friendly, carbon neutral, and non-hazardous, all of which are amazing qualities. The use of biodiesel through blend or direct (usually 20–50% blend is employed) to petrodiesel can help reduce the global's reliance on conventional fuels and surrounding problems with carbon dioxide emissions [3]. Biodiesel could be obtained from a variety of sources, including camelina oil [1], chicken fat [4], microalgae Chlorella pyrenoidosa [5], castor oil [6], waste cooking oil [7], Crotalaria Juncea seeds [8], trisperma (*Aleurites trisperma*), moringa (*Moringa oleifera*), neem (*Azadirachta indica*), trisperma (*Aleurites trisperma*), jatropha (*Jatropha curcas*), and candlenut (*Aleurites moluccana*) [9].

The process of transesterification is very effective for making biodiesel. Due to its low cost, oil from non-edible vegetables is the better option for biodiesel production. Both heterogeneous and homogeneous catalysts can be used in the transesterification process to speed up the reaction. However, it is not recommended to utilize the homogeneous catalyst KOH/NaOH since it has a problem separating and purifying.

M. R. Khan (✉) · H. N. Singh
Department of Mechanical Engineering, National Institute of Technology Manipur, Imphal, Manipur 795004, India
e-mail: mohdrakimudinkhan@gmail.com

B. P. Swain (ed.), *Recent Advances in Materials*, Springer Proceedings in Materials 25,
https://doi.org/10.1007/978-981-99-3844-5_21

The disadvantages of homogeneous catalysis can be overcome by heterogeneous catalysis [10]. When compared to homogeneous ones, heterogeneous catalysts have several advantages, including low cost, low energy consumption, easy separation, and as a result, no liquid effluent generation. Given that natural materials like oyster shells, animal bones, eggshells, and crab shells are primarily constituted of calcium carbonate, which is converted into CaO during the calcination process, it is ideal to produce heterogeneous catalysts from these materials [11]. In the current study, a waste swan eggshell-derived CaO catalyst was used to produce biodiesel from castor seeds. The prepared CaO catalyst was characterized using Fourier-transform infrared spectroscopy (FTIR) and X-Ray Diffraction (XRD) analysis. Response surface methodology (RSM) using the Box Behnken design (BBD) was utilized to optimize the response parameters. Produced castor biodiesel was also characterized using FTIR and *gas chromatography–mass spectrometry* (GCMS).

2 Materials and Methodology

2.1 Materials

Castor beans and used swan eggshells were both collected from the locality of the Indian state of Manipur. Castor seeds were used to make castor oil using a soxhlet apparatus. The membrane from the gathered eggshells was removed after they had been cooked, and the contaminants were then eliminated by washing the shells with distilled water. Figure 1 shows the cleaned swan eggshell.

(a) (b) (c)

Fig. 1 Preparation of CaO catalyst

2.2 Production of Crude Castor Oil

Castor seeds were cleaned and dried before being crushed with a mortar and pestle. Castor seeds that had been crushed were used to produce oil. Castor oil was produced utilizing Soxhlet equipment with n-Hexane as the solvent.

2.3 Preparation of Swan Eggshell CaO Catalyst

The gathered swan eggshells were thoroughly cleaned with the use of distilled water to remove the impurities. It was further left to dry in an oven overnight at the temperature of 90 °C. The eggshell was then processed through a 0.8 mm sieve screen after being ground into powder using a mortar and pestle as shown in Fig. 1b. The powdered shell was put in the electric muffle furnace for calcination at 800–1000 °C for 2–4 h. Figure 1 shows the calcinated swan eggshell catalyst.

2.4 Characterization of Swan Eggshell Catalyst and Castor Biodiesel

The Bruker AXS, Model number D8-Advance; equipped with 1.5406 Cu-K radiations ranging from 10° to 70°, is used to record the diffraction pattern of swan eggshell CaO samples with the use of XRD. The castor biodiesel samples were characterized by utilizing *FTIR* and GCMS.

2.5 Transesterification of Crude Castor Oil

There are several processes involved in transesterification. The transesterification is carried out by mixing an oil sample with the required amount of catalyst and alcohol in a laboratory scale (Maker: Contech) flask in a magnetic stirrer (Maker: Tarsons Digital Spinot) over a hot plate with a constant temperature. An evaporator and centrifugation were used, respectively, to remove the excess alcohol and catalyst. All of the samples were agitated at 500 rpm speed. The transesterification was carried out to obtain biodiesel at fixed parameters oil-to-methanol molar ratio of 1:12, reaction time of 90 min, catalyst concentration of 5%, and reaction temperature kept in 50–70 °C ranges. Then a separation procedure was used to produce biodiesel and glycerol with minimal contaminants. The excess glycerol from the direct biodiesel that could not be recovered was separated. Continued distillation was followed by heating to remove water. The castor biodiesel was eventually acquired.

Table 1 Independent variables used for Box Behnken

Sl. no.	Parameters	Symbol	Levels		
			−1	0	+1
1	Calcination temperature (°C)	B	800	900	1000
2	Calcination time (h)	A	3	4	5
3	Reaction temperature (°C)	C	50	60	70

With more catalysts present, biodiesel yield rises. However, when the catalyst amount reached a particular level, the trend was reversed. The oil-to-methanol molar ratio was comparable. In general, high catalyst concentrations will result in high biodiesel conversion, but excessive catalysts will speed up the free fatty acid esterification and result in the formation of more water more quickly, which will deactivate the acid hydroxyl group because hydration of these groups happens if water is present. Methanol might also dilute the oil, which would then cause the viscosity to drop and aid in the transfer of mass. Nevertheless, when the methanol concentration increased above the ideal level, the catalyst leaching in the methanol grew stronger and reduced the yield of the biodiesel [12].

2.6 RSM Design

With the aid of Design Expert 13 software, the RSM approach, and BBD experimental design, it is possible to investigate and model the impact of variable parameters. In the present study, factors such as calcination time (h), calcination temperature (°C), and reaction temperature (°C) are investigated. Independent parameters used for Box Behnken are shown in Table 1.

3 Results and Discussion

3.1 Response Surface Methodology Analysis

The Design Expert program suggested the quadratic polynomial equation as the best regression equation in statistical modeling for calculating the response output of castor biodiesel based on the findings of the ANOVA. Table 2 shows the experimental design consisting of 17 experimental runs which were designed with help of RSM based on BBD.

Table 3 shows the ANOVA table for the FAME yields quadratic model. The model is found to be significant with the Model F-value of 99.51. An F-value this large may occur owing to noise only 0.01% of the time. Model terms are said to be significant

Table 2 Experimental design using RSM (BBD)

Run order	Calc temp °C	Calc time h	Reaction temp °C	Actual value	Predicted value	std
1	1000	3	50	93.6	93.58	10
2	800	2	60	93.3	93.24	1
3	900	4	70	94.6	94.59	8
4	900	4	50	94.2	94.16	6
5	1000	3	70	93.8	93.75	12
6	800	3	50	92.8	92.85	9
7	900	2	70	94.1	94.14	7
8	1000	2	60	93.9	93.91	3
9	900	3	60	95.1	95.06	16
10	900	3	60	95.2	95.06	13
11	800	3	70	93.1	93.12	11
12	900	3	60	94.9	95.06	17
13	800	4	60	93.5	93.49	2
14	900	3	60	95.0	95.06	14
15	900	2	50	94.1	94.11	5
16	900	3	60	95.1	95.06	15
17	1000	4	60	94.1	94.16	4

if the P-value is comparatively lesser than 0.0500. In this, model terms are *A*, *B*, *C*, and A^2, B^2, and C^2. Model terms are insignificant if the value is greater than 0.1000. The model reduction will enhance our model if it has a larger number of unwanted terms (with the exclusion of those required to keep in a hierarchy). The Lack of Fit F-value of 0.45 indicates it is insignificant relative to the true errors. The significant Lack of Fit F-value has a 73.18% chance of having occurred as a result of noise. As we want the model to get fit, an insignificant lack of fit is required. The final Equation in Terms of Coded Factors is shown below in Eq. 1.

$$
\begin{aligned}
&95.06 + 0.125 * A + 0.3375 * B + 0.1125 * C \\
&+ -3.87315e - 16 * AB + 0.1 * AC + -0.025 * BC \\
&+ -0.2175 * A^2 + -1.1425 * B^2 + -0.5925 * C^2
\end{aligned}
\tag{1}
$$

where *A*—calcination time (h), *B*—calcination temperature (°C), and *C*—reaction time (h).

Table 3 ANOVA for the FAME yield quadratic model

Source	Sum of squares	df	Mean square	*F*-value	*p*-value	Remarks
Model	8.89	9	0.955	99.51	<0.0001	Significant
A-Calcination time	0.125	1	0.125	12.59	<0.0094	
B-Calcination temperature	0.9112	1	0.9112	91.78	<0.0001	
C-Reaction temperature	0.1010	1	0.1012	10.2	<0.0152	
AB	0	1	0	0	1	
AC	0.04	1	0.04	4.03	<0.0847	
BC	0.0025	1	0.0025	0.2518	0.6312	
A^2	0.1992	1	0.1992	20.06	<0.0029	
B^2	5.5	1	5.5	553.56	0.0001	
C^2	1.48	1	1.48	148.88	0.0001	
Residual	0.0695	7	0.0695			
Lack of fit	0.0175	3	0.00155	0.4487	0.7318	Not significant

3.2 *Effects of Input Parameters on the Biodiesel Yield*

Figure 2 shows the 3D model of the biodiesel yield. To illustrate how the independent variables change with the dependent ones, 3D response curves were created. The surface containing the greatest predicted values on the contour diagram was represented by the smallest ellipse. Ellipsoidal outlines are generated when there is a satisfying relationship between the independent variables. The contour diagram's smallest ellipse, which comprised the surface representing the maximum anticipated yield, revealed that every pair of factors has a marginally significant interaction [12].

Figure 2a shows that, with the increase in calcination time of 2–3 h biodiesel yield increases but the yield reduces from 3 to 4 h.

Figure 2b shows that, with the rise in reaction temperature 50–60 °C biodiesel yield rises which is due to an increase in momentum between molecules of triglycerides and the methanol.

Figure 2c shows that, with the rise in calcination temperature from 800 to 900 °C biodiesel yield increases. But with further rises in temperature from 900 to 1000 °C yield of the biodiesel decreases due to ash formation.

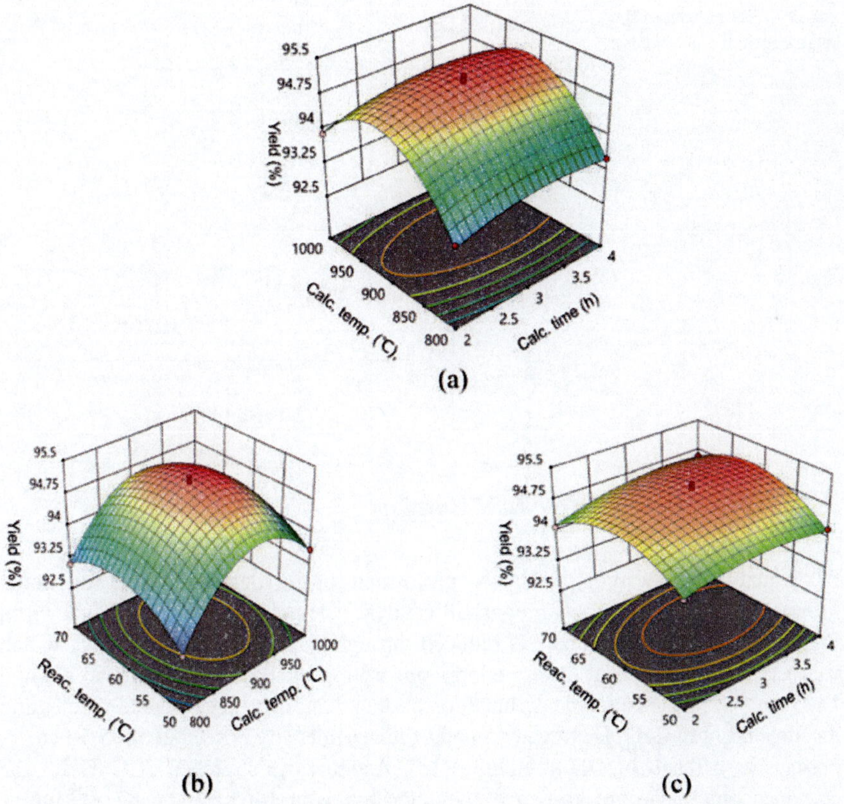

Fig. 2 Response surface plot of RSM **a** Biodiesel yield versus calcination time and calcination temperature, **b** Biodiesel yield versus reaction temperature and calcination temperature, **c** Biodiesel yield versus calcination time and reaction temperature

3.3 XRD Characterization of Calcinated Swan Eggshell

XRD analysis was done to analyze the crystal structure of the swan eggshell CaO. The Bruker AXS; Model number D8-Advance, equipped with 1.5406 Cu-K radiation ranges from 10° to 70° is used to record the diffraction patterns of the synthesized swan eggshell calcium oxide sample by using XRD. Sharp peak formation of the calcinated swan eggshell is a sign that extremely crystal-line materials have formed. The diffraction of the peaks at $2\theta = 18.2116°$, 28.8869°, 29.5814°, 34.2882°, 47.3298°, 50.9818°, 54.5459°, 62.7363°, 64.5045° are shown in Fig. 3. These findings demonstrate strong peaks of XRD reflection with orientations (111), (200), (220), (311), and (222), indicating that the calcinated swan eggshell was properly crystallized throughout the heat treatment procedure. The results match the study reported by other researchers [2, 7].

Fig. 3 XRD pattern of the swan eggshell CaO catalyst

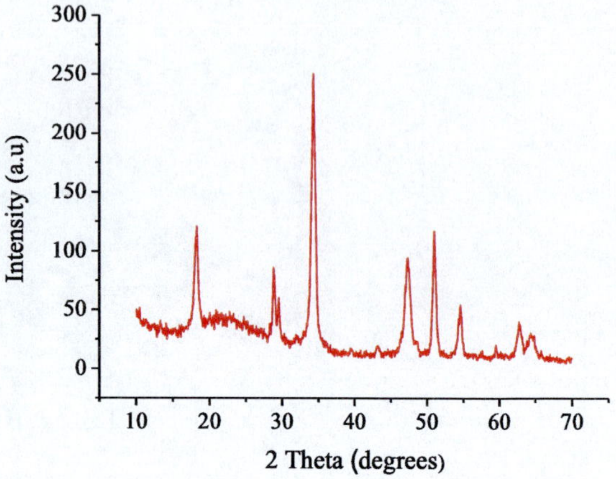

3.4 *FTIR of Swan Eggshell Catalyst*

FTIR analysis of swan eggshells was performed for the identification of functional groups. FTIR spectra of swan eggshell catalyst calcinated at 900 °C are shown in Fig. 4. The mass of carbonate is reduced throughout calcination at 900 °C, which causes the appearance of strong energy absorption bands at 871.09, 1050.77, and 1465 cm^{-1}. The measured peaks have corroborated this drop in carbonate species and its conversion to CaO. However, a band of lower intensity is seen at 3645.50 cm^{-1}, perhaps as a result of OH stretching [13]. Asymmetric stretches of C=O for the carbonaceous group are known by depicted bands at 1465.89 and 871.09 cm^{-1}. This results from ambient CO_2 chemisorption over the CaO surface. The FTIR graph demonstrates that the extra stretches of the Ca-O bonding began at around 395.84 cm^{-1}, which illustrates the production of CaO during the calcination of eggshells at different temperatures and subsequent cooling in the furnace. The Ca-O lattice vibrations of pure calcium oxide explain this [7].

3.5 *FTIR of Castor Biodiesel Sample*

FTIR analysis of castor biodiesel was done to identify functional groups of castor biodiesel samples. Figure 5 shows the infrared spectrometry of the isolated methyl esters. The stretches at 3005.39 cm^{-1} indicate the existence of a C-H bond, and the stretching at 2926.27 cm^{-1} also indicates the existence of CH_2 in the castor biodiesel sample, as can be seen from Fig. 4. At 1741.76 cm^{-1}, the ester C=O can be found, while at 1463.77171 cm^{-1}, CH_2 bending vibrations can be found. C-O ester is found at 1163.77 cm^{-1} and 1168.49 cm^{-1}, while (CO)-O-CH_3 is seen at 1434.7650 cm^{-1}. At 723.04 cm^{-1}, the CH_2 is seen. The current findings of biodiesel are comparable with the finding of another researcher [7].

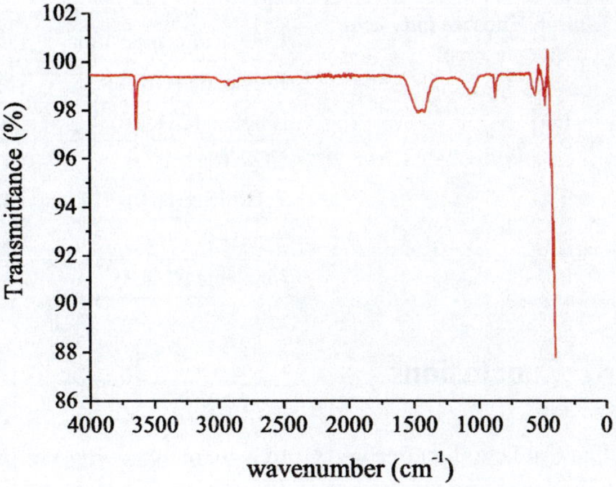

Fig. 4 FTIR graph of swan eggshell CaO catalyst

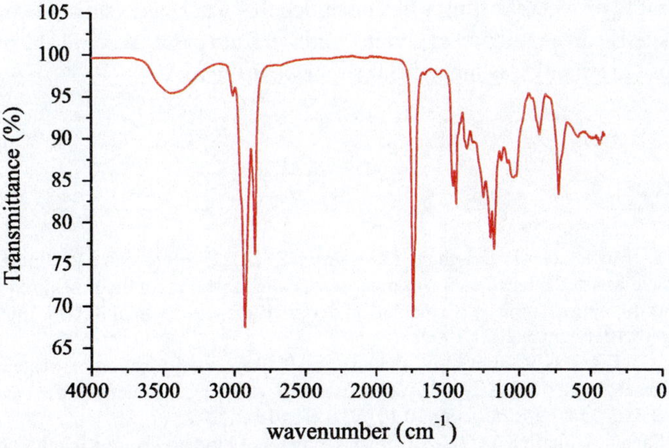

Fig. 5 FTIR graph of the castor biodiesel

3.6 GCMS Characterization of the Biodiesel Sample

Using a GC–MS in the optimal conditions, the free fatty acid methyl ester composition was identified. According to the carbon number, Table 4 displays the free fatty acids composition in the castor biodiesel sample produced using a swan eggshell CaO catalyst. The outcome displays profiles of the sample's fatty acids, including Ricinoleic, Linoleic, Oleic, Palmitic, Stearic, and Linolenic.

Table 4 The free fatty acid of castor biodiesel

Fatty acid composition	Retention time (min)	wt %
Ricinoleic (C18:1)	52.32	85.1
Linoleic (C18:2)	47.58	7.2
Oleic (C18:1)	47.75	5.0
Palmitic (C16:0)	42.49	1.3
Stearic (C18:0)	48.27	1.1
Linolenic (C18:3)	42.27	0.3

4 Conclusions

The CaO catalyst derived from a swan eggshell exhibited excellent catalytic activity and a yield of 95.2% was able to obtain at a calcination temperature of 900 °C, reaction temperature of 60 °C, calcination time of 3 h which is more than the RSM predicted yield of 95.06%. It may be concluded that CaO catalysts made from waste swan egg shells can be utilized to obtain biodiesel from castor seed oil at a reasonable cost because the catalyst is made from a renewable source.

References

1. Rokni K, Mostafaei M, Dehghani M, Kahrizi D (2022) Bioresource technology reports microwave-assisted intensification of transesterification reaction for biodiesel production from camelina oil: optimization by Box-Behnken design. Bioresour Technol Rep 17:100928. https://doi.org/10.1016/j.biteb.2021.100928
2. Jyoti M, Das A, Das V, Bhuyan N, Deka D (2019) Transesteri fi cation of waste cooking oil for biodiesel production catalyzed by Zn substituted waste egg shell derived CaO nanocatalyst. Fuel 242:345–354. https://doi.org/10.1016/j.fuel.2019.01.060
3. Roy T, Sahani S, Sharma YC (2020) Green synthesis of biodiesel from Ricinus communis oil (castor seed oil) using potassium promoted lanthanum oxide catalyst: kinetic, thermodynamic and environmental studies. Fuel 274:117644
4. Ge S, Brindhadevi K, Xia C, Elesawy BH, Elfasakhany A, Unpaprom Y, Doan HV (2021) Egg shell catalyst and chicken waste biodiesel blends for improved performance, combustion and emission characteristics. Fuel 306:121633. https://doi.org/10.1016/j.fuel.2021.121633
5. Ahmad S, Chaudhary S, Pathak VV, Kothari R, Tyagi V (2020) V: Optimization of direct transesterification of Chlorella pyrenoidosa catalyzed by waste egg shell based heterogenous nano—CaO catalyst. Renew Energy. https://doi.org/10.1016/j.renene.2020.06.010
6. Ismail S, Ahmed AS, Anr R, Hamdan S (2016) Biodiesel production from castor oil by using calcium oxide derived from mud clam shell. J Renew Energy 2016:1–8. https://doi.org/10.1155/2016/5274917
7. Singh TS, Verma TN (2019) Taguchi design approach for extraction of methyl ester from waste cooking oil using synthesized CaO as heterogeneous catalyst: Response surface methodology optimization. Energy Convers Manag 182:383–397. https://doi.org/10.1016/j.enconman.2018.12.077
8. Dutta R, Sarkar U, Mukherjee A (2014) Extraction of oil from Crotalaria Juncea seeds in a modified Soxhlet apparatus: physical and chemical characterization of a prospective bio-fuel. Fuel 116:794–802. https://doi.org/10.1016/j.fuel.2013.08.056

9. Martín C, Moure A, Martín G, Carrillo E, Domínguez H, Parajó JC (2010) Fractional characterisation of jatropha, neem, moringa, trisperma, castor and candlenut seeds as potential feedstocks for biodiesel production in Cuba. Biomass Bioenerg 34:533–538. https://doi.org/10.1016/j.biombioe.2009.12.019
10. Nadeem F, Bhatti IA, Ashar A, Yousaf M, Iqbal M, Mohsin M, Nisar J, Tamam N, Alwadai N (2021) Eco-benign biodiesel production from waste cooking oil using eggshell derived MM-CaO catalyst and condition optimization using RSM approach. Arab J Chem 14:103263. https://doi.org/10.1016/j.arabjc.2021.103263
11. Proença BD, Fioroto PO, Heck SC, Duarte VA, Cardozo Filho L, Feihrmann AC, Beneti SC (2021) Obtention of methyl esters from macauba oil using egg shell catalyst. Chem Eng Res Des 169:288–296. https://doi.org/10.1016/j.cherd.2021.03.015
12. Dwivedi G, Sharma MP (2015) Application of Box–Behnken design in optimization of biodiesel yield from Pongamia oil and its stability analysis. Fuel 145:256–262. https://doi.org/10.1016/j.fuel.2014.12.063
13. Ur W, Mohd S, Khan ZA (2021) Environmental technology & innovation valorization of waste chicken egg shells towards synthesis of heterogeneous catalyst for biodiesel production: optimization and statistical analysis. Environ Technol Innov 22:101460. https://doi.org/10.1016/j.eti.2021.101460

Effect of Hydrogen-Diesel Fuel on Combustion Characteristics: By a Numerical Study

S. Mohammed Imran, C. S. E. M. Shyam, Gaddam Dinesh, Namala Harikrishna, Boya Surendra, Upendra Rajak, K. Thirupathi Reddy, Tikendra Nath Verma, and Manoj Arya

1 Introduction

Current research focusses on the use of hydrogen as a substitute for petroleum, since the growth and development of civilization are so reliant on non-renewable sources of energy. In this way, low-heat rejection engines attract the interest of researchers due to their low fuel consumption and reduced pollutants. As is well known, only one-third of the energy produced by the combustion of fuel in internal combustion engines is transformed into productive work. A portion of the remaining energy is used for cooling, and the rest is expelled via the exhaust.

Due to its clear technological importance, modelling turbulent combustion has become an exceedingly difficult, yet crucial task for the scientific community. The com-puting is now completely comparable with trial and concept as an exploration tool for producing multi-scale data that cannot be produced by any other means. In the last decade, CFD has been widely used to strategy a variety of ignition assemblies, and turbulent combustion modelling has garnered a great deal of interest [1, 2].

Addepalli and Mallikarjuna [3] use CFD analysis to determine the result of machine limits on the physiognomies of a GDI engine at three different fuel injection pressures and three different engine speeds (2000, 3000, and 4000 rpm) (200, 300, and 400 bar). They observed that engine speed influenced stormy kinetic energy and tumble ratio more than any other component. Amin and Jafarian [4] presume that

S. M. Imran · C. S. E. M. Shyam · G. Dinesh · N. Harikrishna · B. Surendra · U. Rajak (✉) · K. T. Reddy
Department of Mechanical Engineering, RGM College of Engineering and Technology Nandyal, Nandyal 518501, India
e-mail: upendrarajak86@gmail.com

T. N. Verma (✉) · M. Arya
Department of Mechanical Engineering, Maulana Azad National Institute of Technology Bhopal, Bhopal 462003, India
e-mail: verma.tikks@gmail.com

B. P. Swain (ed.), *Recent Advances in Materials*, Springer Proceedings in Materials 25,
https://doi.org/10.1007/978-981-99-3844-5_22

Table 1 Fuel properties

Properties	Diesel	Hydrogen	5% H2 with diesel	10% H2 with diesel	15% H2 with diesel
Density (kg/m^3)	830	0.085	829.9	829.9	829.9
Viscosity (mm^2/s)	3.4	0.0083	2.8	1.86	1.37
Cetane number	48	5–10	46.1	44.2	42.3
Heat value (MJ/kg)	45.5	120	54.5	63.2	69

further studies used OpenFoam open source software. Using the k-SST turbulence model, they observed that turbulence had no discernable influence on the results of the simulation. Breitung et al. [5] examination evaluated the facility room's maximum flame velocity and over- pressure under the worst-case scenario for an ignition. They found that an early ignition event is anticipated to result in the maximum turbulent flame speed and over pressure of 35–71 m/s and 0.13–0.27 bar, individually [5].

In the present work, a numerical simulation is provided to explain combustion. The effect of adding 5%, 10%, and 15% hydrogen to diesel fuel on the presentation of a four-stroke, diesel engine under varying engine loads was investigated.

2 Fuel Properties and Method

2.1 Fuel Properties

In dual-fuel mode, hydrogen and diesel are the most used fuels. Table 1 lists the characteristics of hydrogen and diesel. Table 1 lists the fuel qualities employed in the present experiment. The following equation may be used to compute the flow rate of hydrogen energy: (1).

$$\text{Hydrogen energy share} = \frac{\text{Mass flow rate of H2} \times \text{LHV of H2}}{\text{Mass flow rate of H2} \times \text{LHV of H2} + \text{Mass flow rate of diesel} \times \text{LHV of diesel}} \tag{1}$$

2.2 Experimental Set-up

In this investigation, all tests were conducted using a single-cylinder, four-stroke, direct-injection diesel engine that generated 3.5 kW of power at 1500 rpm. The practical details and test engine diagram were reported as bore (87.5 mm), stroke

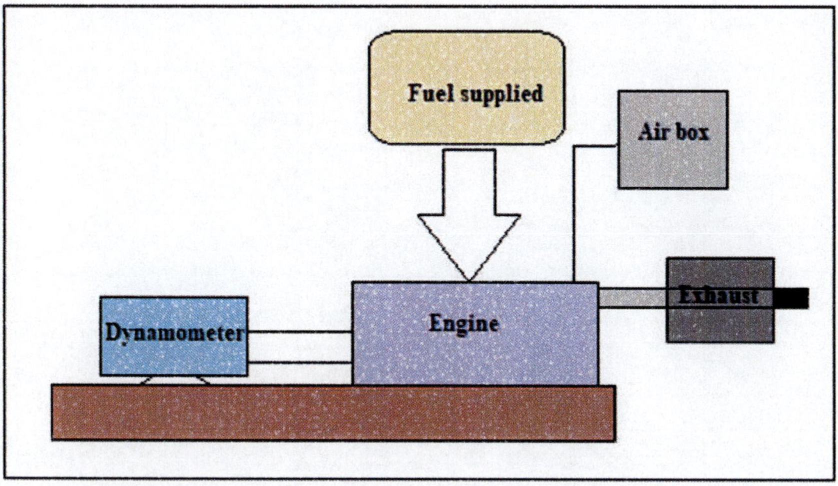

Fig. 1 Engine set-up

(110 mm), swept volume (661.45 cc) and CR (18). Connecting the engine and AC dynamometer to give braking load, through the current injection system, the test mixtures were fed continuously into the combustion chamber at a constant fuel injection pressure and temperature. A solenoid-controlled electronic burette placed between the engine fuel tank and the fuel pump was used to measure fuel consumption, whilst the air box's integrated differential pressure sensor was utilized to evaluate air use (Fig. 1).

3 Results

3.1 Validation of Model

This study looked into the Diesel-RK engine combustion using first law of thermodynamics models. We looked at temperature, pressure, and other significant variables in relation to crank angle or interval. Simulated data were compared with the experimental data from the diesel engine experiment. Figure 2 shows that the experimental and numerical results variation at full load condition, error was fund to be up to 12.0% for peak value of engine brake torque at same operating condition (the operating conditions are the same as in Table 2). Following was the governing equation:

Mass

$$\frac{dm}{dt} = \sum_j \dot{m}_j$$

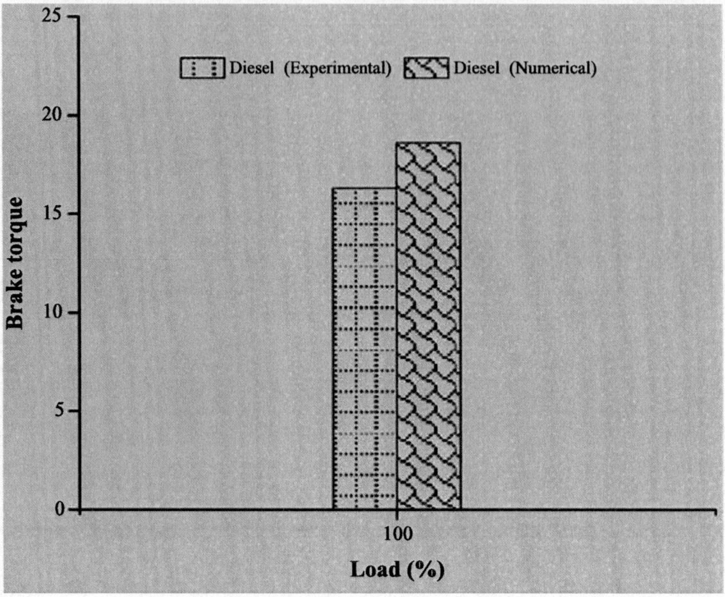

Fig. 2 Validation of brake torque with load

Table 2 Test engine condition

Constraint	Value
Bore/stroke	(87.5 mm*110 mm)
Compression ratio	18
Swept volume	661.45 cc
Power	3.5 kW @1500 rpm

Energy

$$\frac{\mathrm{d}(mu)}{\mathrm{d}t} = -P\frac{\mathrm{d}v}{\mathrm{d}t} + \frac{\mathrm{d}Q_{ht}}{\mathrm{d}t} + \sum_j \dot{\mathrm{m}}_j h_j$$

Species

$$\mathrm{SFC} = \frac{\dot{\mathrm{m}}_j}{P_b}$$

Tolstov's equation (Ignition delay period)

$$\tau = 3.8 \times 10^{-6}\left(1 - 1.6 \times 10^{-4}.n\right)$$

$$\sqrt{\frac{T}{P}}\exp\left(\frac{E_a}{8.312T}-\frac{70}{CN+25}\right)$$

Zeldovich mechanism (NOx emission)

$$[O_2] \leftrightarrow [2O]$$
$$[N_2] + [O] \leftrightarrow [NO] + [N]$$
$$[N] + [O_2] \leftrightarrow [NO] + [O]$$

$$\frac{d[NO]}{d\theta} = \frac{P \times 2.333 \times 10^{-7} \cdot e^{\frac{38020}{T_b}} [N_2]_{e.} [O]_{e.} \left\{1 - \left(\frac{[NO]}{[NO]_e}\right)^2\right\}}{RT_{b.} \left(1 + \frac{2365}{T_b} \cdot e^{\frac{2365}{T_b}} \cdot \frac{[NO]}{[NO]_e}\right)} \cdot \frac{1}{\omega}$$

3.2 *Engine Brake Torque*

The change in engine torque at full load for several fuel samples is shown in Fig. 3. Diesel fuel engine torque was dynamically increased under the addition of hydrogen (5%, 10%, and 10%). It was found that the calorific value of hydrogen fuel is larger than diesel fuel, giving brake torque an almost higher energy share percentage. Torque is one of the most influential aspects of an engine's performance characteristics. As a result of the decreased heat produced during combustion, decreased fuel delivery, and increased mass flow rate, Mustayen et al. [6] research indicated that using biodiesel mix fuel produced lower torque when compared with diesel.

3.3 *Indicated Thermal Efficiency*

Figure 4 displays the variation in stated thermal efficiency at full load for a number of fuel samples. The thermal efficiency of diesel fuel was shown to be reduced when compared with hydrogen additions of 10 and 15%. It was discovered that hydrogen fuel has a higher calorific value than diesel fuel, giving it a nearly higher energy share percentage for specified thermal efficiency. According to prior research conducted by the authors [7], higher hydrogen substitution lowered thermal efficiency and increased nitrogen oxide (NOx) emissions. The engine's efficiency in proportion to the power produced inside the cylinder prior to its transmission to the piston and cylinder is referred to as indicated efficiency [8]. In the present investigation 5% of hydrogen addition to diesel fuel and reduced indicated efficiency.

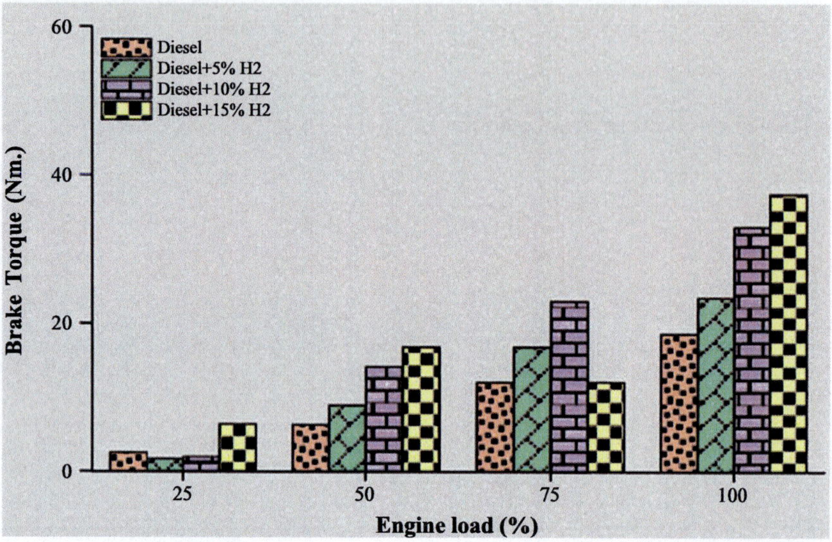

Fig. 3 Brake torque with load

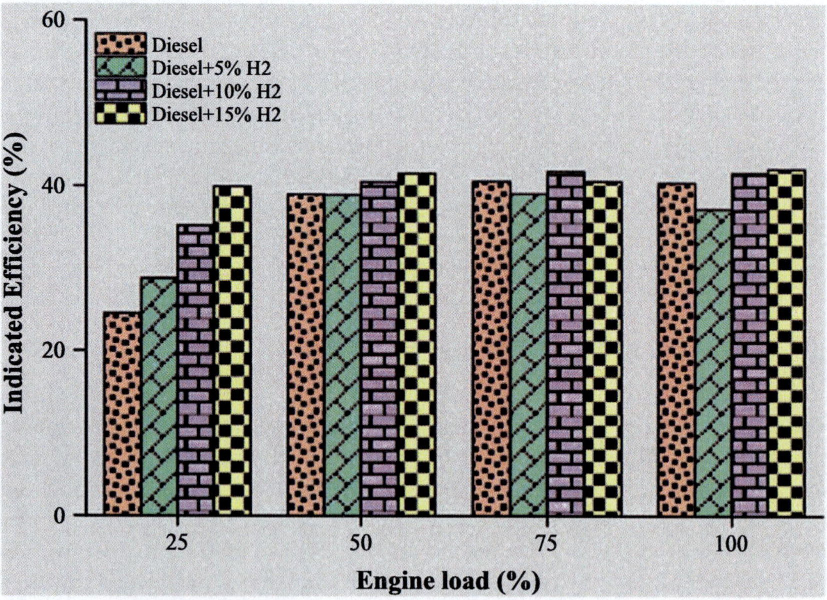

Fig. 4 Indicated efficiency with load

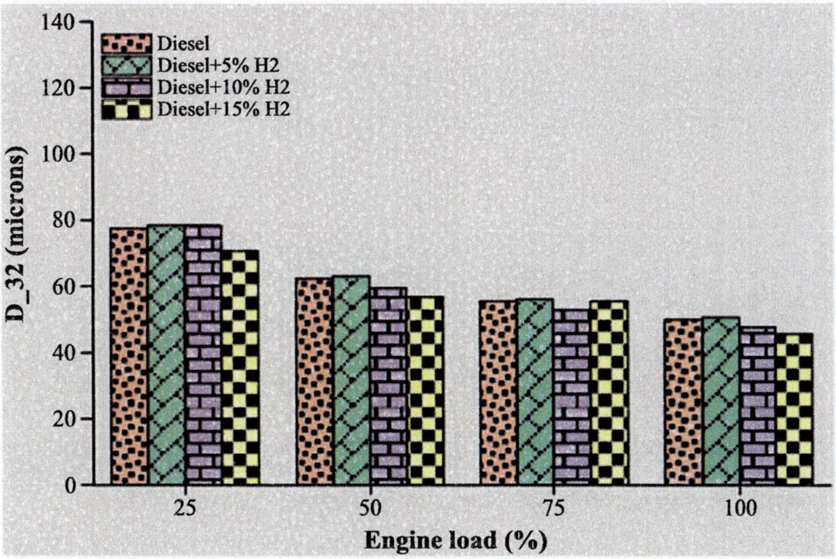

Fig. 5 D32 diameter with load

3.4 D32 Diameter

Figure 5 displays the variation in D32 diameter at full load for a number of fuel samples. The D32 diameter of 15% of hydrogen addition was shown to be reduced when compared with hydrogen additions of 5 and 10%. The diameter of a droplet with a volume to surface area ratio equal to that of a spray is referred to as the D32. Surface tension and viscosity differences can also be attributed to differences in D32. D32 rises as the mass fraction of saturated FAMEs rises [9].

3.5 Smoke Emission

Figure 6 displays the variation in smoke emission at full load for a number of fuel samples. Hydrogen energy share with base fuel was shown to be reduced smoke emission when compared with base fuel (diesel). Hydrogen significantly reduces the amount of smoke that is created. Plain water is the byproduct of hydrogen combustion rather than soot (particulate particles). This is because hydrogen has less carbon than carbon, which lowers particle matter. Additionally, compared with other fuels, hydrogen has a higher diffusivity, which creates a better fuel oxidation for complete combustion and reduced smoke levels [10–12].

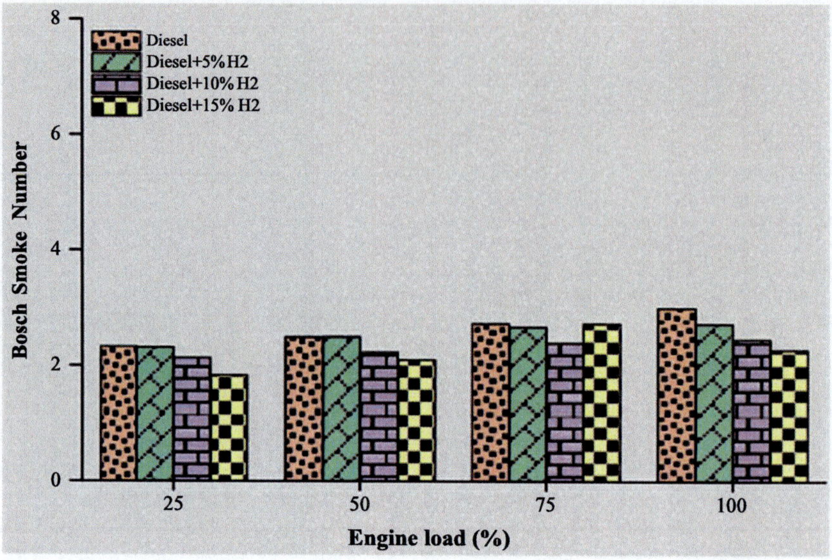

Fig. 6 Bosch smoke number with load

4 Conclusions

The impact of the diesel engine and hydrogen energy share of 5%, 10%, and 15% blends on emission, performance, and combustion as a demonstration of the nature-aspirated single-cylinder diesel engine. The key numerical findings are shown in the list below.

- At 100% load, 5% H2 increased braking torque by 20%. High hydrogen flame velocity results in a more efficient combustion rate, hence enhancing braking torque.
- When the hydrogen (10% and 15%) enrichment is increased, it is discovered that the D32 falls by 4.5% and 8.5% at 100% load, respectively.
- The use of hydrogen drastically reduced the quantity of smoke emissions. Compared with pure diesel, the smoke emissions at 5% H2, 10% H2, and 15% H2 are much lower, at 9.6%, 17.7%, and 24%, respectively.

References

1. Ayed AH, Kusterer K, Funke HH, Keinz J, Bohn D (2017) CFD based exploration of the dry-low-NO_x hydrogen micromix combustion technology at increased energy densities. Propul Power Res 6(1):15–24. https://doi.org/10.1016/j.jppr.2017.01.005

2. Novella R, García A, Pastor JM, Domenech V (2011) The role of detailed chemical kinetics on CFD diesel spray ignition and combustion modelling. Math Comput Model 54(7–8):1706–1719. https://doi.org/10.1016/j.mcm.2010.12.048
3. Addepalli SK, Mallikarjuna JM, Taylor KH (2018) Parametric analysis of a 4-stroke GDI engine using CFD indicated mean effective pressure. Alex Eng J 57(1):23–34. https://doi.org/10.1016/j.aej.2016.10.007
4. Mohammadi MA, Jafarian A (2018) CFD simulation to investigate hydrodynamics of oscillating flow in a beta-type Stirling engine. Energy. https://doi.org/10.1016/j.en-ergy.2018.04.017
5. Breitung W, Halmer G, Kuznetsov M, Xiao J (2018) Science direct analysis oftransient supersonic hydrogen release, dispersion and combustion. Int J Hydrogen Energy 1–11. https://doi.org/10.1016/j.ijhydene.2018.04.158
6. Mustayen AGMB, Rasul MG, Wang X, Negnevitsky M, Hamilton JM (2022) Remote areas and islands power generation: a review on diesel engine performance and emission improvement techniques. Energy Convers Manag 260(October 2021):115614. https://doi.org/10.1016/j.enconman.2022.115614
7. Chakraborty A, Biswas S, Meitei S, Sengupta A (2022) Examining the significance of the ignition characteristics of hydrogen and liquefied-petroleum-gas on the reactivity con- trolled compression ignition and its interspersed profiles induced in an existing diesel engine: a comparative perspective. Energy Convers Manage 268(February):115976. https://doi.org/10.1016/j.enconman.2022.115976
8. Rajak U, Nashine P, Singh TS, Verma TN (2018) Numerical investigation ofperformance, combustion and emission characteristics of various biofuels. Energy Convers Manage 156. https://doi.org/10.1016/j.enconman.2017.11.017
9. Rajak U, Verma TN (2020) Influence of combustion and emission characteristics on a compression ignition engine from a different generation of biodiesel. Eng Sci Technol Int J 23(1):10–20. https://doi.org/10.1016/j.jestch.2019.04.003
10. Seelam N, Gugulothu SK, Venkat R (2022) Science direct exploration of engine characteristics in a CRDI diesel engine enriched with hydrogen in dual fuel mode using toroidal combustion chamber. Int J Hydrogen Energy 47(26):13157–13167. https://doi.org/10.1016/j.ijhydene.2022.02.056
11. Oztürk U, Hazar H, Yılmaz F (2019) Comparative performance and emission characteristics of peanut seed oil methyl ester (PSME) on a thermal isolated diesel engine. Energy 167:260–268. https://doi.org/10.1016/j.energy.2018.10.198
12. Agbulut U, Elibol E, Demirci T, Sarıdemir S, Etem Gürel A, Rajak U, Afzal A, Verma TN (2022) Synthesis of graphene oxide nanoparticles and the influences of their usage as fuel additives on CI engine behaviors. Energy 244:122603. https://doi.org/10.1016/j.en-ergy.2021.122603

Gamma Spectroscopy Studies of Rice Samples Grown in Imphal Valley of Manipur, India

Karanjit Leiphrakpam and Mamata Maisnam

1 Introduction

Natural and man-made (anthropogenic) radionuclides are the sources of radioactivity in the environment. Natural radionuclides include radioisotopes like K^{40}, U^{238} and Th^{232} and their decay products. These naturally occurring radioactive materials shortly known as NORM have long half-lives. Naturally occurring radionuclides may be divided into primordial radionuclides which are found on Earth and have existed in their present form ever since the creation of Earth, and cosmogenic radionuclides which are continuously produced due to the interaction of cosmic rays. Not only K^{40}, but naturally occurring long half-lived radioisotopes such as U^{238} and Th^{232} also contribute to the earth's natural radioactivity. Human activities such as the mining of coal, extraction of crude oil and natural gases, production of geothermal energy, water and wastewater treatment, use of phosphate fertilizers, mining of uranium, thorium, etc. increase the level of NORM [1]. These radioactive materials are present in all environmental samples like soil, air, water, vegetation, food and even building materials.

Humans are constantly exposed to natural ionizing radiation. These include terrestrial as well as cosmic (extraterrestrial) origin. Cosmic or extraterrestrial radiation is mainly due to cosmic rays. High-energy cosmic rays are constantly bombarding the Earth's atmosphere. Through nuclear reactions with nitrogen, oxygen, and other atmospheric nuclei, as well as through other processes, these high-energy cosmic ray interactions with the atmosphere result in the production of several radionuclides. Through this process, many sub-atomic particles like muons, electrons, mesons, etc. are also produced. The production of these radionuclides and sub-atomic particles varies with altitude and latitude, however, the production is constant with time relatively [2]. Of all the total cosmogenically-produced radionuclides, about 70% are

K. Leiphrakpam (✉) · M. Maisnam
Department of Physics, National Institute of Technology Manipur, Langol, Imphal 795004, India
e-mail: karanjitleiphrakpam1@gmail.com

B. P. Swain (ed.), *Recent Advances in Materials*, Springer Proceedings in Materials 25,
https://doi.org/10.1007/978-981-99-3844-5_23

formed in the stratosphere and the remaining are formed in the troposphere [3]. Important cosmogenic radionuclides include H^3, Be^7, C^{14} and Na^{22} [4]. In India, the ambient dose rate due to cosmic rays is estimated as 31.96 $nGyh^{-1}$ Reference [5]. At sea level worldwide estimated value of ambient dose from cosmic rays is about 31 $nGyh^{-1}$. Reference [6] Terrestrial sources of ionizing radiations originate from the radionuclides that were formed during the formation of our planet and their daughter nuclides. These terrestrial sources of naturally occurring radionuclides consists of singly occurring radionuclides like K^{40}, Rb^{87} and the daughter nuclides in the radioactive decay series of uranium series (U^{238}), thorium series (Th^{232}) and actinium series (U^{235}).

K^{40} is the most prominent and it is widely distributed in rocks and soil among the singly occurring radionuclide. K^{40} has a half-life of 1.3×10^9 years. Of total potassium available, the abundance of K^{40} is about 0.0012% and hence in 1 kg of the soil contains about 2.36 mg of K^{40} which is radioactive. Potassium is an essential element for most life forms. The body's potassium level is in strict homeostasis and is unaffected by environmental changes. K^{40} also significantly contributes to the gamma radiation field in the environment. The radioactivity of K^{40} is almost constant under all conditions, and hence no fractionalization of K^{40} takes place in the environment. Rubedium-87 is another singly occurring radionuclide and is of less importance since the weight ratio of potassium to rubidium in the soil is roughly 90:1.

Uranium-238 (U^{238}) is the most abundant isotope of natural uranium found in all rocks and soil in different concentrations. Both uranium and thorium are oxidized metals. Since their oxides are of very low density they are concentrated in the Earth's crust [7]. Among the decay products of U^{238}, Radium-226 (Ra^{226}) and its decay products are of great importance in natural radiation exposure. It is estimated that about 98% of the external doses due to U^{238} are from Ra^{226} and its daughter nuclide [8]. The concentration of Ra^{226} in the soil is not evenly distributed. Even within a country, one could get a different amount of concentration of Ra^{226}. This is because of the heterogeneous nature of the soil. UNSCEAR 2008 reported that the concentration of Ra^{226} in the soil varies from 0.5 to 1000 $Bqkg^{-1}$ [6]. Both calcium and radium have similar chemical properties and hence enter our body through the food chain and gets accumulated in the bones.

Th^{232} and its decay products are another important source of natural radiation and are largely distributed in soil and rocks in various concentrations. U^{238} and Th^{232} and their decay products generate the bulk of natural radioactivity in the environment. About four times that of uranium content Th^{232} is contained in the igneous rocks. The specific activity of Th^{232} is 4.07 Bq kg^{-1} while that of U^{238} is 12.21 Bq kg^{-1} and the radioactivity due to these two nuclides is nearly 1:1. Th^{232} has a half-life of about 1.4×10^{10} years through an alpha emission decay to a beta emitter Ra^{228} having a half-life of 5.76 years. Ra^{228} has considerable significance in natural radioactivity studies. Even though Ra^{228} occurs in the soil and water in a ratio of about 1:1 to that of Ra^{226}, however, there is little knowledge of its presence in foods or tissues of humans [4].

Soil is a significant source of radiation exposure in the environment. Hence soil is an important source of radiological contamination. The radioactive materials are transferred to humans through ingestion and inhalation. Human beings require food for life support. So, we can predict any radiological hazards from our foodstuffs by knowing the contents of radioactivity in the foodstuffs. Hence, it is very important to identify the radiological hazards from the food we eat, sand, soil and other materials used in the construction of buildings [9].

The radioactivity in the soil is mainly from the presence of K^{40}, U^{238} and Th^{232} and their progeny, and anthropogenic sources. Due to the production of γ–rays and the inhalation of radon and its progeny, they cause radiological hazards that are both internal and external [1–10]. It is crucial to measure external gamma doses from environmental sources because of how they affect overall doses and how individual doses vary depending on the route. These doses depend on the amount (concentration) of K^{40}, and U^{238} and Th^{232} present in the geology of rocks and soils of the area [11].

Rice is the staple food in many nations including India. Among the Indian states, Manipur leads in rice consumption both in rural and urban areas with monthly per capita consumption of 15.56 kg and 15.15 kg, respectively. Nowadays, it has become one of the serious topics for health consciousness in the world, so it is very essential to study the presence of radioactivity in the foods that we consume. Many researchers conducted the determination of different radionuclides concentrations in food samples and dose assessments from the consumption of foodstuff by the population [1–13]. The results of the analysis aided in the development of a baseline of radiation exposure to the public at large from the consumption of rice from that specific geological region. So the number of radionuclides transferred through the consumption of foods (rice) accumulated in our body can be known by measuring the concentration of radionuclides [12, 13]. In this paper, activity of U^{238} (Ra^{226}), Th^{232} and K^{40} are measured and studied.

2 Experiment

Rice grains were fetched from different areas of Imphal Valley. The collected rice samples are kept in an air-tight container. Only six rice samples were chosen in this present study. Impurities like stones, leaves, roots, etc. were taken out from each of the samples collected. The samples were ground into a fine powder and sieved through a 0.5 mm sieve to achieve uniform-size particles. The sieved samples were heated using a hot air oven at around 65 °C for around 1 day (~20 h) to remove moisture content. Each of these samples was again packed inside different air-tight containers of the same size. The lid of the container is tightened using glued tape to make sure that there is no leakage of gas from the container. The samples are kept for 30 days to achieve secular equilibrium between Ra^{226}, Th^{232} and their daughter nuclides, before counting gamma rays using NaI(Tl) scintillation detector [14].

The measurements of natural radioactivity in each of the rice samples were carried out by using NaI(Tl) (Ortec® digiBASE™ model 818/2) detector from Ortec. The detector is provided with a lead shield of a thickness of about 10 cm. The detector is connected to a computer and the spectrum was analyzed using Maestro multichannel analysis software from Ortec. Using standard gamma source Cs^{137} and Co^{60} point source, the energy calibration of the detector was carried out. The detector has an energy resolution (FWHM) of 6.7% at the peak of 662 keV for Cs^{137}. Each sample was counted for about 36,000 s. The background was also counted at the same time. The background count was subtracted from each sample's gross count to get net counts. Each of the samples was assumed to achieve secular equilibrium in the U^{238} and Th^{232} decay series. The activity of Ra^{226} was calculated from the gamma-ray lines of Bi^{214}, while the activity of Th^{232} was calculated from the gamma-ray lines of Ac^{228} and Tl^{208}. The activity of K^{40} was directly calculated from the single γ- ray line at 1460.8 keV [15].

3 Results and Discussion

3.1 Activity Concentration

The activity concentration of natural radionuclides U^{238}, Th^{232} as well as K^{40} was investigated from the collected samples. The radioactivity concentrations in each of the rice samples were calculated using the formula [16].

$$A = \frac{N_{CPS} \times 1000}{\in \times I \times w(g)} \tag{1}$$

where A is the activity in $BqKg^{-1}$, N_{CPS} is the net count per second, ε is the gamma efficiency evaluated as a function of transition energy, I is the emission probability of the gamma-ray energy and w is the weight of samples in grams. The activity concentration of Ra^{226}, Th^{232} and K^{40} for the rice samples has been measured and given in Table 1. From Table 1, it is observed that, the activity concentration of Ra^{226} ranged from 4.53 $Bqkg^{-1}$ (R4) to 9.45 $Bqkg^{-1}$ (R1), for Th^{232} concentration ranged from 5.77 $Bqkg^{-1}$ (R4) to 14.14 $Bqkg^{-1}$ (R2) and K^{40} concentration ranged from 14.56 (R1) $Bqkg^{-1}$ to 61.47 (R6) $Bqkg^{-1}$.

The concentration of the present study of Ra^{226}, Th^{232} and K^{40} are all lower than the world average values of 32 $Bqkg^{-1}$for Ra^{226}, 45 $Bqkg^{-1}$1 for Th^{232} and 412 $Bqkg^{-1}$for K^{40} [17].

Table 1 Activity concentration of Ra^{226}, Th^{232} and K^{40} in rice samples in $Bqkg^{-1}$

Sample ID	Location	Ra^{226}	Th^{232}	K^{40}	Ra_{eq}
R1	Khonghampat	9.45	7.31	14.56	21.02
R2	Kakching	6.37	14.14	19.75	28.11
R3	Thoubal	5.35	11.04	36.79	23.64
R4	Nongpok Sekmai	4.53	5.77	26.58	8.94
R5	Kakching	7.60	6.74	19.11	18.71
R6	Thoubal	7.87	7.73	61.47	23.67

3.2 *Radiological Parameters*

3.2.1 Radium Equivalent Activity (Ra_{eq})

Radium equivalent activity (Ra_{eq}) is the most commonly used radiation hazard index. The gamma radiation hazards are caused by specific radionuclides of Ra^{226}, Th^{232} and K^{40}. Radium equivalent activity (Ra_{eq}) is the weighted sum of the activities of the Ra^{226}, Th^{232} and K^{40} assuming that their respective gamma-ray dose rates of 370 $Bqkg^{-1}$ of Ra^{226}, 259 $Bqkg^{-1}$ of Th^{232} and 4810 $Bqkg^{-1}$ of K^{40} are equal [18, 19] and it is given by [20].

$$Ra_{eq}\left(Bqkg^{-1}\right) = A_{Ra} + 1.42\,A_{Th} + 0.077\,A_{K} \tag{2}$$

where the activity concentration of Ra^{226}, Th^{232} and K^{40} is represented by A_{Ra}, and A_{Th} and K^{40}, respectively. To keep the annual radiation dose less than 1.5 mGy y^{-1} the value of Ra_{eq} is limited to 370 $Bqkg^{-1}$ [21]. The Ra_{eq} for rice samples is given in Table 1. The calculated values of Ra_{eq} for all the rice samples varied from 8.94 $Bqkg^{-1}$ (R4) to 28.11 $Bqkg^{-1}$ (R2). The evaluated values of Ra_{eq} are all less than the highest permitted value, i.e. 370 $Bqkg^{-1}$ [21].

3.2.2 External and Internal Hazard Indices

External hazard index (H_{ex}) and internal hazard index (H_{in}) were determined to limit the external γ-radiation dose from the samples. The H_{ex} and H_{in} are calculated by using the Eqs. (3, 4) [22, 23].

$$H_{ex} = \frac{A_{Ra}}{370} + \frac{A_{Th}}{259} + \frac{A_{K}}{4810} \leq 1 \tag{3}$$

$$H_{in} = \frac{A_{Ra}}{185} + \frac{A_{Th}}{259} + \frac{A_{K}}{4810} \leq 1 \tag{4}$$

The values of the external hazard index H_{ex} vary from 0.024 (R4) to 0.076 (R2). The values of H_{ex} for each of the samples are lower than critical value 1. Respiratory organs get damaged due to inhalation of radon. Therefore, the internal hazard index H_{in} is used to quantify internal exposure to radon and its daughter. The evaluated values of H_{in} are from 0.036 (R4) to 0.093 (R2). The calculated H_{in} values are all less than the recommended limit, unity (Fig. 1 and Table 2).

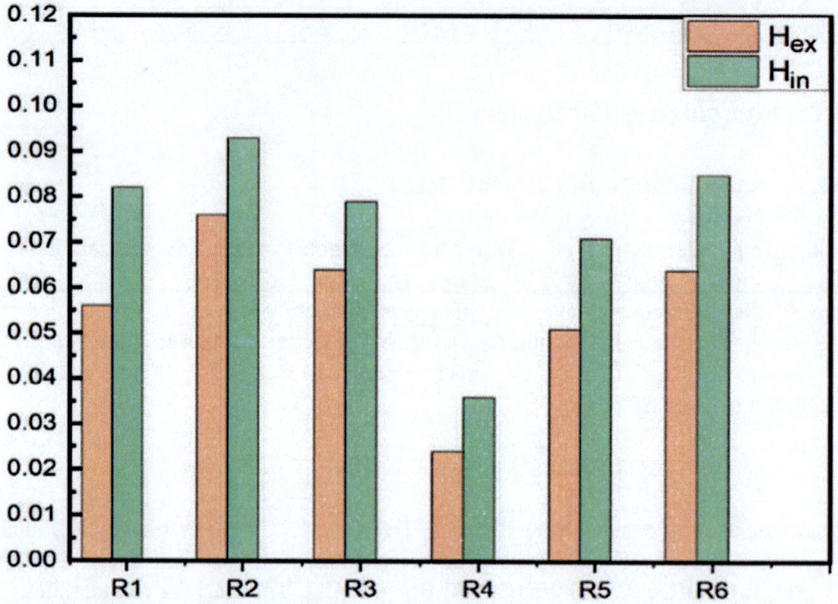

Fig. 1 Bar graph showing external hazard index (H_{ex}) and internal hazard index (H_{in}) of rice samples studied

Table 2 External hazard index (H_{ex}) and internal hazard index (H_{in}) of rice samples studied

Sample ID	Location	H_{ex}	H_{in}
R1	Khonghampat	0.056	0.082
R2	Kakching	0.076	0.093
R3	Thoubal	0.064	0.079
R4	Nongpok Sekmai	0.024	0.036
R5	Kakching	0.051	0.071
R6	Thoubal	0.064	0.085

4 Conclusion

To determine the health risk, it is crucial to determine the activity concentration. The activity concentration of Ra^{226}, Th^{232} and K^{40} for the rice samples under study are less than the permissible maximum values of Ra^{226}, Th^{232} and K^{40} reported by UNSCEAR. From the result of this present work, the values of Ra_{eq}, internal hazard index H_{in} and external hazard index H_{ex} are all less than their corresponding allowed limits recommended by UNSCEAR. From this present study, it was found that the six rice samples under study, which are consumed mostly in Imphal valley, are radiologically safe for consumption.

References

1. Nahar A, Asaduzzaman K, Islam MM, Rahaman MM, Begum M (2018) Assessment of natural radioactivity in rice and their associated population dose estimation. Radiat Eff Defects Solids 1–9
2. Perkins RW, Nielsen JM (1965) Cosmic ray produced radionuclides in the environment. Health Phys 11:1297–1304
3. National Council of Radiation Protection (NCRP) (1998) Measurement of radon and radon daughters in air. United States, NCRP Rep. No. 97, Bethesda, MD, Maryland
4. Eisenbud M (1987) Environmental radioactivity from natural, industrial and military sources, 3rd edn. Academic Press Inc., California
5. Nambi KSV, Bapat VN, David M, Sundaram VK, Santa CM, Soman SD (1986) Natural background radiation and population dose distribution in India, HPD, BARC, India
6. United Nations Scientific Committee on effects of Radiation (UNSCEAR-2008) Report to the general assembly, vol 1, Annex B
7. Osburn WS (1965) Primordial radionuclides: their distribution, movement, and possible effects within terrestrial ecosystem. Health Phys 11:1275–1295
8. Mishra UC, Sadasivan S (1971) Natural radioactivity levels in Indian soils. J Sci Ind Res 30:59–62
9. Ezzulddin SK, Ahmed AH, Samad AI, Othaman SQ (2017) Radioactivity measurement of nuts and seeds available in Erbil city markets. AIP Conf Proc 1888:020022
10. Alrefae T, Nageswaran TN (2013) Radioactivity of long lived gamma emitters in rice consumed in Kuwait, 24–27
11. Mehra R, Singh M (2011) Measurement of radioactivity of ^{238}U, ^{226}Ra, ^{232}Th and ^{40}K in soil of different geological origins in northern India. J Environ Prot 960–966
12. Laith N, Nada T, Fouzey K (2015) Measuring radioactivity level in various types of rice using NaI (Tl) detector. Am J Eng Res 126–132
13. Ramachandran TV, Mishra UC (1989) Measurement of natural radioactivity levels in Indian foodstuffs by gamma spectrometry. Int J Radiat Appl Instrum, Part A. Appl Radiat Isot 723–726
14. Ibrahiem NM, El Ghani AHA, Shawky SM, Ashraf EM, Farouk MA (1993) Measurement of radioactivity levels in soil in the Nile delta and middle Egypt. Health Phys 620–627
15. Alharbi T (2020) Establishment of natural radioactivity baseline, mapping, and radiological hazard assessment in soils. Arab J Geosci 13:415
16. Ashikun N, Khandoker Asaduzzaman M, Moinul Islam M, Mashiur R, Mahfuza B (2018) Assessment of natural radioactivity in rice and their associated population dose estimation. Radiat Eff Defects Solids 173:11–12
17. UNSCEAR (2008) Sources and effects of ionizing radiation. Report to the General Assembly. New York, United Nation

18. Kessaratikoon P, Awaekechi S (2008) Natural radioactivity measurement in soil samples collected from municipal area of Hat Yai district in Songkhla province, Thailand. KMITL Sci Technol J 52–58
19. Luikuku AS, Karunakara N, Nthoiwaa GP (2014) Assessment of γ-radiation levels and associated dose rates from surface soils in the eastern part of Botswana. Int J Low Radiat 344–354
20. Stranden ERLING (1976) Some aspects on radioactivity of building materials. Physica Norvegica 163–167
21. Singh S, Rani A, Mahajan RK (2005) ^{226}Ra, ^{232}Th and ^{40}K analysis in soil samples from some areas of Punjab and Himachal Pradesh, India using gamma ray spectrometry. Radiat Meas 431–439
22. Beretka J, Mathew PJ (1985) Natural radioactivity of Australian building materials, waste and by products in. Health Phys 87–95
23. Xinwei L (2005) Natural radioactivity in some building materials of Xi'an, China. Radiat Meas 94–97

Mechanical and Wear Behavior of A713–Al_2O_3–$CaCO_3$ Aluminum Hybrid Metal Matrix Composites

Sunil Kumar Pulluru, Anil Kumar Birru, and Goutam Sutradhar

1 Introduction

Lightweight materials are attractive to the automobile and aerospace sectors. Aluminum has many advantages over other materials because of its superior corrosion resistance, relative strength-to-weight ratio, and ability to be recycled. However, aluminium matrix is poor in wear resistance. Reinforcement particles like Al_2O_3, ZrO_2, WC, SiC, B4C, and Mg_2Si are added to enhance the microstructural and wear characteristics of the AMC because of their intrinsic properties [1–5]. The density and hardness were increased significantly with the dispersion of reinforcement. It leads to difficulty in machining. To overcome this problem, secondary reinforcement particles are added to the matrix material [6–9]. It was noticed that the wear properties of the monolithic 2014A1 are low compared with 2014A1 reinforced with SiC [10]. The average coefficient of friction for unreinforced matrix material is quite high than AMC [11]. It was observed that the percentage of SiC content was increased from 15 to 25, the wear property of the A356-Al-SiCp composite increases gradually [12]. The weight loss is high for pure alloy compared with the composite sample during sliding conditions. The mass loss is proportional to the sliding distance. It may increase with the sliding distance. It was observed that the mass loss is significantly reduced with the increase wt.% of Al_2O_3 particles [13–15]. Because of Al_2O_3, the plastic deformation reduced under wear test. This leads to reduce the contact area between the tribo-pair which aids in low friction. Hard reinforced particles meet the counter surface during the initial sliding condition. These aspirants were removed from the surface under severe rubbing conditions [16]. The Gr particles have solid lubricant behavior which is easily eroded during the polishing for the specimen, and it will reduce the friction between the tribo-pair. The solid lubricant makes the reinforced particles move easily and re-arrange themselves in the matrix alloy

S. K. Pulluru · A. K. Birru (✉) · G. Sutradhar
Department of Mechanical, National Institute of Technology, Manipur, Imphal 795004, India
e-mail: birruresearch@gmail.com

B. P. Swain (ed.), *Recent Advances in Materials*, Springer Proceedings in Materials 25,
https://doi.org/10.1007/978-981-99-3844-5_24

[17–19]. $CaCO_3$ particles possess good tribological properties. $CaCO_3$ nanoparticles were used in lubricating the grease because of their anti-wear properties. The wear scar diameter and friction coefficient gradually increase with the concentration of $CaCO_3$ nanoparticles. This is due to the formation of agglomeration, which can be yielded easily during sliding conditions. Very few research has been conducted on the fabrication of aluminum composites using $CaCO_3$ as reinforcement. Prabhu et al. [19] investigated wear resistance performance of $CaCO_3$ on metal matrix composites [20, 21].

Stir casting method was preferred because, it includes a stirring operation that helps to distribute the particles uniformly throughout the molten metal. Secondary processes including rolling, forging, and extrusion are preferred to distribute the particles uniformly and to eliminate the casting defects [22–25].

The tensile strength and wear properties of as-cast A713 and A713–Al_2O_3–$CaCO_3$ cold rolled hybrid composite material were studied in this work. This study provides the basic information required to enhance the mechanical and tribological properties of A713 aluminum alloy in engineering applications.

2 Fabrication of Composites

Stir casting process was employed to prepare unreinforced and hybrid composite specimen. The composition of primary alloy and hybrid composite alloy was displayed in Table 1. The EDS examination shows 100% composition and confirmed that particles are successfully reinforced through stir casting process. After many pilot experiments, the volumetric ratio of 15 wt-% $CaCO_3$ and 15 wt-% Al_2O_3 have satisfactory results. The AMC reinforced with 15 wt-% $CaCO_3$ and 15 wt-% Al_2O_3 have been selected for final examination. The substrate is heated to 800 °C and the reinforced particles are heated to a temperature of 300 °C to improve wettability and to reduce the temperature difference between matrix and reinforced particles. These particles are added into the molten particle with a uniform continued stirring motion of 5 min for uniform distribution of particles. It was cooled at atmospheric temperature and the cast alloys are machined into rectangular cross-sections. These samples are applied for secondary processes such as cold rolling. During the rolling operation, the thickness of the specimen is reduced by 25% in each pass. The thickness is reduced from 11 to 6 mm in three passes. As-cast samples are fabricated without the addition of reinforced particles. There was no secondary process applied to the as-cast samples.

2.1 *Study of Wear and Tensile Behavior*

The mechanical and wear behavior of an unreinforced A713 alloy and A713–Al_2O_3–$CaCO_3$ cold rolled hybrid composite alloy were studied. The tensile test and wear

Table 1 Composition of A713 (as-cast) alloy and A713–Al_2O_3–$CaCO_3$ rolled hybrid composite

Sample	C (wt.%)	O (wt.%)	Ca (wt.%)	Zn (wt.%)	Si (wt.%)	Al (wt.%)	Total
A713	19.64	13.66	0.07	9.93	0	56.7	100
A713–Al_2O_3–$CaCO_3$	11.27	9.17	0.37	7.17	1.53	70.49	100

test are conducted, and the results are compared and discussed briefly. The average of three trials is used to calculate the results of each experiment. The following sections provide a brief discussion of the test methodologies used.

2.2 *Tensile Test*

According to ASTM E8 Standards, the specimens are prepared. The specimen measures 100 mm in entire length and the gauge measures 32 mm and 10 mm in length and width. The tensile test is performed using the universal testing machine (UTM).

2.3 *Wear Test*

Wear tests were performed by using a pin-on-disk type testing machine for unreinforced A713 alloy and A713–Al_2O_3–$CaCO_3$ cold rolled hybrid composite alloy. The experiment is conducted under the dry sliding condition at room temperature. The samples were machined to get the dimension of 30 mm length and 6 mm diameter. The specimen surfaces were polished with an automatic polisher before wear tests. The pins are maintained in contact with the rotating steel disk under a load of 1N and a sliding distance of 500 m. The wear rate and weight loss are considered as a response to determining the wear resistance of the material. The specimen's weight has been recorded both before and after each test, and weight loss has been computed. A digital balance with a minimum count of 0.1 mg was used to measure the weight. The disk is cleaned with acetone to eliminate contaminants after each test.

3 Results and Discussion

3.1 Study Tensile Test

Figure 1. presents the stress strain relation and ultimate tensile strength for A713 and A713–Al_2O_3–$CaCO_3$ cold rolled hybrid composite. The results revealed that the ultimate tensile for hybrid composite were increased by 52.63%. This is due to two major reasons: (a) Due to the presence of reinforced particles (Al_2O_3 and $CaCO_3$) (b) Due to the cold rolling process. The hard reinforced particles induced into soft matrix material tend to produce dislocations [26]. The density of dislocations was increased due to the mismatch of coefficient of thermal expansion for the reinforced and matrix material, which tends to increase the tensile strength. The particles Al_2O_3 and $CaCO_3$ dispersed in the matrix, which locks the motion of the dislocations. On the other side, work hardening takes place due to the rolling operation. The grains are re-arranged and elongated toward the rolling direction. In this process, the grains are crushed and broken down into small pieces which tends to reduce the grain size and increase the density of dislocation [27–29]. The synergetic effect of reinforcement particles and rolling operation for the A713–Al_2O_3–$CaCO_3$ hybrid composite material leads to improve the tensile strength to a high extent. Table 2 displayed the tensile strength of rolled hybrid and unreinforced specimen.

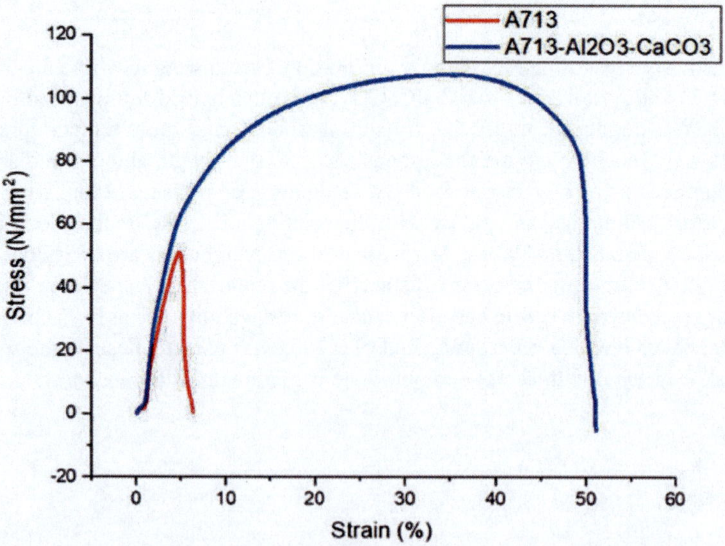

Fig. 1 Stress-strain curve of as-cast and rolled A713–Al_2O_3–$CaCO_3$ hybrid composite

Table 2 Tensile strength of hybrid and as-cast specimen

Sample	Ultimate tensile strength (N/mm^2)
A713	19.64
A713–Al_2O_3–$CaCO_3$	11.27

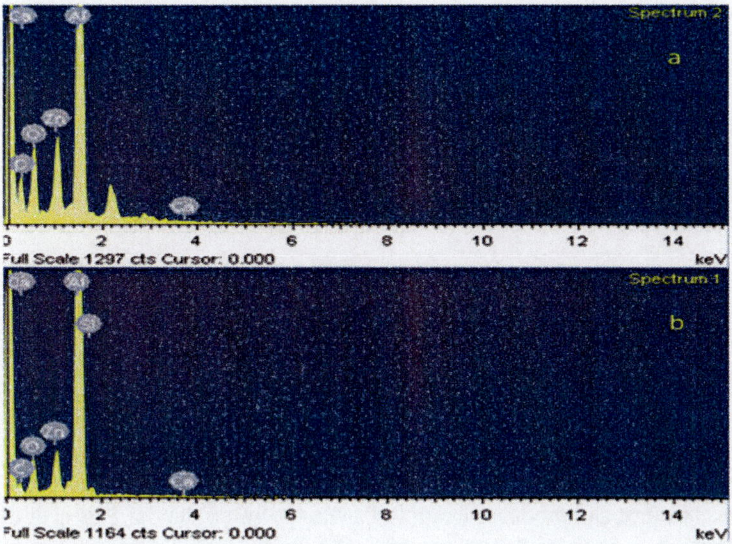

Fig. 2 EDS analysis indicating different phase peaks in **a** Unreinforced A713 alloy **b** A713–Al_2O_3–$CaCO_3$ rolled hybrid composite alloy

3.2 EDS Analysis

Figure 2 shows the EDS spectrum of A713 and A713–Al_2O_3–$CaCO_3$ rolled hybrid composite material. The composite specimen detects Si content, and the wt-% of Al and Ca elements has increased in the fabricated specimen.

3.3 Wear Rate

The mass loss under wear test can be evaluated by measuring the weight before and after wear test using an electronic balance. Figure 3 ensures that the wear rate for A713 matrix material is higher than A713–Al_2O_3–$CaCO_3$ hybrid composite during dry sliding condition. In A713–Al_2O_3–$CaCO_3$ hybrid composite, hard particles such as Al_2O_3 act as load-bearing elements. It enhances the strength of the composite. The composite is more resistant to plastic deformation at the subsurface and is more resistant to wear. As a result, the A713–Al_2O_3–$CaCO_3$ composite's sample experiences

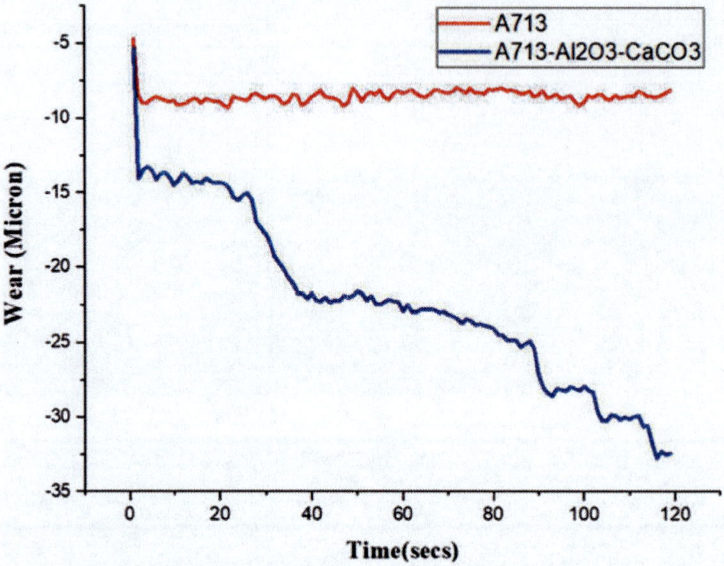

Fig. 3 Wear rate of rolled hybrid and as-cast specimen under the dry sliding condition

less wear surface and produces a reduced amount of wear debris. This is due to the self-lubricant and anti-wear property of $CaCO_3$ particles that aids in enhancing the wear resistance and reducing the COF. Suitable solid lubricant materials have major effect in reducing coefficient of friction in the oxide environment. The coefficient of friction on the aluminum surface tends to reduce because of presence of generation of aluminum hydroxides ($Al(OH)_3$). $CaCO_3$ tends to create tribolayer between the sample and counter surface during the sliding condition. The tribolayer can reduce the resistance and friction between the counter surfaces. The $CaCO_3$ particles can support the formation of a stable tribolayer.

The inclusion of $CaCO_3$ particles increases the thickness of tribolayer that leads to reduce the friction coefficient between the tribo-pair. Also, it can increase the wear surface area protected by the tribolayer, which tends to reduce the wear loss of the surface. Figure 4 shows the wear rate and weight loss of the specimen. It shows that the wear rate and weight loss were significantly low for hybrid composite than as-cast specimen. The addition of $CaCO_3$ may result in a reduction in porosity, which increases the lubrication of the wear surface. As a result, adding $CaCO_3$ can increase the wear resistance.

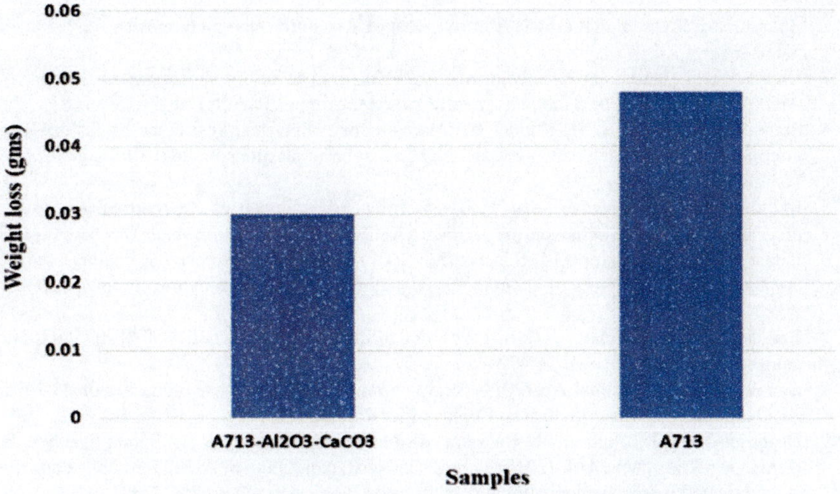

Fig. 4 Weight loss of rolled hybrid and as-cast specimen under the dry sliding condition

4 Conclusion

In this study, tensile properties and wear behavior of A713 and A713–Al_2O_3–$CaCO_3$ hybrid composite have been evaluated.

1. A713-15 wt.%, Al_2O_3-15 wt.% $CaCO_3$ rolled hybrid composite experiences lowest wear rate. It exhibits highest reduction of 60% in weight loss and enhances the tensile strength of 52.63% than A713. This is due to the synergetic effect of reinforcement and cold rolling process.
2. The secondary reinforcement of $CaCO_3$ plays a vital role in enhancing the wear resistance due to its anti-wear and self-lubricant property. It can generate the tribolayer between the counter surface which tends to reduce the friction coefficient between the contact area. It can improve the wear behavior under dry sliding.
3. The tamping effect of Al_2O_3 and $CaCO_3$ can generate nucleation sites, which tends to formation of grain refinement in the matrix. This leads to reduce the porosity and improve the tensile strength and wear property. The grains are elongated and uniformly distributed with the rolling process, which increases the interfacial bonding between the particles and matrix material.

References

1. Pramanik (Feb 2016) Explicit understanding of reinforcement effects on wear resistance of metal matrix composites. Trans Nonferrous Met Soc China 26(2):348–358

2. Bhansali KJ, Mehrabian R (1982) Abrasive wear of aluminum-matrix composites. Phys Metall 34:30–34
3. Umanath K, Palanikumar K, Selvamani ST (2013) Analysis of dry sliding wear behaviour of Al6061/SiC/Al_2O_3 hybrid metal matrix composites. Compos Part B: Eng 53:159–168
4. Song B, Dong S, Liao H, Coddet C. Microstructure and wear resistance of FeAl/Al_2O_3 intermetallic composite coating prepared by atmospheric plasma spraying. Surf CoatTechnol 268:24–29
5. Balasubramaniam I, Maheswaran R (2015) Effect of inclusion of SiC particulates on the mechanical resistance behaviour of stir-cast AA6063/SiC composites. Mater Des 65:511–520
6. Urena A, Rams J, Campo M, Sánchez M (2009) Effect of reinforcement coatings on the dry sliding wear behaviour of aluminium/SiC particles/carbon fibres hybrid composites. Wear 266:1128–1136
7. Baradeswaran A, Perumal AE (2014) Wear and mechanical characteristics of Al 7075/graphite composites. Compos Part B 56:472–476
8. Baradeswaran A, Perumal AE (2014) Study on mechanical and wear properties of Al 7075/ Al_2O_3/graphite hybrid composites. Compos Part B 56:464–471
9. Deaquino-Lara R, Soltani N, Bahrami A, Gutierrez- Castanada E, Garcia-Sanchez E, Hernandez-Rodriguez MAL (2015) Tribological characterization of Al7075-graphite composites fabricated by mechanical alloying and hot extrusion. Mater Des 67:224–331
10. Alpas AT, Zhang J (1994) Effect of microstructure (particulate size and volume fraction) and counterface material on the sliding wear resistance of particulate-reinforced aluminum matrix composites. Metall Mater Trans A 25:969–983
11. Suresha S, Sridhara BK (2012) Friction characteristics of aluminium silicon carbide graphite hybrid composites. Mater Des 34:576–583
12. Pramila Bai BN, Ramasesh BS, Surappa MK (1992) Dry sliding wear of A356–Al–SiCp composites. Wear 157:295–304
13. Kok M (2006) Abrasive wear of Al_2O_3 particle reinforced 2024 aluminium alloy composites fabricated by vortex method. Composites: Part A 37:457–464
14. Kok M (2005) Production and mechanical properties of Al_2O_3 particle-reinforced 2024 aluminium alloy composites. J Mater Process Technol 161:381–387
15. Hosking FM, Portillo FF, Wunderlin R, Mehrabian R (1982) Composite of Al alloys: fabrication and wear behaviour. J Mater Sci 17:477–498
16. Sannino AP, Rack HJ (1995) Dry sliding wear of discontinuously reinforced aluminium composites: review and discussion. Wear 189:1–19
17. Ravindran P, Manisekar K, Vinoth Kumar S, Rathika P (2013) Investigation of microstructure and mechanical properties of hybrid aluminum nano composites with the additions of solid lubricants. Mater Des 51:448–456
18. Sunqing Q, Junxiu D (Feb 2000) Wear and friction behaviour of $CaCO_3$ nanoparticles used as additives in lubricating oils. Lubr Sci 12(2):205–212
19. Prabhu MS, Perumal AE, Arulvel S, Issac RF (2019) Friction and wear measurements of friction stir processed aluminium alloy 6082/$CaCO_3$ composite. S0263–2241(19):30379–3
20. Jokhio MH, Panhwar MI, Unar MA (Jan 2011) Manufacturing of aluminum composite material using stir casting process. J Eng Technol 30(1)
21. Lei Y, Yan H, Wei Z, Xiong J, Zhang P, Wan J, Wang Z (2021) Effect of hot extrusion on microstructure and tribological behavior of Al_2O_{3p} reinforced 7075 aluminum-matrix composites. J Central S Univ 28:2269–2284
22. Himanshukala K, Mer KS, Kumar S (2014) A review on mechanical and tribological behaviors of stir cast aluminum matrix composites. Procedia Mater Sci 6:1951–1960
23. Sajjadih SA, Ezatpour HR, Beygi H (15 Nov, 2011) Microstructure and mechanical properties of Al–Al_2O_3 micro and nano composites fabricated by stir casting. Mater Sci Eng: A 528(29–30):8765–8771
24. Radhika N, Balaji TV, Palanippan S. Studies on mechanical properties and tribological behaviour of LM25/SiC/Al_2O_3 composites. J Eng Sci Technol 10(2):134–144

25. Wang B, Chen X-H, Pan F-S, Mao J-J, Fang Y (2015) Effects of cold rolling and heat treatment on microstructure and mechanical properties of AA 5052 aluminum alloy. Trans Nonferrous Met Soc China 25:2481–2489
26. Wang T, Huang Y, Ma Y, Wu L, Yan H, Liu C, Liu Y, Liu B, Liu W (2021) Microstructure and mechanical properties of powder metallurgy 2024 aluminum alloy during cold rolling. J Mater Res Technol 15:3337–3348
27. Benamirouche S, Abdi A, Hemmouche L, Mejias A, El Amine M, Belouchrani DC, Amrouche A, Trari M (2022) Multiscale study of cold-rolling deformation on mechanical and corrosion behaviors of AA2024-T4 aluminum alloy. J Indian Chem Soc 99:100307
28. Kishan V, Devaraju A, Lakshmi KP (2017) Influence of volume percentage of NanoTiB2 particles on tribological & mechanical behaviour of 6061-T6 Al alloy nano-surface composite layer prepared via friction stir process. Defence Technol 13(1):16–21
29. Palanikumar K, AshokGandhi R, Raghunath BK, Jayaseelan V (2019) Role of calcium carbonate ($CaCO_3$) in improving wear resistance of polypropylene (PP) components used in automobiles. Mater Today: Proc 16:1363–1371

In Silico Analysis of Ferrocenyl-Analogs as the Potential Drugs Against Aggressive UK-Based Strain of SARS-CoV-2 Novel Coronavirus

Maynak Pal, Abhishek Panwar, Sharmila Wahengbam, Dulal Musib, and Mithun Roy

1 Introduction

The COVID-19 pandemic has had a significant motility with over 243 million cases and 4.9 M deaths reported since November 2019 [1]. Although no cure has been found yet, vaccination is considered the most effective way to prevent infection [2]. To prevent the disease, WHO suggests the use of antiviral drugs to slow the spread of the virus [3–5]. A new strain of the virus called the SARS-CoV-2 novel coronavirus variant (VARIANT UNDER INVESTIGATION 2020/12/01) has emerged and is rapidly spreading globally, especially in south-east England.

The SARS-CoV-2 novel Coronavirus virus is known to resist antiviral drugs and have an aggressive immune response, but the long-term effects on patients and the efficacy of vaccines against the mutated variant have yet to be fully analyzed [6]. A study by Wang et al. found that vaccines may only be effective against the mutated variant for one-third of the population. Therefore, it is crucial to find a drug that can effectively combat this strain of the virus [7].

M. Pal · A. Panwar · S. Wahengbam · D. Musib · M. Roy (✉)
Department of Chemistry, National Institute of Technology Manipur, Langol 795004, India
e-mail: MithunRoy@mithunroy.nitmanipur.ac.in

M. Pal
e-mail: MaynakPal@mithunroy.nitmanipur.ac.in

A. Panwar
e-mail: AbhishekPanwar@mithunroy.nitmanipur.ac.in

S. Wahengbam
e-mail: SharmilaWahengbam@mithunroy.nitmanipur.ac.in

D. Musib
e-mail: DulalMusib@mithunroy.nitmanipur.ac.in

B. P. Swain (ed.), *Recent Advances in Materials*, Springer Proceedings in Materials 25,
https://doi.org/10.1007/978-981-99-3844-5_25

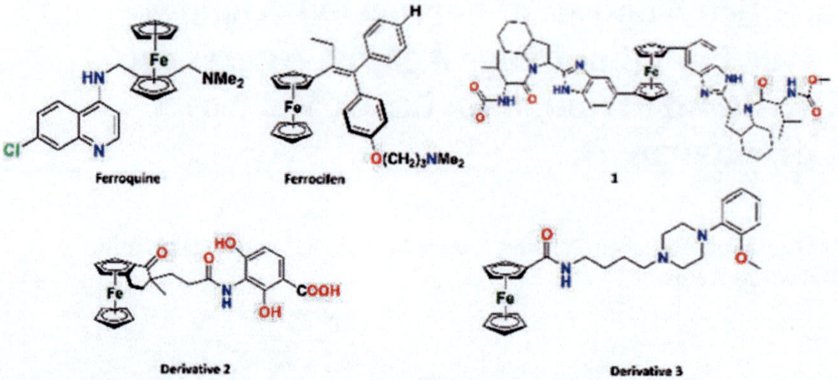

Scheme 1 Schematic illustration of the ferrocene derivatives selected for molecular docking against the S protein of the newly emerged UK strain of severe acute respiratory syndrome Coronavirus-2

The SARS-CoV-2 novel Coronavirus virus contains 23 mutations, including 6 synonymous mutations, 13 irregular mutations, and 4 amino acid deletions. The majority of these mutations are located in vital structural proteins such as the nucleocapsid (N) protein, spike (S) protein, and the ORF8 genetic protein. The S protein, in particular, has been affected by mutations such as N501Y, A570D, S982A, and D1118H [8]. The N501Y mutation, which is commonly found in other variants and located in the RBD of the S protein, is associated with increased transmissibility and more fatalities. The N501Y mutation is a crucial target for discovering a drug against the SARS-CoV-2 novel coronavirus variant [9, 10].

Transition metal complexes have recently gained attention as potential candidates for antiviral applications due to their wide range of properties and ability to disrupt viral proteins [11–15]. Ferrocene complexes, in particular, have been identified as they may inhibit SARS-CoV-2 novel Coronavirus spike protein [14, 15]. One of these complexes, Ferroquine, has already been established as an anticancer and antiviral drug, and many other ferrocene-derived compounds are being explored for their potential against the SARS-CoV-2 novel Coronavirus variant [16–23] (Scheme 1).

2 Results and Discussion

Ferroquine, a medication that has been approved by the FDA and exhibits both antitumor and anti-HIV properties, has been found to be more effective than chloroquine complexes according to Phase III clinical trials. This research endeavors to analyze the inhibition ability of Ferroquine against the S protein receptor domain of the mutated strain of SARS-CoV-2 novel Coronavirus virus. The docking simulation of Ferroquine was performed on the ANGIOTENSIN-CONVERTING ENZYME-S protein mutant (VARIANT UNDER INVESTIGATION 202012/01) and demonstrated remarkable binding effectiveness, with a binding affinity ~−7.1 kcal/mol and an IC ~ 6.3 micro molar. This prevents the S protein of the virus from binding to the host's ANGIOTENSIN-CONVERTING ENZYME-receptor protein. Ferroquine complexes engage with the ACE-2 protein through interactions that are both electrostatic and structural, with its rigidity and charge separation creating opportunities for stacking interactions and electrostatic bonds. The N501Y mutation has an impact on the inhibition capacity of Ferroquine to the ANGIOTENSIN-CONVERTING ENZYME protein, as it blocks the interaction with the ANGIOTENSIN-CONVERTING ENZYME site, enabling Ferroquine to bind to the RBD through intramolecular bonding, resulting in a higher binding affinity.

Previous studies has shown that a complex of Ferroquine, referred to as Complex 1, has the ability to inhibit the spike and RdRp proteins of SARS-CoV-2 novel Coronavirus. Our computer simulations demonstrate that Complex 1 can form several non-covalent bonds with proteins, and it could be a potential candidate for intramuscular injections. Our docking simulation analysis reveals that Complex 1 binds to the spike protein of the SARS-CoV-2 novel Coronavirus mutant strain with a binding affinity of approximately 10 kcal/mol and an IC value of 0.3 micro molar. This makes Complex 1 a promising option for suppressing the mutant strain (Fig. 1).

In addition to its antiviral and antitumor properties, Complex 2 has the ability to form non-covalent bonds with proteins. Our docking simulation study suggests that Complex 2 binds to the spike protein of the mutant strain of SARS-CoV-2 novel Coronavirus with a binding affinity of approximately -8 kcal/mol and an IC value of 1.6 micro molar, thanks to its strong interactions that involve five hydrogen bonds. As a result, Complex 2 has potential as an inhibitor of the SARS-CoV-2 novel coronavirus mutant spike protein.

Our docking simulation research has revealed that Complex 3 possesses antiviral properties and is capable of binding to the mutated form of the S protein with a binding affinity of approximately −8 kcal/mol and an IC of approximately 1 micro molar.

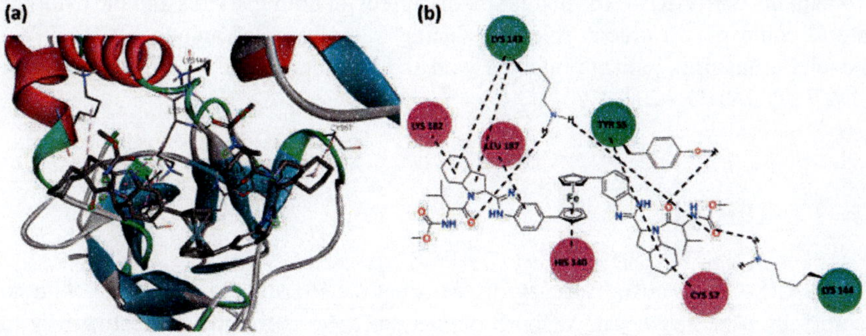

Fig. 1 **a** Optimal docking configuration showing the presence of non-covalent interactions between compound 1 and the NF-κβ protein. **b** Schematic illustration that depicts all the non-covalent bonding interactions

This binding is stronger compared with other proposed drugs and is augmented by hydrogen bonds and π-lone pair interactions, making Complex 3 a promising candidate for inhibiting the mutated spike protein.

Our findings indicate that of all the complexes studied, Complex 1 had the most notable impact on inhibiting the mutated spike protein of SARS-CoV-2 novel Coronavirus, with a binding affinity of approximately −10.22 kcal/mol. These results motivated us to explore its ability to inhibit the Nuclear factor kappa B and INTERLEUKIN-6 proteins, which control the cytokine storm responsible for inflammation in individuals infected with SARS-CoV-2 novel Coronavirus.

Our study of ferrocenyl Complex 1 through docking simulation revealed a binding affinity of approximately −8.28 kcal/mol and an IC of approximately 0.856 micro molar, as indicated in Fig. 1. The strength of the binding is largely attributed to the combination of hydrophobic interactions, hydrogen bonding and electrostatic interactions with protein residues. In 2021, the occurrence of the SARS-CoV-2 novel coronavirus VARIANT UNDER INVESTIGATION 202012/01 mutant variant resulted in widespread cases globally and has been noted to have reduced responsiveness to vaccines.

Our computational research on the inhibitory effects of ferrocenyl complexes with antiviral and antitumor properties against SARS-CoV-2 novel Coronavirus mutations revealed that Complex 1 has strong potential as a therapeutic option. The in silico study showed that it exhibits a binding affinity of −10.22 kcal/mol toward the mutant spike protein of the virus. In addition, the complex demonstrated strong binding toward the modulator proteins INTERLEUKIN-6 and Nuclear factor kappa B, with binding energies of −9.07 and −8.28 kcal/mol, respectively. These findings suggest that Complex 1 has the potential to inhibit SARS-CoV-2 novel Coronavirus replication and modulate the cytokine storm that can lead to inflammation in patients with ARDS.

The strong binding affinity of Complex 1 toward the human INTERLEUKIN-6 and Nuclear factor kappa B proteins, with a binding affinity of −9.07 kcal/mol and an IC ~ 0.222 micro molar, makes it a promising modulator of inflammation for COVID-19 patients with ARDS. Its dual inhibitory effect on both the virus and the cytokine storm could play a crucial role in reducing mortality rates among SARS-CoV-2 novel Coronavirus patients infected with the mutant variant VARIANT UNDER INVESTIGATION 202012/01.

3 Conclusion

The SARS-CoV-2 novel Coronavirus virus has caused a global pandemic, with no cure fully established yet. Vaccination has emerged as the most effective way to prevent infection, but the new mutated variant (VARIANT UNDER INVESTIGATION 2020/12/01) is posing a threat. The N501Y mutation in the receptor binding domain of the S protein is associated with increased transmissibility and drug resistance, making it a crucial target for drug development. Ferrocene complexes, in particular Ferroquine, have garnered attention as potential inhibitors of the SARS-CoV-2 novel Coronavirus virus. Ferroquine, a FDA-approved drug with antitumor and anti-HIV effects, has been shown to be more effective than chloroquine complexes in clinical trials and has the potential to inhibit both the viral spike protein and the INTERLEUKIN-6 and Nuclear factor kappa B proteins. Our studies found that Complex 1 is more potent than ferroquine in terms of inhibiting viral spike protein and the INTERLEUKIN-6 and Nuclear factor kappa B proteins. However, further studies are necessary to fully understand the potential of Ferroquine as a solution for combating the SARS-CoV-2 novel Coronavirus virus.

Acknowledgements We sincerely thank the Science and Engineering Research Board (SERB) (CRG/2021/004337) for financial support, MHRD India for scholarship support, and NIT Manipur for providing infrastructure. We also thank TEQIP-III (NPIU), NIT Manipur for providing computational infrastructure that facilitated our in-silico studies.

References

1. W. H. Organization, weekly operational update on COVID-19. https://www.who.int/publications/m/item/weekly-epidemiological-update-on-covid-19-31-march-2021
2. W. H. Organization. https://www.who.int/emergencies/diseases/novel-coronavirus-2019/covid-19-vaccines
3. Ahsan W, Javed S, Bratty MA, Alhazmi HA, Najmi A (2020) Treatment of SARS-CoV-2 novel coronavirus: how far have we reached? Drug DiscovTher 14:67–72
4. Richardson P, Griffin I, Tucker C, Smith D, Oechsle O, Phelan A, Stebbing J (2020) Current challenges of the 2019-COVID pandemic: where we started, where we are, and where do we go? Lancet 395:30–31
5. Kouznetsov VV (2020) Eur J Med Chem 203:112647
6. Cyranoski D (2020) Why emergency COVID vaccine approvals could pose a dilemma. Nature 588:18–19
7. Wang et al (2021) mRNA vaccine-elicited antibodies to SARS-CoV-2 novel coronavirus and circulating variants. Nature 592:616–622
8. Singh J, Ehtesham NZ, Rahman SA, Hasnain SE (2021) Structure-function investigation of a new VARIANT UNDER INVESTIGATION-202012/01 SARS-CoV-2 novel coronavirus variant. Biorixv. https://doi.org/10.1101/2021.01.01.425028
9. W. H. Organization, SARS-CoV-2 novel coronavirus Variants. https://www.who.int/csr/don/31-december-2020-sars-cov2-variants/en
10. Tian F, Tong B, Sun L et al (2021) N501Y mutation of spike protein in SARS-CoV-2 novel coronavirus strengthens its binding to receptor ANGIOTENSIN-CONVERTING ENZYME. Elife 10:69091
11. Nagendra Prasad HS, Ananda AP, Najundaswamy S, Nagashree S, Mallesha L, Dayananda BP, Jayanth HS, Mallu P (2021) Design, synthesis and docking simulation studies of novel piperazine metal complexes as potential antibacterial candidate against MRSA. J Mol Struct 1232:130047
12. Pahonțu E, Proks M, Shova S, Lupascu G, Llies CD, Barbuceanu F-S, Socea L-L, Badea M, Paunescu V, Istrati D, Gulea A, Draganescu D, Pirvu DEC (2019) Synthesis, characterization,

docking simulation studies and in vitro screening of new metal complexes with Schiff base as antimicrobial and antiproliferative agents. Appl Organomet Chem 33:5185
13. Bursal E, Turkan F, Buldurun K, Turan N, Aras A, Colak N, Murahari M, Yergeri CM (2020) Transition metal complexes of a multidentate Schiff base ligand containing pyridine: synthesis, characterization, enzyme inhibitions, antioxidant properties, and docking simulation studies. Biometals 34:393–406
14. Pal M, Musib D, Zade AJ, Chowdhury N, Roy M (2021) Computational studies of selected transition metal complexes as potential drug candidates against the SARS-CoV-2 Virus. ChemistrySelect 6:7429
15. Pal M, Musib D, Roy M (2021) Transition metal complexes as potential tools against SARS-CoV-2 novel coronavirus: an in silico approach. New J Chem 45:1924–1933
16. Ornelas C (2011) Application of ferrocene and its complexes in cancer research. New J Chem 35:1973–1985
17. Larik FA, Saeed A, Fattah TA, Muqadar U, Channar PA (2016) Recent advances in the synthesis, biological activities and various applications of ferrocene Complexes. Appl Organomet Chem 31:e3664
18. Santos MM, Bastos P, Catela I, Zalewska K, Branco LC (2017) Recent advances of metallocenes for medicinal chemistry. Mini Rev Med Chem 17:771–784
19. Peter S, Aderibigbe BA (2019) Ferrocene-based compounds with antimalaria/anticancer activity. Molecules 24:3604
20. Patra M, Gasser G (2017) The medicinal chemistry of ferrocene and its complexes. Nat Rev Chem 1:0066
21. Xiao J, Sun Z, Kong F, Gao F (2020) Current scenario of ferrocene-containing hybrids for antimalarial activity. Eur J Med Chem 185:111791
22. Chirullo B, Sgarbanti R, Limongi D, Shytaj L, Alvarez D, Das B, Boe A, Fonseca SD, Chomont N, Liotta L, Petricoin E, Norelli S, Pelosi E, Garaci E, Savarino A, Palamara AT (2013) Cell Death Dis 12:e944
23. Song P, Li W, Xie J, Hou Y, You C (2020) Cytokine storm induced by SARS-CoV-2 novel coronavirus. Clin Chim Acta 509:280–287

Effective Adsorption of Arsenic from Synthetic Water Sample Using Activated Sunflower Seed Shells

Aniket Dahasahastra, Lairenlakpam Helena, and P. Albino Kumar

1 Introduction

Arsenic (*As*) is a naturally occurring as well as an anthropogenic contaminant present in groundwater. Higher concentrations of *As*, due to its acute toxicity, have already been linked to multiple carcinogenic as well as non-carcinogenic characteristics in humans. Sustained exposure to *As* primarily leads to respiratory, reproductive, cardiovascular, hematological and neurological problems in humans. The inorganic form of *As* is a well-known cause of bladder, skin, and lung cancers [1]. According to IS:10500–2012, the acceptable limit for *As* in drinking water is 10 μg/L [2], while the permissible limit of 50 μg/L is adopted in the absence of an alternate source. The predominant species of *As* present in groundwater is *As (III)* and *As (V)* [3, 4]. Antagonistic effects of *As* directly depend on its dosages and exposure to the human body, and also include dermatological effects including melanosis and keratosis [5].

These crystal rocks become disturbed, broken, and crushed as a result of various man-made activities like drilling and tunnelling, and are essentially left exposed to the Earth around them. *As* contained in these cracked rock formations dissolves in the percolating water as it passes through the substrata, eventually mixing with the groundwater source and contaminating it. Despite the existing treatment options, there is always a need to find ways to deliver more effective care at a more affordable price. One of the most actively researched areas for water treatment researchers today is the lack of a workable treatment method for *As* removal with a comparably higher degree of removal efficiency, and the method to be used to remove *As* from drinking water [6].

A. Dahasahastra
Department of Civil Engineering, Visvesvaraya National Institute of Technology, Nagpur, India

L. Helena · P. A. Kumar (✉)
Department of Civil Engineering, National Institute of Technology, Manipur, Imphal, India
e-mail: albinoiit@gmail.com

B. P. Swain (ed.), *Recent Advances in Materials*, Springer Proceedings in Materials 25,
https://doi.org/10.1007/978-981-99-3844-5_26

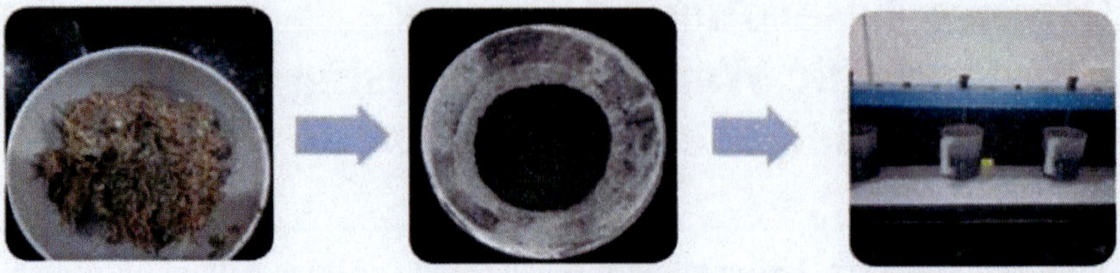

Fig. 1 Process flowchart

The oxidation states of *As* are determined by the pH of the water sample. Either a one-step therapy or a two-step treatment is afterwards used. While a two-step treatment involves oxidation before the removal of *As*, a one-step treatment just involves direct *As* removal and has a very low removal efficiency. The latter procedure is laborious and time-consuming, but it has a better removal efficiency. This turns out to be a significant drawback because oxidation is a sluggish process that can take days or weeks to complete [7, 8]. Furthermore, in some cases, use of ion-selective barriers were also employed for the removal of *As* but they were observed to be too costly, because membranes used for *As* removal are very difficult to recover [9]. Coagulation-flocculation has also been employed in certain literature for the removal of *As* from drinking water, as it converts dissolved *As* into an insoluble solid which gets precipitated and subsequently removed [10]. This method of coagulation however requires a high initial investment and skilled supervision to optimize the dosing with a proper understanding of the complex chemistry involved.

Hence, considering the financial feasibility and ease of operation, adsorption techniques are the most commonly adopted treatment strategy for *As* removal [11]. However, the efficiency of this technique depends primarily on the physical and chemical properties of the adsorbent [12, 13]. The effectiveness of a natural adsorbent's ability to efficiently and affordably remove *As* from a synthetic water sample was examined in this study. The adsorbent was created from the shell of a sunflower seed, and mathematical techniques such as isotherms and kinetics were used to examine the overall efficacy of the technology used, as shown in Fig. 1.

2 Materials and Methods

2.1 Materials

Analytical-grade chemicals were used in the process of adsorbent preparation. Arsenic trioxide, nitric acid, and sodium hydroxide were procured from Romi Chemicals, Manipur. Sunflower seed shells were collected from the nearby market of the National Institute of Technology, Manipur campus. Digital pH meter (ESICO 1012), hot air oven (KOMAI) and muffle furnace (Thermotech TIC 4000) were all procured through Advance Inspection and Testing Lab, New Delhi.

2.2 Adsorbent Preparation

The collected sunflower seed shells were ground into small fibrous pieces, followed by washing under running tap water to remove any residual dust adhering to it. This ground powder was dried overnight at 50 °C in a hot air oven for roughly 12 h. Chemical activation of the powdered precursor was done with nitric acid (1N) by pre-soaking the ground and dried material for 24 h. The resulting chemical-impregnated material was kept inside the muffle furnace. The furnace was heated at a heating rate of 10 °C/min to the final carbonization temperature of 550 °C, after which the material was kept inside the furnace for 20 min at a constant temperature of 550 °C.

The resulting material was washed with warm distilled water to remove any residual organic and mineral matter. It was finally washed with cold distilled water till the wash water achieves a neutral pH. The processed material was then filtered through Whatman quantitative filter paper grade 41, and the filtrate is finally dried for 24 h at 105 °C inside a hot air oven. Twenty-two-micron sieve was used for sieving the prepared adsorbent and the quantity retained on the pan was used for characterization. The same was stored in an air-tight container.

2.3 Preparation of Standard As (III) Solution

Stock *As (III)* solution was prepared by dissolving 1.32 g of As_2O_3 (Arsenic Trioxide) in a known quantity of distilled water containing 4 g of NaOH (sodium hydroxide) and later diluted to 1 L such that 1 ml stock solution contains 1 mg of *As (III)*. Intermediate *As (III)* solution was then prepared by diluting previously prepared 10 ml stock *As (III)* solution with 985 ml distilled water containing 5 ml concentrated HNO_3 (nitric acid) such that 1 ml intermediate solution contains 10 μg of *As (III)*. Finally standard *As (III)* solution was synthesized by diluting 10 ml of previously prepared intermediate *As (III)* solution with 985 ml distilled water, containing the same amount of concentrated nitric acid used for sample preservation. (1 ml = 0.1 μg of *As (III)*). The intermediate *As (III)* and standard *As (III)* solutions were stored at room temperature and used for analysis.

2.4 Experimental Analysis

Jar test apparatus was used in this study to achieve batch process as explained earlier. Multiple beakers each containing 500 ml of standard *As (III)* solution were taken and natural sunflower seed shells adsorbent was added to each of them in a pre-determined quantity. In this study, two sets of experiments with initial concentrations of 5 ppm and 3 ppm were performed [14, 15]. The pH of the solution was regularly monitored at 10 min interval and maintained to a specified value using either 1N NaOH or 1N HNO_3. 2 g/L dose of adsorbent was added to each sample. The solutions were thoroughly agitated at 300 rpm for 180 min. After completion of each batch process, the residual *As (III)* concentration in the solution was found using a commercially available *As* detection kit [16–19]. *As (III)* was analyzed using PerkinElmer atomic adsorption spectrometer (AAS) model AAnalyst 400.

2.5 *Estimation of* As (III)

The amount of *As (III)* adsorbed per unit amount of adsorbent at equilibrium (q_e) was then determined as follows:

$$q_e = \frac{Co - Ce}{D} \times V \tag{1}$$

where

Co initial concentration of adsorbate (mg/L),
Ce final concentration of adsorbate remaining in solution at equilibrium,
V volume of water to be treated (L), and
D amount of adsorbent added in (gm).

3 Results and Discussions

Results are obtained after batch processes were analyzed to optimize the pH, dosage, and kinetic studies as well as isotherms which signify the mathematical expression for adsorption. The efficiency of the adopted natural adsorbent in the adsorption of *As (III)* from the synthetic water sample was determined as explained ahead.

3.1 *Characterization of Adsorbent*

FTIR was used to confirm that the adsorption of *As (III)* is due to the adsorbent used in this study. Readings were taken pre- and post-adsorption. It is evident from Fig. 2a and b that the adsorption of *As (III)* on activated sunflower seeds occurs for a range of wave numbers between 1100 and 1600 cm^{-1}. Similar findings were observed by Sbeibley and Fowler, 1966, who in their study found that the As_2O_3 (Arsenous Oxide) spectrum lies in the same range of wave numbers as observed in this study.

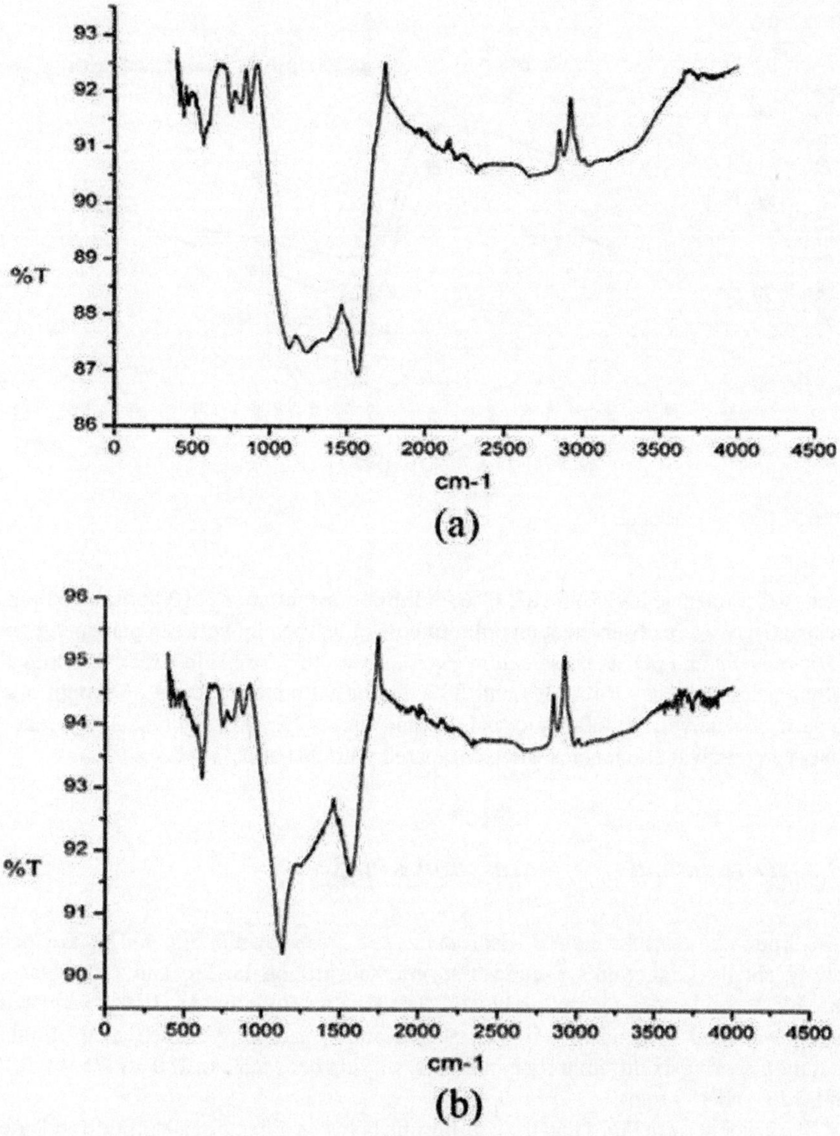

Fig. 2 FTIR spectroscopy for *As (III)*: **a** Before adsorption **b** After adsorption

3.2 *Optimization of pH*

As (III) removal efficiency significantly varies with the pH of test samples. Analysis was performed with the help of jar test apparatus to determine optimized pH for effective *As (III)* adsorption on natural sunflower seeds adsorbent. Highest *As (III)*

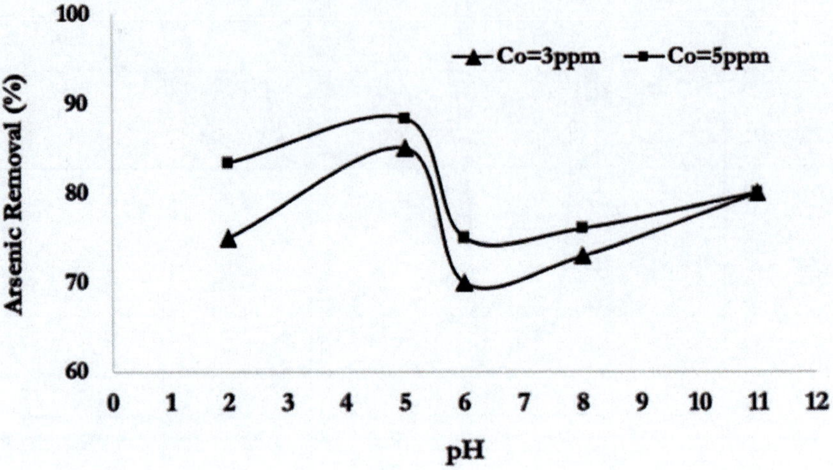

Fig. 3 Optimization of pH

removal efficiencies, 85% and 88.33% for initial concentrations of 3 ppm and 5 ppm, respectively, were observed at an optimum pH of 5. The plot between percentage *As (III)* removal and pH, as presented in Fig. 3, shows that for pH less than 5, removal efficiencies drop down to 73% and 83% for initial concentrations of 3 ppm and 5 ppm, respectively. Similarly, for pH of 6 and above, there is a significant difference between removal efficiencies when compared with that at pH 5.

3.3 *Determination of Adsorption Kinetics*

Adsorption kinetics for natural adsorbent at 295 K is shown in Fig. 4. This kinetics study entails Lagergren's pseudo-first-order adsorption kinetic and Lagergren's pseudo-second-order adsorption kinetic. The kinetics study for *As (III)* on activated sunflower seed shell adsorbent was carried out by sampling 5 ml treated sample solution each at six different time intervals, varying between 5 to 180 min (5, 15, 30, 60, 120, and 180 min).

It is evident from Fig. 4 that the equilibrium between the rate of adsorption and rate of desorption was achieved at 120 min after which further agitation lead to desorption and there was an increase in the *As (III)* concentration for the 10 ppm set. On the other hand, the equilibrium was achieved at 180 min for 5 ppm initial *As (III)* concentration, which signifies the relation between initial concentration and removal rate. It can therefore be safely deduced that with an increase in the adsorbate concentration, the adsorption rate increases and equilibrium can be achieved comparatively earlier.

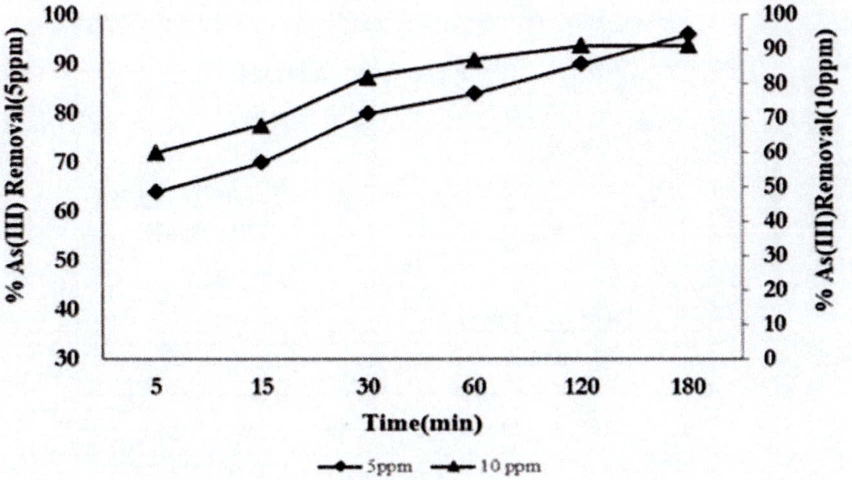

Fig. 4 General adsorption kinetics

Lagergren's pseudo-first-order plot as shown in Fig. 5 depicts the relation between the natural log of difference between instantaneous removal and removal at equilibrium and time. A significantly high coefficient of correlation (R^2) of 0.9665 and 0.9667 was obtained for both 5 ppm and 10 ppm initial *As (III)* concentrations, respectively. Similarly, Lagergren's pseudo-second-order plot between the ratio of time to instantaneous removal rate, as shown in Fig. 6, depicts the R^2 of 0.9978 and 0.9998 for initial *As (III)* concentrations of 5 and 10 ppm, respectively. An in-depth study of the plot revealed that the second-order kinetics was strongly followed during the process of adsorption of *As (III)* on naturally activated sunflower seed adsorbent irrespective of the initial concentration.

3.4 *Intraparticle Diffusion*

The equation shown below [20] served as the foundation for the intraparticle diffusion model that was given in Fig. 7. When intraparticle diffusion is the only rate-limiting process, the line drawn between qt and t1/2 crosses the origin. But, if film diffusion is also occurring, an intercept *C* is given, giving an idea of the boundary layer's thickness.

$$q_t = k_i \times t^{\frac{1}{2}} + C \quad (2)$$

where q_t is the amount of solute on the surface of the sorbent at time t (mg/g), k_i is the intraparticle diffusion rate constant (mg/g.min$^{1/2}$).

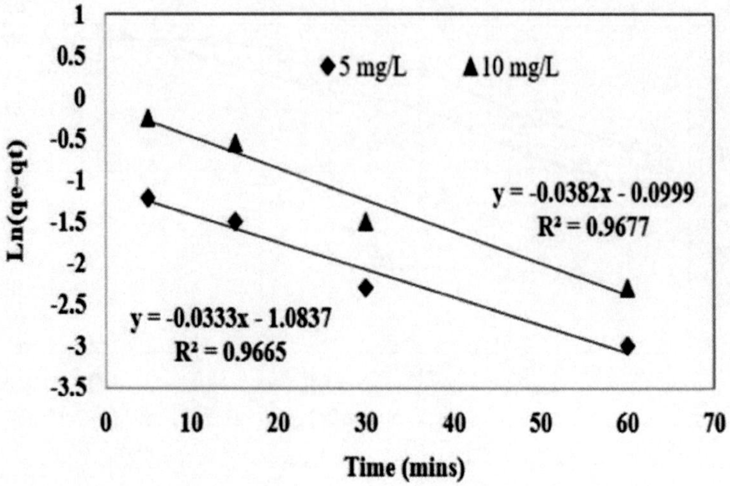

Fig. 5 Lagergren's pseudo-first-order adsorption kinetics

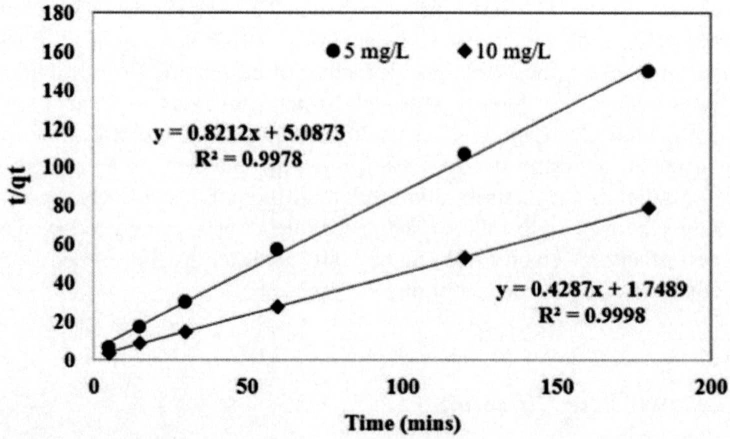

Fig. 6 Lagergren's pseudo-second-order adsorption kinetics

Figure 7 makes it clear that the 10-ppm diffusion plot's R2 value, 0.8085, is not sufficient. Hence, it was necessary to divide the graph into two linear halves. As seen in Fig. 8, the boundary layer diffusion was initially displayed linearly, and then intraparticle diffusion was displayed linearly. This demonstrates that both of the reported processes involved film diffusion in addition to intraparticle diffusion for the adsorption processes.

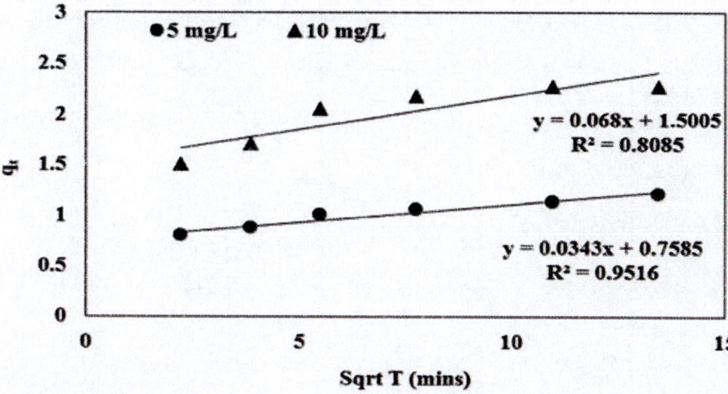

Fig. 7 Intraparticle diffusion model for the adsorption of arsenic on natural adsorbent

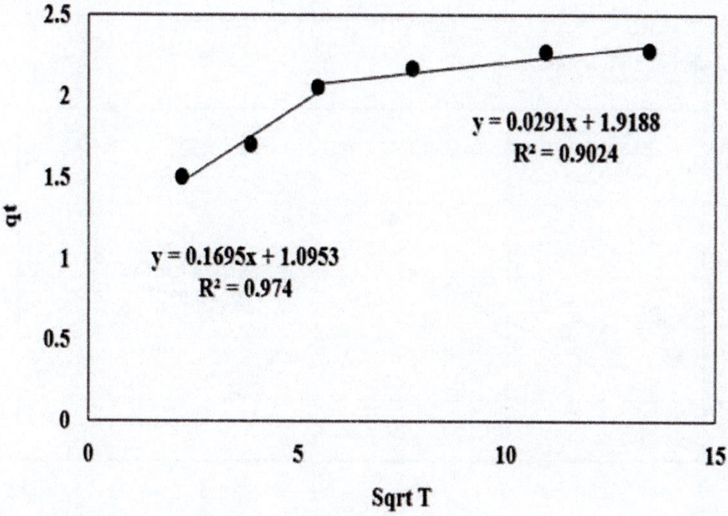

Fig. 8 Intraparticle diffusion model for adsorption of *As (III)*

3.5 Adsorption Isotherms

After optimization of pH and reaction kinetics, an experiment for determining adsorption isotherm was carried out for six different dosages of adsorbent (0, 0.5, 2, 4, 6, and 8 g/L). The Langmuir and Freundlich adsorption models were applied to the observed results from the adsorption isotherms experiment. The same is presented ahead in Figs. 9 and 10 (Table 1).

It is evident from Table 2 that the plot for Langmuir isotherm, i.e. plot between $1/C_e$ and $1/q_e$, has a comparatively higher R^2 value equal to 0.994 when compared

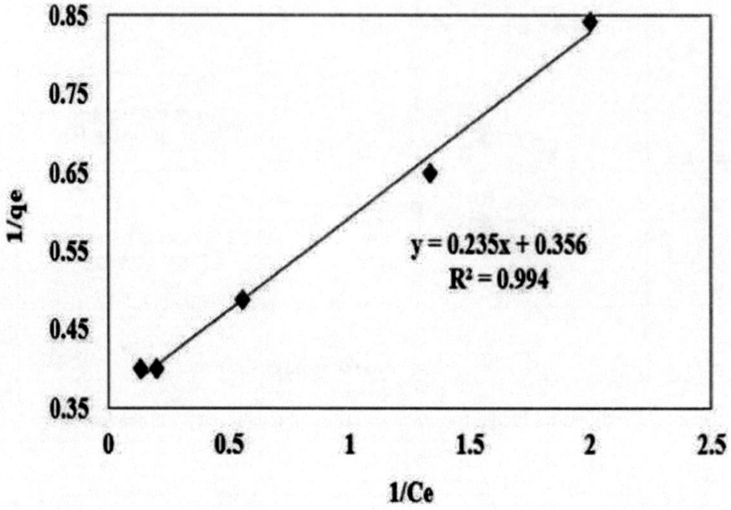

Fig. 9 Langmuir adsorption isotherms

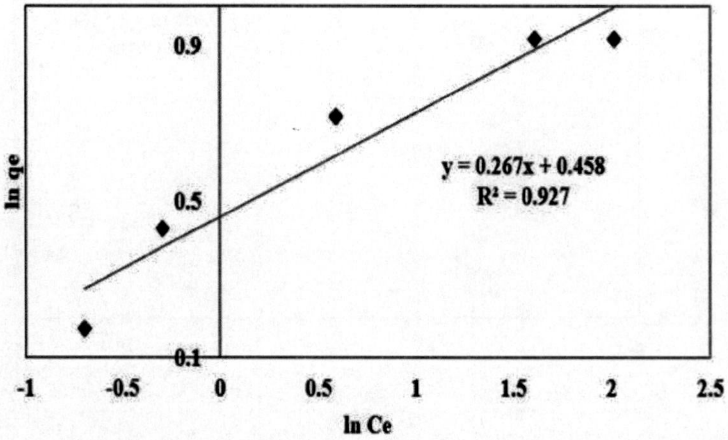

Fig. 10 Freundlich adsorption isotherms

Table 1 Comparison of isotherm models

SN	Isotherm model	Initial concentration (ppm)	Constants		R^2
1	Langmuir isotherm	10	a	2.8090	0.994
			b	1.5149	
2	Freundlich isotherm	10	ln (k_f)	0.4580	0.927
			1/n	0.2670	

with that of Freundlich isotherm, i.e. plot between *ln* (C_e) and $ln(q_e)$ for which R^2 value was 0.927. It is hence safe to assume that Langmuir isotherm is more practically applicable for *As (III)* removal by activated sunflower seeds used as a natural adsorbent.

4 Summary and Conclusions

The main goal of this study was to advance adsorption as an effective method for removing As (III) from synthetic water solution. As (III) was removed from a sample of synthetic water using an easily accessible and inexpensive natural adsorbent. Prior to the investigation, the sunflower seeds utilized as natural adsorbents were chemically activated. Batch procedures were carried out, and FTIR was used to characterize the adsorbent. The sample pH and adsorbent dose were improved by the use of numerous experimental runs. For both the starting concentrations of 3 ppm and 5 ppm, it was found that the highest As (III) removal efficiencies of about 85–90% were recorded at pH 5. Also, the outcomes of kinetic tests showed that the rate of adsorption was directly proportional to and reliant upon the initial concentration of adsorbate in the sample. It was found that the adsorption process wasn't just restricted to intraparticle diffusion but also had a significant function for film diffusion. The idea that chemical adsorption was primarily responsible for controlling the process is further supported by the technical applicability of Langmuir isotherm. The results of this investigation clearly demonstrate that critical As (III) pollution in the water sample may be successfully treated using activated sunflower seed shells as natural adsorbents.

References

1. Chakrabarti D, Singh SK, Rashid MH, Rahman MM (2019) Arsenic: occurrence in groundwater. In: Encyclopedia environmental health, 2nd edn. Elsevier Inc., pp 153–168
2. BIS (2012) Indian standards drinking water specifications IS 10500:2012
3. Korte NE, Fernando Q (1991) A review of arsenic (III) in groundwater. Crit Rev Environ Control 21:1–39
4. Wang Y, Jiao JJ (2014) Multivariate statistical analyses on the enrichment of arsenic with different oxidation states in the quaternary sediments of the pearl river delta, China. J Geochem Explor 138:72–80
5. Mandal BK, Suzuki KT (2002) Arsenic round the world: a review. Talanta 58:201–235
6. Singh R, Singh S, Parihar P, Singh VP, Prasad SM (2015) Arsenic contamination, consequences and remediation techniques: a review. Ecotoxicol Environ Saf 112:247–270
7. Johnston SG, Bennett WW, Burton ED, Hockmann K, Dawson N, Karimian N (2018) Rapid arsenic(V)-reduction by fire in schwertmannite-rich soil enhances arsenic mobilisation. Geochim Cosmochim Acta 227:1–18
8. Ungureanu G, Santos S, Boaventura R, Botelho C (2015) Arsenic and antimony in water and wastewater: overview of removal techniques with special reference to latest advances in adsorption. J Environ Manage 151:326–342

9. Shih MC (2005) An overview of arsenic removal by pressure-driven membrane processes. Desalination 172:85–97
10. Mondal P, Bhowmick S, Chatterjee D, Figoli A, Van der Bruggen B (2013) Remediation of inorganic arsenic in groundwater for safe water supply: a critical assessment of technological solutions. Chemosphere 92:157–170
11. Jang Y-C, Somanna Y, Kim H (2016) Source, distribution, toxicity and remediation of arsenic in the environment—a review. Int J Appl Environ Sci ISSN 11:973–6077
12. Chen S, Qin C, Wang T, Chen F, Li X, Hou H, Zhou M (2019) Study on the adsorption of dyestuffs with different properties by sludge-rice husk biochar: adsorption capacity, isotherm, kinetic, mthermodynamics and mechanism. J Mol Liq 285:62–74
13. Gong X-J, Li Y-S, Dong Y-Q, Li W-G (2020) Arsenic adsorption by innovative iron/calcium in-situ-impregnated mesoporous activated carbons from low-temperature water and effects of the presence of humic acids, 250:126275
14. Kong Y, Li M, Zhou Y, Pan R, Han Z, Ma J, Chen Z, Shen J (2022) Carbothermal synthesis of nano-iron-carbon composites for arsenate removal from high-arsenic acid wastewater, 10(2):107140
15. Ye C, Deng J, Huai L, Cai A, Ling X, Guo H, Wang Q (2022) Multifunctional capacity of CoMnFe-LDH/LDO activated peroxymonosulfate for p-arsanilic acid removal and inorganic arsenic immobilization: performance and surface-bound radical mechanism. Sci Toal Environ 806, Part 1
16. Reddy RR, Rodriguez GD, Webster TM, Abedin MJ, Karim MR, Raskin L, Hayes KF (2020) Evaluation of arsenic field test kits for drinking water: recommendations for improvement and implications for arsenic affected regions such as Bangladesh. Water Res 170:115325
17. Van Geen A, Cheng Z, Seddique AA, Hoque MA, Gelman A, Graziano JH, Ahsan H, Parvez F, Ahmed KM (2005) Reliability of a commercial kit to test groundwater for arsenic in Bangladesh. Environ Sci Technol 39:299–303
18. Das J, Sarkar P, Panda J, Pal P (2014) Low-cost field test kits for arsenic detection in water. J Environ Sci Heal—Part A Toxic/Hazardous Subst Environ Eng 49:108–115
19. Zhong XL, Wen SH, Wang Y, Luo YX, Li ZM, Liang RP, Zhang L, Qiu JD (2019) Colorimetric and electrochemical arsenate assays by exploiting the peroxidase-like activity of FeOOH nanorods. Microchim Acta 186:1–10
20. Nethaji S, Sivasamy A, Mandal AB (2013) Adsorption isotherms, kinetics and mechanism for the adsorption of cationic and anionic dyes onto carbonaceous particles prepared from Juglans regia shell biomass. Int J Environ Sci Technol 10:231–242
21. Ahsan H, Chen Y, Parvez F, Zablotska L, Argos M, Hussain I, Momotaj H, Levy D, Cheng Z, Slavkovich V, Van Geen A, Howe GR, Graziano JH (2006) Arsenic exposure from drinking water and risk of premalignant skin lesions in Bangladesh: baseline results from the health effects of arsenic longitudinal study. Am J Epidemiol 163:1138–1148
22. Boyle RW, Jonasson IR (1973) The geochemistry of arsenic and its use as an indicator. J Geochem Explor. 2:251–296
23. Lv S, Li C, Mi J, Meng H (2020) A functional activated carbon for efficient adsorption of phenol derived from pyrolysis of rice husk, KOH-activation and EDTA-4Na-modification. Appl Surf Sci 510:145425
24. Nejad MS, Sheibani H (2022) Super-efficient removal of arsenic and mercury ions from wastewater by nanoporous biochar-supported poly 2-aminothiophenol. J Environ Chem Eng 10(3):2107363
25. Kalaruban M, Loganathan P, Nguyen TV, Nur T, Johir MAH, Nguyen TH, Trinh MV, Vigneswaran S (2019) Iron-impregnated granular activated carbon for arsenic removal: application to practical column filters, 239:235–243

A Review on the Performance of the Textured Hydrodynamic Journal Bearing

Deepak Byotra and Sanjay Sharma

1 Introduction

With the rapid growth in machine tools in the recent years and massive use of high-speed machineries, new modified techniques are being explored to increase the performance of hydrodynamic bearings. Tribological properties have good impact on resources and energy efficiency in cylinder liners, etc. in journal bearing. The allied components of any machines play an important part in the smooth running of journal bearing. Automotive engines in the past require the journal bearings to work and operate at high-speed temperature and severe conditions. To meet this requirement, plain bearings with circumferential microgrooves have been developed in the past. These bearings are employed to enhance the load-bearing capacity and reduction in the friction coefficient. In the recent decades, the use of journal bearing increases considerably as it is capable to perform better in heavy loading conditions in various industries like turbo machines, engines, etc. The various shapes of bearings like axial groove, waviness, elliptical find their applications in different machineries to get the improved performance like load-carrying capacity and reduction in coefficient of friction. Good bearing design, lubrication, geometry could improve the bearing life and reduction in the machine failure. Many parameters like surface roughness, Newtonian, non-Newtonian flow, laminar, turbulent flow are considered for investigation by many researchers in their study in the past. The use of different fluids as lubricants at numerous places has its own relevance and importance. Lubrication plays an important role to minimize the friction for smooth, relative motion, and reduction in wear and friction. The micro-irregularities can develop sufficient fluid film pressure which thereby increases the load capacity of bearing. Texturing on the surface has been getting the more attention and becoming an important aspect

D. Byotra (✉) · S. Sharma
School of Mechanical Engineering, Shri Mata Vaishno Devi University, Katra, Jammu and Kashmir 182320, India
e-mail: deepak.byotra@smvdu.ac.in

B. P. Swain (ed.), *Recent Advances in Materials*, Springer Proceedings in Materials 25,
https://doi.org/10.1007/978-981-99-3844-5_27

in the fabrication of hydrodynamic bearing. Surface texturing of journal bearing has become the one of the aspects to enhance the journal bearing performance. The texturing increases the thickness of the film between two surfaces and thereby reduces the friction coefficient. A controlled amount of porosity on tribological surfaces has also been found to reduce friction in the sliding surface and surface texturing using boundary conditions for lubrication in the form of micro-dimples, which create a reservoir and retain the lubricants, increasing the pressure and load-carrying capacity of the hydrodynamic journal bearing. This created the interest in scientists and engineers to perform the experimental and theoretical studies for the enhancement of the performance to greater extent. At present, many scholars have studied the various aspects of the texturing in hydrodynamic journal bearing, and this paper summarizes the traditional shapes of textures along with the new shapes, design, geometry, etc. in journal as shown in Figs. 1, 2 and 3 and puts forward the ideas of new shapes which are not yet explored for the extensive use of bearings.

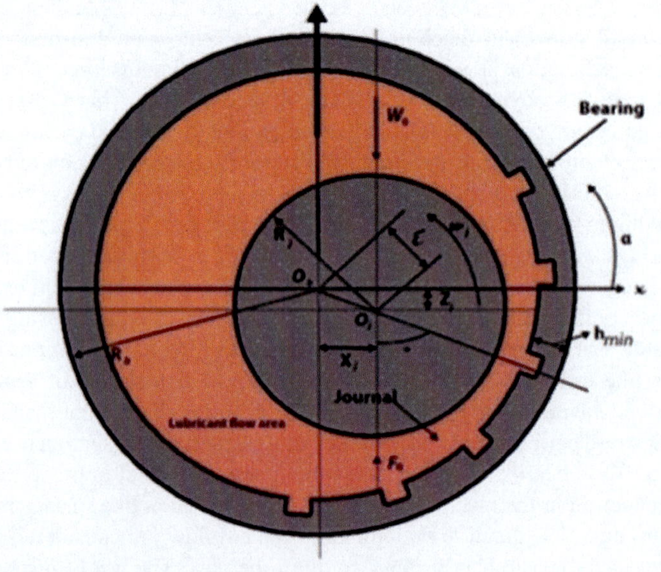

Fig. 1 Sectional view of the hydrodynamic journal bearing

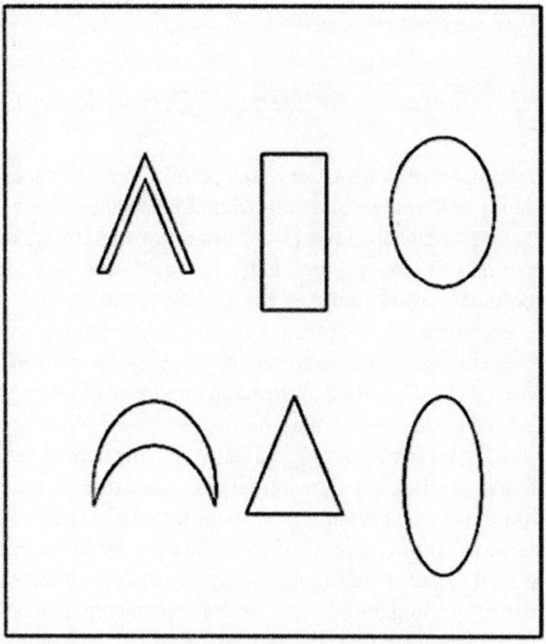

Fig. 2 Various shapes of textures

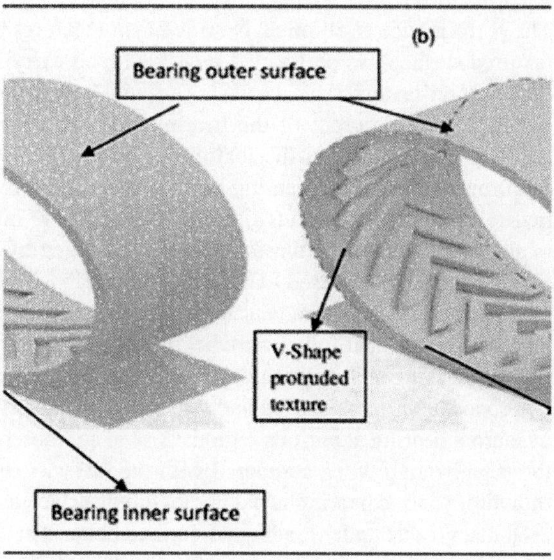

Fig. 3 V-shape texture

2 Literature Review

2.1 Literature Related to Effect of Various Shapes of Textures

The impact of various textures profiles and geometries on journal bearing performance and wedge effect in triangular shape texture increases the capacity of the bearing to sustain load [1]. Comparing the effects of micro-grooving and surface texturing in journal bearings, it was found that the microgrooving lowers the coefficient of friction than the surface texturing [2].

Investigation on textured and untextured surfaces of hybrid journal bearing and results showed that textured bearing has improved stability than the untextured bearing [3]. The effect of the ideal texture shape on bearing performance was studied, and it was discovered that the convex shape, which is parallel to the sliding surface, has a higher load-carrying capacity than the concave shape [4]. The effect in spherical dimples/textures as shown in Fig. 5 on bearing surface and observed enhancement in film thickness, pressure, and axial flow in bearing characteristics [5]. The effect of shape of texture and position on bearing performance and found that enhancement in load capacity and reduction in coefficient of friction [6, 7]. The bearing performance by using dimple textures and enhancement in LCC found to be maximum with partial texturing than the full texturing [8]. The study on the impact of geometrical abnormalities indicated that they had an impact on the misalignment conditions [9]. The performance of elliptical bearing having textures is observed and found that the textured surface has performed increased load-carrying capacity at higher density than the non-textured surface [10]. The study on the impact in cone texture as shown in Fig. 6 on the output of the bearing showed the enhancement in load-carrying capacity [11]. Study on the textures' effect as depicted in Fig. 4 on steady-state performance in journal bearing produced good results in increased bearing performance by applying various models [12, 13]. The influence of various properties of the lubricants on textured surface is observed by using thermal fluid methodology and found increased LCC in the output [14]. The comparison between full and partial surfaces of texture bearing is made and observed that partial texture bearing performs better than full texture bearing [15]. The performance of bearing is investigated by using texture position, size and found reduced friction and increased LCC by proposing proper position and size [7]. The impact of textures in the performance of journal bearing at various regions is seen and observed increased performance in the negative half when compared with the full wave texture [16]. The performance of hydrodynamic bearing is having textured surface by adopting CFD approach and seen that groove under loading performed better than the smooth surfaces [17]. The comparison in the impact of dimple depth, density, and profile on surface of the bearing using CFD is observed and found that the surface with dimple surface in lubrication increased the functioning of the bearing when compared with smooth surface [18]. The impact of indentation, textured surface has shown improvement in the LCC and frictional force due to the formation of small reservoirs for lubricants [19]. The hydrodynamic impact of different shapes of texture, geometry, size on the bearing performance is studied and observed enhancement in performance cubic shape texture in comparison to the other geometries [20]. The impact on tribological performance of surface texture by various different optimization methods is observed and seen that the approach is advantageous in understanding the relationship between the partially textured surfaces and other bearing geometry [21]. The morphology in cavitations in the textured and untextured surfaces in the thrust bearing is studied by applying the Reynolds and JFO models and found that the JFO predictions are more accurate than the Reynolds model [22]. The effect of microgrooves on the surface is investigated by applying computational fluid dynamics method in 2D and 3D cases under low and high loading and observed good performance at 3D than the 2D case [23]. The impact of partially textured surface bearing performance is seen by using Reynolds equation and observed good results in terms of LCC and pressure in the film [24]. The study on the lead-free materials and texturing on the bearing surface is conducted by using pin disc setup and found improvement in the potential in terms of wear rate and friction when compared with the non-textured surfaces [25]. The impact of various textured shapes to optimize the curvature direction on the bearing performance is seen and found that convex shape of texture performs better than the texture having concave profile [4]. The impact of micro-textures spherical in shape on the concave and convex profiles on the bearing performance is investigated by adopting geometric, dynamic models and observed that in convex textures, the LCC of the bearing is increased and in concave texture the LCC is decreased [26]. The study on the two-lobe bearing having cylindrical textures is conducted by using Reynolds equation and FEM in Newtonian lubrication and observed an improvement in bearing performance with non-circularity index and further enhances with any further increase in non-circularity [27]. By using the one-dimensional Reynolds equation to analyse the effects of partial slip texture sliders in magneto-hydrodynamic (MHD) lubrication, it was discovered that journal bearings with partial texturing slip using the MHD approach enhance the bearing's LCC [28]. Analyzed the different models to understand the impact of various shapes, density and size of textures on the bearing performance and provided a comprehensive report on flow of fluid cavitations zone etc. [29]. The effect of regularly spaced rectangular textured surface of parallel bearing is seen by applying mass-conserving analysis and found reduced friction and improvement in load-carrying capacity in comparison with the plain bearing [30]. A model is developed to see the surface roughness effect in the form texturing by solving Reynolds equation and observed that developed model reduces the friction and optimizes the groove characteristics [31]. The impact of surface texturing by designing the texture shape and size using the finite element simulation and FDM under various lubrication techniques is studied and found maximum improvement in friction reduction in full film lubrication regime [32]. The topographic surface effect of film lubrication using different shapes of textures is studied and found that the chevron shape textures are more effective than groove in increasing hydrodynamic lubrication case [33]. The effects of partial slip texture slider were investigated using the one-dimensional Reynolds equation in magnetohydrodynamic (MHD) lubrication, and it was discovered that journal bearings with partial texturing slip using the MHD approach enhance the LCC of the bearing [34]. The performance of the cylindrical texturing on the different regions of the bearing surface is seen and observed the improvement in static characteristics and dynamic characteristics in the upstream zone [35]. The impact of surface texturing on hydrodynamic bearing having inlets oil holes is observed and compared it with the untextured bearing and found that the textured surface performed better in comparison with the non-textured surface of the bearing at constant speed and oil supply [36]. A model is developed for circumferentially grooved journal bearing to predict the static and dynamic characteristics and obtained satisfied result for all parameters expect the flow rate [37]. A model is developed to see the effect of texture on pressure in bearing dynamic performance and observed that the stiffness and other factors in textured bearing performed better than the untextured bearing [38]. The effect of dimples of various size, depth, etc. on the friction characteristics of the bushes is observed by using etching and machining techniques and observed that dimples on the whole surface have good impact than on the half surface of the bushing [39]. The impact of cavitations and various texturing characteristics on the performance of the contact surfaces is studied and observed that the parallel cavitations showed a good response in comparison to the smooth surfaces and optimal values are found in the partially textured surfaces [40]. The surface texturing performance individually and with slip is investigated and observed that no-slipping and micro-texturing improve steady-state and dynamics characteristics of bearing [41]. The combined impact of rectangular texturing in surface and boundary slip is studied by using computational fluid dynamic analysis and Navier slip model in journal bearing and found that increased texturing and friction reduction has observed at high eccentric ratio [42]. The impact of location of partial texturing in bearing by using Navier equation and computational dynamic fluid method on the performance of LCC and COF is observed and found that the bearing performance is decreased at the maximum pressure area but found to be increased in the fluid divergent area [43]. The journal bearing performance is observed by comparing the texturing surface with the non-texturing surface by finite element method (FEM) and found that the surface texturing with definite geometry and other parameters performed better than the non-textured surfaces [44]. The influence of micro-dimples of different shapes, sizes, etc. on COF on hardened steel is seen and observed that surface with micro-dimples has low friction in comparison to plain surface [45]. A model is developed for texturing on the surface with different patterns using mapping technique and observed very good results in terms of enhancement in performance in the workpiece in comparison with the conventional practices [46]. The methodology is developed to improve the grooving geometry of hydrodynamic bearings by using spline function to make it flexible and obtained good results in increasing the stiffness of gas thrust bearing when compared with the conventional geometries [47]. The square-shaped textures' geometry is studied to see its effect on load capacity, friction coefficient, etc., in flat bearing and found that optimal values' parameters remains almost the same [48]. The impact of hydrodynamic lubrication on micro-pocketed journal bearings was assessed, and it was observed that a change in pressure in the pocket near the bearing's intake was what generated the increased lubrication layer and decreased friction [49]. The effect of micro-dimples formed by laser surfaces texturing on partial and full surfaces of the thrust bearing is studied and found optimum parameters for load-carrying capacity by using this technique [50]. The hydrodynamic impact of micro-pore textures was examined by solving Reynolds' equation, and it was discovered that the friction coefficient is reduced to the minimum when pores are texturized optimally [51]. The impact of texturing and slip and no-slip in bearing performance operating under different operating conditions is observed, and using various governing equations and on the basis of comparison, they observed that selection of textures/slip geometry is crucial; also slip and no-slip pattern improved the bearing performance [52]. The comparison in the noise level between compound grooved and simple grooved textures is observed by applying the computation fluid dynamic method in the journal bearing and found that compound grooved bearing performed better than the bearing having simple groove [53]. The influence of textured surfaces in various bearings and their effects on variables like friction coefficient, sliding velocity, and Hersey number are studied. It was shown that textured bearings out performed smooth bearings in terms of hydrodynamic lubrication [54]. The textures' performance formed by laser-etched technique on crank bearing using test rig of journal bearing is investigated and found reduction in friction and shear rate even at the higher loading region bearing having worn surface [55]. The performance of spherical shaped micro-textured hybrid journal bearing is examined by using the Reynolds's equation, FEM and found that the textured increases the performance of the bearing and reduced untextured bearing [56]. The impact of full, partially textures upon the dynamic characteristics of the journal bearing is seen using different geometries, location, and comparison with the untextured surface and observed that textured surface having proper geometries has better performance than the surface without texturing [57]. The level of acoustic power micro-grooved bearing performance is studied by adopting the computation fluid dynamic method and software Swanson Analysis Systems Inc (ANSYS) and used to solve numerically the various problems using finite element method and observed that dimple can enhance the noise level depending upon the size and geometry [58]. The bi-triangular textures' steel bearing's performance in terms of wear and friction coefficient under dry conditions is seen, and results reveal that the high-density textures reduced the wear in comparison with the smooth surface, but coefficient of friction is high at high speeds [59]. The surface texturing impact on journal bearing performance is investigated and observed that response of friction response in bearing is in line with transverse to the sliding directions [60]. The impact of groove, lubricant, etc. on slider bearing performance is observed and found that the groove formation reduced the wear friction [61]. The impact of textured with contaminated lubricating oils on bearing characteristics is studied and observed that the textures' presence reduced the chances of bearing failure [62]. The mathematical model is developed to study the various shapes of textures compared it with the available experimental data and found an error in the coefficient of friction (COF) values [63]. The parameters of shape, texturing, etc., numerically and experimentally through mass-conserving technique are studied and observed that numerical approach cannot supplement the experimentally approach with different shapes of textures [64]. The different texturings in sliding bearing by optimizing partial textures' parameters are observed and proposed surface treatment of texturing on the surface for future development [65].

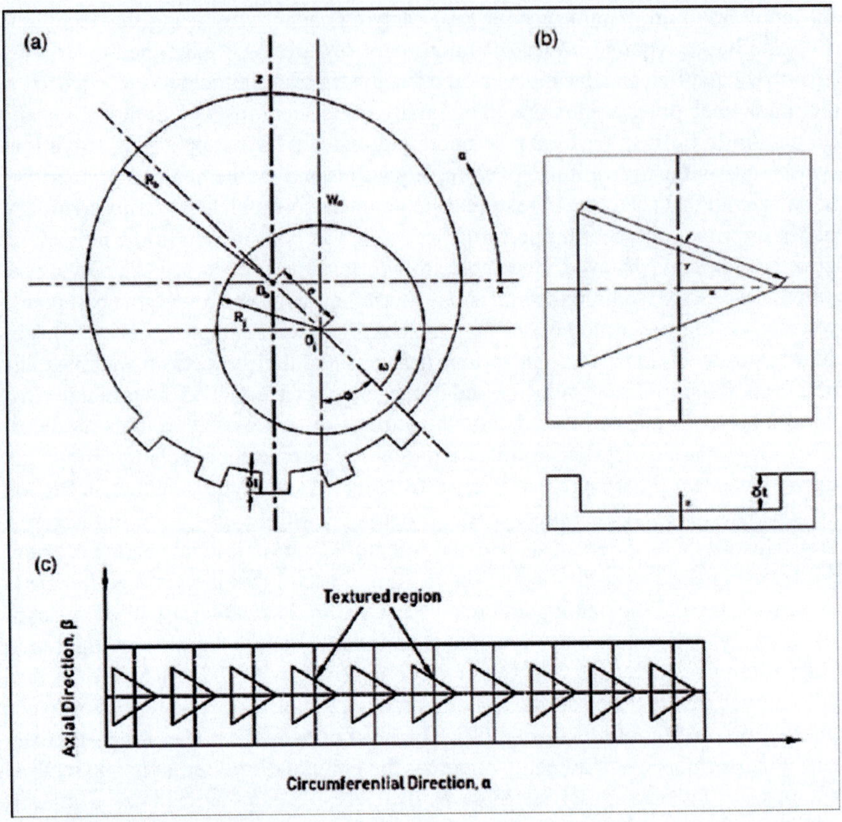

Fig. 4 Chevron-shape texture

2.2 *Literature Related to Effect of Performance*

The influence of textured surface on full and partial regions of the bearing is studied and found the reduction in acceleration amplitude in case of textured bearing when compared with untextured journal bearing [66]. The flexibility of the texture and its impact on journal bearing performance are investigated, and it is discovered that textured surfaces have higher film pressure than smooth ones [67]. The impact of different types of surface texture in the performance of bearing is studied and found more enhancements in the negative half texture than the textures at full wave [68]. The evaluation of partial and full textures' bearing performance is studied and observed enhancement in partially texture surface when compared with full textures [69]. Various studies have observed the impact of textures on bearing performance and observed enhancement in load-carrying capacity of the bearing [70–72]. In the positive pressure zone of the journal bearing, Gauss-Seidel and the linear complementarity problem (LCP) technique were used, and good results were obtained when

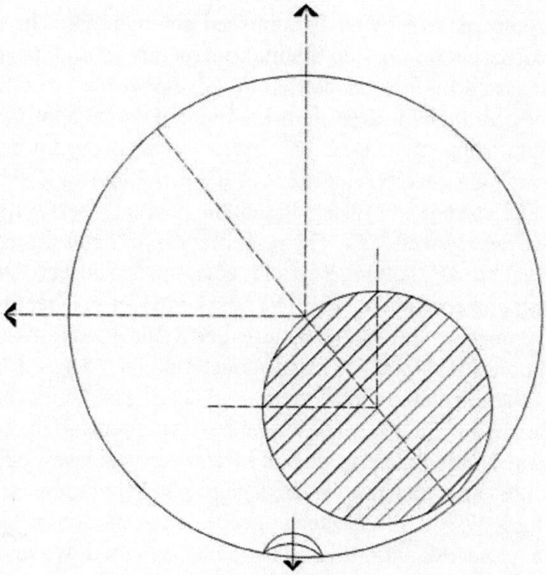

Fig. 5 Micro-grooved-shape texture

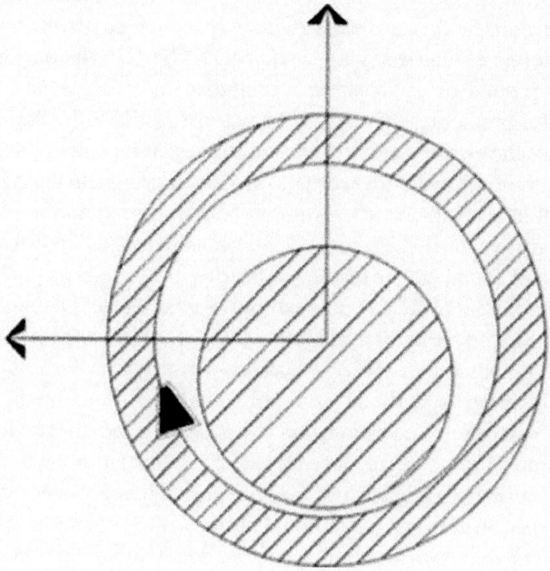

Fig. 6 Cone-shape texture

compared to previously published findings [73]. The cavitational model with mass-conserving model on the micro-structure journal bearing is studied and found that mass-conserving model has an edge over the cavitational model [74]. The impact of nanofluids in bearing using computational approach is observed and showed an improvement in LCC [75]. When comparing the precision medium duty (PMD) method's results with those of fixed and multigrid mesh in terms of computational efficiency, it was observed that the bearing's performance with texturing performed extremely well [76]. The performance of textured surface and slip of the bearing is evaluated by solving Reynolds equation and observed the enhanced load capacity and friction coefficient [77]. The misaligned journal bearing is studied by using different geometries and bottom textures and found good response at high eccentricity ratio for both used parameters [1]. A model is predicted based on development of pressure and variation on the shafts' textures of bearing and found good performance of bearing in terms of LCC [78]. The influence of textured surface for the full, half, and sinusoidal waves' roughness is studied by using power law model on the bearing performance and found that full wave roughness helps in increase of LCC and reduced friction force [79]. The trajectories of the center motion of the journal are studied by using nonlinear equations to observe the static and dynamic performance and found that stability margin in nonlinear is better than the linear analysis [80]. The lubrication with irregularities on the micro-surface is observed by comparing the analytical results with the results obtained through experiments which are in agreements in terms of load-carrying capacity (LCC) [81]. The development of pressure in various textured on shaft surface is studied by using neural network to develop a model for prediction of various characteristics like LCC capacity, etc., of the bearing [82]. Finite element analysis was used to examine the impact of hydrodynamic lubrication in cylindrical bore bearings on load carrying and the rupture's portion and discovered that bearing characteristics rely on the direction of loading [83]. The circumferential groove journal bearing (CGJB) considering cavitations and dynamic stability is seen and found better results in bearing performance [84]. The geometry with alloyed tool steel (SKD11) pin textured in bearing steel using cross-hatch pattern in paraffin lubricant is studied and found that the pattern considered for this study improved friction parameters [85]. The static performance with non-Newtonian fluids of journal bearings is studied by adopting control volume method and Elrod algorithm solved by Reynolds equation, and results have good effects on load capacities and side flow rates [86]. The impact of wear and deformation on the micro-grooved surface of the journal bearing is studied and reported that grooves avoid the seizure conditions and help in lowering the temperature of the bearing as the flow of oil increases [87]. The micro–macro multiscale method with mixed lubrication using 3D Navier stokes equation is studied and found that in micro-cavitations the flow pressure increases at particular location of the bearing and Reynolds equation is employed in macro-method at contact load in deformed surface around the pocket by many simulations to have increased pressure and flow factors [88]. The gasoline engine bearing performance bearing is evaluated by using two different lubricants and found that mathematical model-predicted friction is close to the data obtained experimentally [89]. The effect of thermo-hydrodynamic solutions in axial grove using turbulent flow conditions is studied by solving the governing equations on bearing performance and observed that the pressure and temperature are affected by the parameters of bearing [90]. The effect of local radius along with various motions on grooved thrust bearing is seen and observed that oscillating motion can have self-excited bearing [91]. The pressure distribution using governing Reynolds differential equation and finite difference method in a groove thrust bearing is studied and observed that gas flow, stiffness, etc. are calculated analytically and presented with the experimental data [92]. The characteristics' behavior of different thrust bearings having tilted pads under some operating conditions is studied and observed low friction, metal contact under fluid film lubrication at bearing area [93]. The face geometry due to discontinue thickness of thrust bearing by using Reynolds's equation and Newton–Raphson theory is investigated and found that some of the tools showed good response than the earlier used grooved seals [94]. The thrust bearing performance in unidirectional, bi-directional, etc. by applying the laser surface texturing is evaluated and observed improved performance in comparison with the untextured surface [95]. The coustical properties by applying boundary condition reported by Jakobsson–Floberg–Olsson (JFO) and also by implementing the cavitations' algorithm using frequency analysis in journal bearings are studied and found good results in coustical frequency in its pure form as super-harmonics [96]. The nonlinear bearing characteristics in terms of stiffness, frequency response, etc. are observed and compared it with the linear characteristics through simulation and found the good results [97]. The bearing's performance under the influence of high speed, load, etc. is reviewed and observed that the materials and lubricants should be in line with the proposed features of the bearing [98]. The bearing performance with vegetable, mineral oils with different textures and geometry is investigated and observed that the combination of vegetable oil with different textures can enhance the bearing performance with different combination and also reduced the impact on the environmental conditions [99]. The v-shape textured as shown in Fig. 3 is studied with the other textures' shapes and found good results at particular region at particular eccentricity ratio and depth of texture [100]. Recently, some literatures are also found which have described about the performance of active magnetic bearings under the influence of various faults in the rotor system [101–104]. After studying the literature, it has been observed that the improvement in the performance of the journal bearing can be enhanced analytically with formation and optimization of textures in the different regions of the bearing surface by selecting the particulars operating and geometric parameters as reported by [12]. The dimensionless form of governing Reynolds equation for fluid flow is adopted for distribution of pressure by using various methods like finite element method [FEM], finite difference method [FDM], etc. and selected the suitable boundary conditions. The performance parameters like LCC, COF, etc. can be calculated by using the computational procedure as shown in the flowchart. All these modules in the flowchart as shown in Fig. 7 are executed by iteration process which stops execution after achieving the convergence criteria as reported by ref. [12] and attaining the journal equilibrium position so that by using the nodal pressure, the performance parameters can be computed.

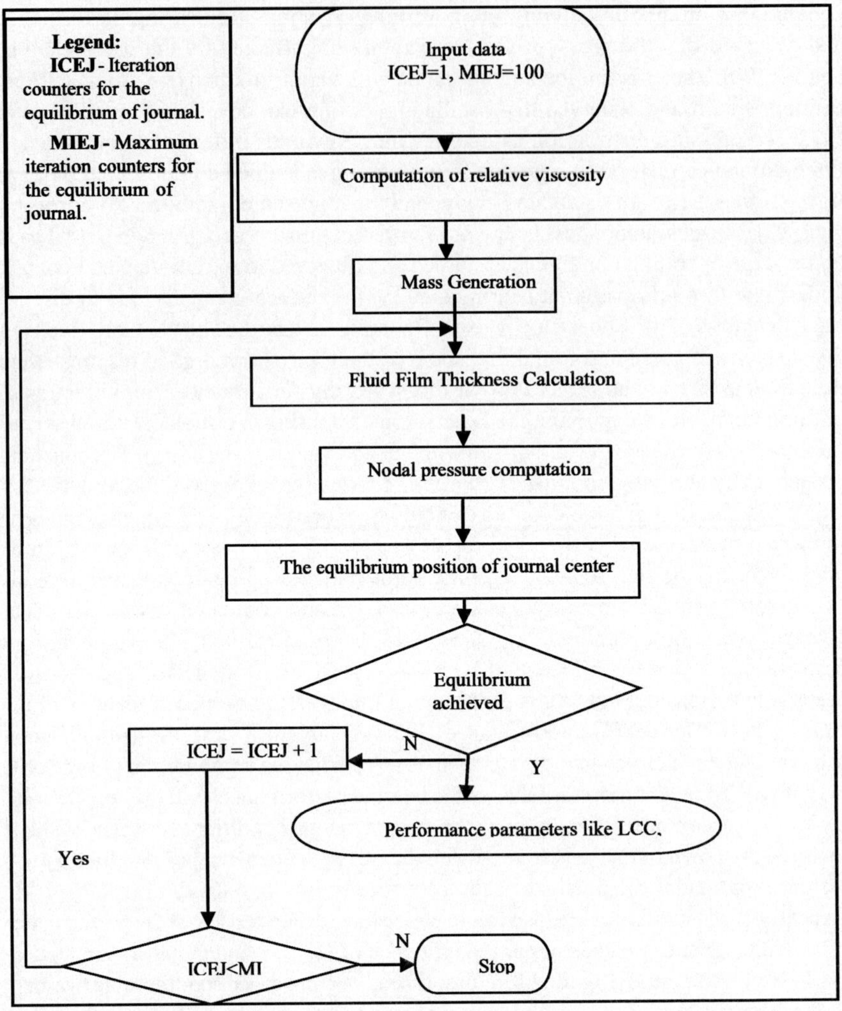

Fig. 7 Flowchart for solution scheme

3 Conclusions

It has been observed that many researchers have done investigation on different conventional shape textures like triangular, chevron, rectangular, square, ellipse, etc. to see their impact in terms of static characteristics as well as dynamic characteristics on the performance of the bearings. It is observed in the study that the size, depth, and orientation and geometry of the textures are the important factors in improving the bearing performance. Because of the textures' ability to operate as a reservoir, pressure builds up in that area, improving the area's ability to support loads. It has also been found in the study that the performances characteristics like load-bearing capacity, frictional force, etc. in different regions of the journal bearings vary and pressure increasing region of the bearing performs better in comparison to the other regions, and this region is explored extensively by many researchers. The comparison of different shapes of textures in journal bearing is also found in this study. It is beneficial for researchers to examine other traditional texture forms and compare them to other texture shapes in light of the ongoing development in texturing shapes in order to assess and improve the performance of the journal bearing. This approach not only breaks the constraint of getting optimal shape, size but also achieves the density and geometrical parameters of the specific bearing. Future step is planned to use this approach in the next research to explore the possibility of any increased performance in the functioning of journal bearing.

References

1. Manser B, Belaidi I, Hamrani A, Khelladi S, Bakir F (2020) Texture shape effects on hydrodynamic journal bearing performances using mass-conserving numerical approach. Tribol-Mater, Surf Interfaces 14(1):33–50
2. Kango S, Sharma RK, Pandey RK (2014) Comparative analysis of textured and grooved hydrodynamic journal bearing. Proc Inst Mech Eng, Part J: J Eng Tribol 228(1):82–95
3. Khatri CB, Sharma SC (2016) 2 Influence of textured surface on the performance of non-recessed hybrid journal bearing operating with non-Newtonian lubricant. Tribol Int 95:221–235
4. Saleh AM, Crosby W, El Fahham IM, Elhadary M (2020) The effect of liner surface texture on journal bearing performance under thermo-hydrodynamic conditions. Ind Lubr Tribol
5. Tala-Ighil N, Maspeyrot P, Fillon M, Bounif A (2007) Effects of surface texture on journal-bearing characteristics under steady-state operating conditions. Proc Inst Mech Eng, Part J: J Eng Tribol 221(6):623–633
6. Lampaert SG, Quinci F, Van Ostayen RA (2020) Rheological texture in a journal bearing with magnetorheological fluids. J Magn Magn Mater 499:166218
7. Yu R, Li P, Chen W (2016) Study of grease lubricated journal bearing with partial surface texture. Ind Lubr Tribol
8. Brizmer V, Kligerman Y (2012) A laser surface textured journal bearing. J Tribol 134(3)
9. Rajput AK, Yadav SK, Sharma SC (2017) Effect of geometrical irregularities on the performance of a misaligned hybrid journal bearing compensated with membrane restrictor. Tribol Int 115:619–627
10. Ganji TS, Kakoty SK, Pandey RK (2014) Analysis on micro elliptical textured journal bearing. Int J Curr Eng Technol 2(2):648–650
11. Shinde A, Pawar P, Shaikh P, Wangikar S, Salunkhe S, Dhamgaye V (2018) Experimental and numerical analysis of conical shape hydrodynamic journal bearing with partial texturing. Procedia Manufact 20:300–310
12. Sharma S, Jamwal G, Awasthi RK (2019) Enhancement of steady state performance of hydrodynamic journal bearing using chevron-shaped surface texture. Proc Inst Mech Eng, Part J: J Eng Tribol 233(12):1833–1843
13. Tala-Ighil N, Fillon M, Maspeyrot P (2011) Effect of textured area on the performances of a hydrodynamic journal bearing. Tribol Int 4(3):211–219
14. Zhang Y, Chen G, Wang L (2019) Effects of thermal and elastic deformations on lubricating properties of the textured journal bearing. Adv Mech Eng 11(10):1687814019883790

15. Niu Y, Hao X, Xia A, Wang L, Liu Q, Li L, He N (2022) Effects of textured surfaces on the properties of hydrodynamic bearing. Int J Adv Manuf Technol 118(5):1589–1596; diluted magnetic semiconductor: a density functional theory explanation. Phys Lett A 1:2–5 (2017). https://doi.org/10.1016/j.physleta.2017.09.018
16. Kango S, Singh D, Sharma RK (2012) Numerical investigation on the influence of surface texture on the performance of hydrodynamic journal bearing. Meccanica 47(2):469–482
17. Cupillard S, Cervantes M, Glavatskih S (2008) A CFD study of a finite textured journal bearing. In IAHR symposium on hydraulic machinery and systems 27/10/2008
18. Yong H, Balendra R (Aug, 2009) CFD analysis on the lubrication behaviours of journal bearing with dimples. In: 2009 international conference on mechatronics and automation. IEEE, pp 1279–1284
19. Morris N, Rahnejat H, Rahmani R (2011) Tribology of partial pad journal bearings with textured surfaces
20. Tala-Ighil N, Maspeyrot P, Fillon M, Bounif A (2007) Hydrodynamic effects of texture geometries on journal bearing surfaces. In: 10th international conference on tribology ROTRIB'07, Bucharest, Romania from 8 to 10 November
21. Rahmani R, Rahnejat H (2018) Enhanced performance of optimised partially textured load bearing surfaces. Tribol Int 117:272–282
22. Zhang J, Meng Y (2012) Direct observation of cavitation phenomenon and hydrodynamic lubrication analysis of textured surfaces. Tribol Lett 46(2):147–158
23. Cupillard S (2009) Thermohydrodynamics of sliding contacts with textured surfaces (Doctoral dissertation, Luleå tekniska universitet)
24. Hamdavi S, Ya HH, Rao N (2016) Effect of surface texturing on hydrodynamic performance of journal bearing. ARPN J Eng Appl Sci 11(1):172–176
25. Sharma VK, Singh RC, Chaudhary R (2020) Improvement in the performance of journal bearings by using lead-free bearing material and surface texturing. Int J Surf Sci Eng 14(3):238–256
26. Wang J, Zhang J, Lin J, Ma L (2018) Study on lubrication performance of journal bearing with multiple texture distributions. Appl Sci 8(2):244
27. Dashti Rahmatabadi A, Zare Mehrjardi M (2018) The effect of shell texturing on the performance of noncircular hydrodynamic two lobe journal bearings. Modares Mech Eng 18(2):293–304
28. Rao TVVLN, Rani AMA, Mohamed NM, Ya HH, Awang M, Hashim FM (2018) Analysis of magnetohydrodynamic partial slip texture slider and journal bearing. In: Proceedings of Asia international conference on tribology, pp 90–91
29. Gropper D, Wang L, Harvey TJ (2016) Hydrodynamic lubrication of textured surfaces: a review of modeling techniques and key findings. Tribol Int 94:509–529
30. Fowell MT, Medina S, Olver AV, Spikes HA, Pegg IG (2012) Parametric study of texturing in convergent bearings. Tribol Int 52:7–16
31. Mezghani S, Demirci I, Zahouani H, El Mansori M (2012) The effect of groove texture patterns on piston-ring pack friction. Precis Eng 36(2):210–217
32. Podgornik B, Vilhena LM, Sedlaček M, Rek Z, Žun I (2012) Effectiveness and design of surface texturing for different lubrication regimes. Meccanica 47(7):1613–1622
33. Costa HL, Hutchings IM (2007) Hydrodynamic lubrication of textured steel surfaces under reciprocating sliding conditions. Tribol Int 40(8):1227–1238
34. Kulkarni HD, Rasal AB, Bidkar OH, Mali VH, Atkale SA, Wangikar SS, Shinde AB (2019) Fabrication of micro-textures on conical shape hydrodynamic journal bearing. Int J Trends Eng Technol 36(1):37–41
35. Singh N, Awasthi RK (2020) Influence of dimple location and depth on the performance characteristics of the hydrodynamic journal bearing system. Proc Inst Mech Eng, Part J: J Eng Tribol 234(9):1500–1513
36. Gupta KK, Kumar R, Kumar H, Sharma M (2013) Study on effect of surface texture on the performance of hydrodynamic journal bearing. Int J Eng Adv Technol 3(1):49–54

37. Dowson D, Ruddy AV, Sharp RS, Taylor CM (1985) An analysis of the circumferentially grooved journal bearing with consideration of lubricant film reformation. Proc Inst Mech Eng C J Mech Eng Sci 199(1):27–34
38. Murthy AN, Etsion I, Talke FE (2007) Analysis of surface textured air bearing sliders with rarefaction effects. Tribol Lett 28(3):251–261
39. Lu X, Khonsari MM (2007) An experimental investigation of dimple effect on the stribeck curve of journal bearings. Tribol Lett 27(2):169–176
40. Dobrica MB, Fillon M, Pascovici MD, Cicone T (2010) Optimizing surface texture for hydrodynamic lubricated contacts using a mass-conserving numerical approach. Proc Inst Mech Eng, Part J: J Eng Tribol 224(8):737–750
41. Arif M, Shukla DK, Kango S, Sharma N (2020) Implication of surface texture and slip on hydrodynamic fluid film bearings: a comprehensive survey. Tribol Online 15(4):265–282
42. Tauviqirrahman M, Jamari J, Wibowo BS, Fauzan HM, Muchammad M (2019) Multiphase computational fluid dynamics analysis of hydrodynamic journal bearing under the combined influence of texture and slip. Lubricants 7(11):97
43. Liang X, Liu Z, Wang H, Zhou X, Zhou X (2016) Hydrodynamic lubrication of partial textured sliding journal bearing based on three-dimensional CFD. Ind Lubr Tribol
44. Tala-Ighil N, Fillon M (2015) Surface texturing effect comparative analysis in the hydrodynamic journal bearings. Mech Ind 16(3):302
45. Wakuda M, Yamauchi Y, Kanzaki S, Yasuda Y (2003) Effect of surface texturing on friction reduction between ceramic and steel materials under lubricated sliding contact. Wear 254(3–4):356–363
46. Oliveira JFGD, Bottene AC, Franca TV (2010) A novel dressing technique for texturing of ground surfaces. CIRP Ann 1:361–364
47. Hashimoto H, Ochiai M (2008) Optimization of groove geometry for thrust air bearing to maximize bearing stiffness
48. Rahmani R, Shirvani A, Shirvani H (2007) Optimization of partially textured parallel thrust bearings with square-shaped micro-dimples. Tribol Trans 50(3):401–406
49. Fowell M, Olver AV, Gosman AD, Spikes HA, Pegg I (2007) Entrainment and inlet suction: two mechanisms of hydrodynamic lubrication in textured bearings
50. Brizmer V, Kligerman Y, Etsion I (2003) A laser surface textured parallel thrust bearing. Tribol Trans 46(3):397–403
51. Ronen A, Etsion I, Kligerman Y (2001) Friction-reducing surface-texturing in reciprocating automotive components. Tribol Trans 44(3):359–366
52. Aurelian F, Patrick M, Mohamed H (2011) Wall slip effects in (elasto) hydrodynamic journal bearings. Tribol Int 44(7–8):868–877
53. Meng FM, Zhang W (2018) Effects of compound groove texture on noise of journal bearing. J Tribol 140(3)
54. Galda L, Sep J, Olszewski A, Zochowski T (2019) Experimental investigation into surface texture effect on journal bearings performance. Tribol Int 136:372–384
55. Vlădescu SC, Fowell M, Mattsson L, Reddyhoff T (2019) The effects of laser surface texture applied to internal combustion engine journal bearing shells–an experimental study. Tribol Int 134:317–327
56. Tomar AK, Sharma SC (2020) An investigation into surface texture effect on hole-entry hybrid spherical journal bearing performance. Tribol Int 151:106417
57. Matele S, Pandey KN (2018) Effect of surface texturing on the dynamic characteristics of hydrodynamic journal bearing comprising concepts of green tribology. Proc Inst Mech Eng, Part J: J Eng Tribol 232(11):1365–1376
58. Meng F, Wei Z, Minggang D, Gao G (2016) Study of acoustic performance of textured journal bearing. Proc Inst Mech Eng, Part J: J Eng Tribol 230(2):156–169
59. Kumar M, Ranjan V, Tyagi R (2020) Effect of shape, density, and an array of dimples on the friction and wear performance of laser textured bearing steel under dry sliding. J Mater Eng Perform 29(5):2827–2838

60. Usman A, Park CW (2018) Numerical optimization of surface texture for improved tribological performance of journal bearing at varying operating conditions. Ind Lubr Tribol
61. Sep J, Pawlus P, Galda L (2013) The effect of helical groove geometry on journal abrasive wear. Arch Civil Mech Eng 13(2):150–157
62. Dadouche A, Conlon MJ (2016) Operational performance of textured journal bearings lubricated with a contaminated fluid. Tribol Int 93:377–389
63. Li D, Yang X, Wu Y, Cheng J, Wang S, Wan Z, Liu W, Xia G (2022) Theoretical analysis and experimental research of surface texture hydrodynamic lubrication. Chin J Mech Eng 35(1):1–15
64. Codrignani A, Frohnapfel B, Magagnato F, Schreiber P, Schneider J, Gumbsch P (2018) Numerical and experimental investigation of texture shape and position in the macroscopic contact. Tribol Int 122:46–57
65. Song F, Yang X, Dong W, Zhu Y, Wang Z, Wu M (2022) Research and prospect of textured sliding bearing. Int J Adv Manuf Technol 1–25
66. Dong J, Wang X, Zhang J, Xiang X, Nie Z, Shen J (2017) An experimental research on the vibration of surface-textured journal bearings. Shock Vibr
67. Yu R, Chen W, Li P (2016) The analysis of elastohydrodynamic lubrication in the textured journal bearing. Proc Inst Mech Eng, Part J: J Eng Tribol 230(10):1197–1208
68. Kango S, Singh D, Sharma RK (2012) Numerical investigation on the influence of surface texture on the performance of hydro-dynamic journal bearing. Meccanica 47:469–482
69. Niu Y, Hao X, Xia A, Wang L, Liu Q, Li L, He N (2022) Effects of textured surfaces on the properties of hydrodynamic bearing. Int J Adv Manufact Technol 118:1589–1596
70. Sharma S, Jamwal G, Awasthi RK (2019) Enhancement of steady state performance of hydrodynamic journal bearing using chevron-shaped surface texture. Proc Inst Mech Eng, Part J: J Eng Tribol 233:1833–1843
71. Shinde A, Pawar P, Shaikh P, Wangikar S, Salunkhe S, Dhamgaye V (2018) Experimental and numerical analysis of conical shape hydrodynamic journal bearing with partial texturing. Procedia Manuf 20:300–310
72. Buscaglia GC, Ciuperca I, Jai M (2005) The effect of periodic textures on the static characteristics of thrust bearings
73. Chandrawat HN, Sinhasan R (1987) A comparison between two numerical techniques for hydrodynamic journal bearing problems. Wear 119:77–87
74. Ausas R, Ragot P, Leiva J, Jai M, Bayada G, Buscaglia GC (2007) The impact of the cavitation model in the analysis of micro-textured lubricated journal bearings
75. Sadabadi H, Sanati Nezhad A (2020) Nanofluids for performance improvement of heavy machinery journal bearings: a simulation study. Nanomaterials 10:2120
76. Syed NR, Kakoty SK (2022) Computational efficiency improvement of dimple textured hydrodynamic journal bearing using progressive mesh densification method. J Tribol 144
77. Rao TVVLN, Rani AMA, Nagarajan T, Hashim FM (2012) Analysis of slider and journal bearing using partially textured slip surface. Tribol Int 56:121–12
78. Sinanoğlu C, Nair F, Karamış MB (2005) Effects of shaft surface texture on journal bearing pressure distribution. J Mater Process Technol 168(2):344–353
79. Kango S, Sharma RK (2010) Studies on the influence of surface texture on the performance of hydrodynamic journal bearing using power law model. Int J Surf Sci Eng 4(4–6):505–524
80. Chandrawat HM, Sinhasan R (1988) A study of steady state and transient performance characteristics of a flexible shell journal bearing. Tribol Int 21(3):137–148
81. Hamilton DB, Walowit JA, Allen CM (1966) A theory of lubrication by micro-irregularities. Trans ASME J Basic Eng 88:177–85
82. Sinanoğlu C, Nair F, Karamış MB (2005) Effects of shaft surface texture on journal bearing pressure distribution. J Mater Process Technol 168(2):344–353
83. Gethin DT, El Deihi MKI (1987) Effect of loading direction on the performance of a twin-axial groove cylindrical-bore bearing. Tribol Int 20(4):179–185
84. Lundholm G (1969) The circumferential groove journal bearing considering cavitation and dynamic stability (Real time operation for circumferential groove journal bearing with cavitation and dynamic stability)

85. Suh MS, Chae YH, Kim SS, Hinoki T, Kohyama A (2010) Effect of geometrical parameters in micro-grooved crosshatch pattern under lubricated sliding friction. Tribol Int 43(8):1508–1517
86. Li WL, Weng CI, Lue JI (1996) Surface roughness effects in journal bearings with non-Newtonian lubricants. Tribol Trans 39(4):819–826
87. Kumada Y, Hashizume K, Kimura Y (1996) Performance of plain bearings with circumferential microgrooves. Tribol Trans 39(1):81–86
88. De Kraker A, van Ostayen RA, Van Beek A, Rixen DJ (2007) A multiscale method modeling surface texture effects
89. Mufti RA, Priest M (2009) Theoretical and experimental evaluation of engine bearing performance. Proc Inst Mech Eng, Part J: J Eng Tribol 223(4):629–644
90. Solghar AA, Nassab SG (2011) Thermohydrodynamic behaviors of finite journal bearings with cavitation. Mech Ind 12(1):5–15
91. Malanoski SB, Pan CHT (1965) The static and dynamic characteristics of the spiral-grooved thrust bearing
92. James DD, Potter AF (1967) Numerical analysis of the gas-lubricated spiral-groove thrust bearing-compressor
93. Fogg A (1946) Fluid film lubrication of parallel thrust surfaces. Proc Inst Mech Eng 155(1):49–67
94. Bonneau D, Huitric J, Tournerie B (1993) Finite element analysis of grooved gas thrust bearings and grooved gas face seals
95. Etsion I, Halperin G, Brizmer V, Kligerman Y (2004) Experimental investigation of laser surface textured parallel thrust bearings. Tribol Lett 17(2):295–300
96. Rho BH, Kim KW (2003) Acoustical properties of hydrodynamic journal bearings. Tribol Int 36(1):61–66
97. Choy FK, Braun AM, Hu Y (1992) Nonlinear transient and frequency response analysis of a hydrodynamic journal bearing
98. Singh P, Sharma N (2014) Optimum design of journal bearing through surface texturing—a review. Int J Res Advent Technol 2(1)
99. Rasep Z, Yazid MM, Samion S (2021) Lubrication of textured journal bearing by using vegetable oil: a review of approaches, challenges, and opportunities. Renew Sustain Energy Rev 146:111191, 37
100. Sharma S, Jamwal G, Kumar Awasthi R (2021) The effect of V-shape protruded and dimple textured on the load-carrying capacity and coefficient of friction of hydrodynamic journal bearing. Proc Inst Mech Eng, Part J: J Eng Tribol 235(5):997–1011
101. Kumar P, Tiwari R (2020) Development of a novel approach for quantitative estimation of rotor unbalance and misalignment in a rotor system levitated by active magnetic bearings. Iran J Sci Technol-Trans Mech Eng 45:769–786
102. Kumar P, Tiwari R (2020) Dynamic analysis and identification of unbalance and misalignment in a rigid rotor with two offset discs levitated by active magnetic bearings: a novel trial misalignment approach. Propul Power Res 10:58–82
103. Kumar P, Tiwari R (2021) Finite element modelling, analysis and identification using novel trial misalignment approach in an unbalanced and misaligned flexible rotor system levitated by active magnetic bearings. Mech Syst Signal Process 152:107454
104. Tiwari R, Kumar P (2022) An innovative virtual trial misalignment approach for identification of unbalance, sensor and active magnetic bearing misalignment along with its stiffness parameters in a magnetically levitated flexible rotor system. Mech Syst Signal Process 167:108540

Computational Analysis of Axisymmetric Supersonic Jet Impingement on Flat and Incline Plate

Thanggoulen Hmangte and Dushyant Singh

1 Introduction

Supersonic-impinging jets have been the subject of various investigations, most of which were motivated by practical applications. For instance, during a rocket launch, multistage separation, lunar and planetary landing, and take-off, vertical take-off and landing (VTOL), short take-off and landing (STOL), etc. The supersonic jet is accompanied by a very complex flow field such as shock-shock interactions, shock-boundary interaction, highly non-uniform downstream flow from the nozzle exit, and a recirculation region just under the main plate shock. The flow structure depends on the nozzle exit Mach number, pressure ratios, and nozzle exit to impingement distance. Though understanding and analyzing the physics behind a supersonic jet is extremely difficult as the nature of the flow changes continuously by changing any one of the abovementioned parameters, the application it has in many fields prompts one to learn more about supersonic jet impingement. Jiang et al. [1] provide a thorough analysis of jet impingement.

2 Literature Review and Objective

Perpendicular impingement has been the focus of the majority of earlier studies on supersonic jet impingement on a flat plate. A few well-known examples are the experimental research carried out by Henderson [2], Carling and Hunt [3], and Kalghatgi and Hunt [4]. These experiments helped to establish the existence of stagnation

T. Hmangte (✉)
Department of Mechanical Engineering, National Institute of Technology, Manipur, India
e-mail: thanggoulenhmangte@gmail.com

D. Singh
Department of Mechanical Engineering, National Institute of Technology, Jalandhar, India

B. P. Swain (ed.), *Recent Advances in Materials*, Springer Proceedings in Materials 25,
https://doi.org/10.1007/978-981-99-3844-5_28

bubbles, the near wall jet, and the concept of jet structure as significant phenomena. A recirculation region also known as a stagnation bubble, which forms close to the plate surface, affects the stability of the jet as well as the peak pressure on the surface of deflectors. Lamont and Hunt's experiment [5] with under-expanded jets impinging on an inclined flat plate was the first significant contribution to the in-depth study of jet impingement on an inclined plate. The geometrical parameters considered have a plate angle of $\theta = 30–90°$, nozzle exit to an impinging distance of $z/d_n z/d_n = 0.5–4$, and pressure ratios, P.R of 1.2 and 2, respectively. They measured the pressure variations on the surface plate using pressure taps. By the shadowgraph technique, they were able to observe the flow fields as well. The results showed that plate angles significantly affect the distributions of pressure on the plate surface. Recent developments in high-performance computing have made it possible to numerically simulate these complex impinging flow fields. Using the PSP/TSP technique, Nakai et al. [6] studied jet impingement in an experimental condition, and the outcomes were consistent with the findings of the study conducted numerically. The study showed that provided the impinging distance, the orientation of plate angle, pressure ratios of the nozzle, and free jet shock length various types of flow can be simulated without the experimental work. The same author [7] illustrates the advantages and limitations of the suggested classification. Asymmetric stagnation bubble generation, the size and position of the maximum wall pressure, the impact of plate inclination, and other aspects are all studied numerically by Kim and Chang [8]. For moderate plate inclinations, a fairly reasonable prediction of the impinging jet is made by comparison with the available experimental results [5]. Goto et al. [9] studied the mechanism of maximum peak pressure. Wu et al. [10], numerically analyzed various types of shock formation during impingement of supersonic jet on a plate. The results of the simulation indicate the site of the peak pressure on the plate along with shadowgraph images which were in good agreement with the available experimental data. By using a commercial CFD tool to solve a 3-D RANS equation, Dharavath and Chakraborty [11] were able to simulate the experiment carried out by Nakai et al. [6] for various plate angles. Experimental results and computed pressure on the plate and numerical Schlieren pictures agree quite well. The jet structure is accurately predicted by each turbulence model employed in the study. The findings showed that RANS simulation can visualize the flow pattern and capture finer aspects of the jet structure. Singhal et al. [12] numerically study the effect of deflector plates with and without water-cooled conditions. The deflector has an angle of $\theta = 35°$ inclination and is 18 m from the nozzle. The water-cooled situation was studied using the discrete phase model, which predicts lower temperatures that can be further lowered by increasing the number of holes in the deflector that the coolant can pass through. Oh et al. [13] investigation of the Korea Space Launch Vehicle-first-stage II rocket engine used five distinct single-sided flame deflectors. The nozzle exit has Mach number 3.25, $z/d_n z/d_n = 6.3$ and the deflector has a 30° inclination. Model 5 which has a straight outlet and open top is the optimal design as it allows a larger gas emission area. Zhou et al. [14] simulate an $X - 1$ launch vehicle with a single-sided flame deflector having impinging angle ranges from $\theta = 20–30°$ along with an uplift angle of $\theta = 5°$. The diameter of nozzle exit is 1.2 m and the impingement point is 8.4 m away from the nozzle exit. The optimal deflector design for a four-nozzle rocket engine having an angle of $\theta = 25^\circ$ impinging angle and appropriate uplift angle is required as it reduces the recirculation of the exhaust gases.

In the present study, assessment of computational results for supersonic jet impingement on a flat plate using various turbulence models viz., the standard $k - \varepsilon$, RNG $k - \varepsilon$, realizable $k - \varepsilon$, SST $k - \omega$, and BSL $k - \omega$ was carried out. Lamont and Hunt's [5] experimental condition was simulated using ANSYS—FLUENT. To choose an appropriate turbulence model, the numerically computed results are compared with the experimental data.

3 Numerical Model

In the present investigation, an incompressible, steady, and three-dimensional turbulent flow is assumed. For the simulation, commercial ANSYS FLUENT (20) [15] software was used to solve the continuity, momentum, and energy equations. It uses a finite volume approach to solve the 3-D Reynolds Averaged Navier Stokes (RANS) problem on a structured grid.

The following are the suitable governing equations for the turbulent flow of incompressible gas as provided in [16]:

$$\frac{\partial \mathrm{u}_i}{\partial x_i} = 0\frac{\partial \mathrm{u}_i}{\partial x_i} = 0 \tag{1}$$

$$\rho \mathrm{u}_j \frac{\partial \mathrm{u}_i}{\partial x_j} = -\frac{\partial p}{\partial x_i} + \frac{\partial}{\partial x_j}\left[\mu\left(\frac{\partial \mathrm{u}_i}{\partial x_j} + \frac{\partial \mathrm{u}_j}{\partial x_i}\right) - \rho\overline{\mathrm{u}_i'\mathrm{u}_j'}\right] \tag{2}$$

$$\rho \mathrm{u}_j \frac{\partial T}{\partial x_j} = \frac{\partial}{\partial x_j}\left[\frac{\mu}{pr}\frac{\partial T}{\partial x_j} - \rho\overline{T'\mathrm{u}_j'}\right]\rho \mathrm{u}_j \frac{\partial T}{\partial x_j} = \frac{\partial}{\partial x_j}\left[\frac{\mu}{pr}\frac{\partial T}{\partial x_j} - \rho\overline{T'\mathrm{u}_j'}\right] \tag{3}$$

in which u_i is the component of average velocity, T is temperature, p is the pressure, $\rho\rho$ is density, $\mu\mu$ is the viscosity. Fluctuation quantities $\rho\overline{\mathrm{u}_i'\mathrm{u}_j'}\rho\overline{\mathrm{u}_i'\mathrm{u}_j'}$ is Reynolds-stress tensor, $\mathrm{u}_i'\mathrm{u}_i'$ and T' T' represent velocity fluctuation in the i-direction and temperature fluctuation, respectively. To quantify the turbulent or Reynolds stress, turbulence closure is necessary because the above equations are not a closed set. Each turbulence model utilized in the current investigation has the same comprehensive description as [17–21].

4 Boundary Condition

See Figs. 1 and 2.

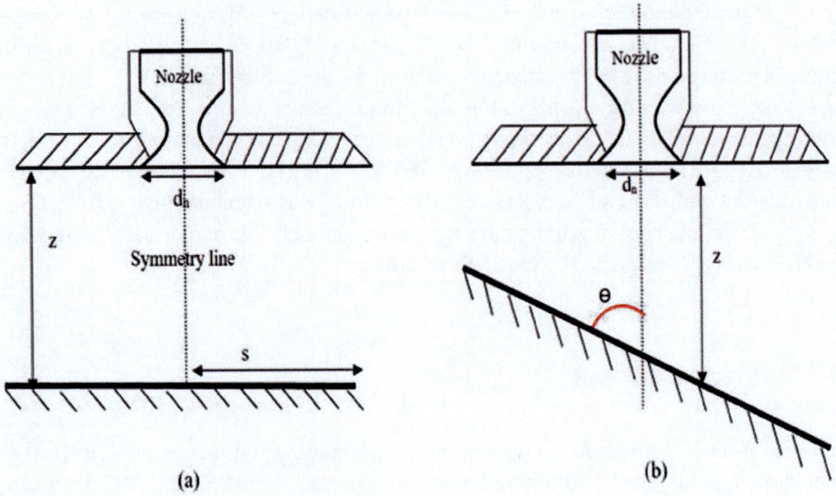

Fig. 1 Domain of the present numerical study **a** Perpendicular plate and **b** Incline flat plate *Source* Author

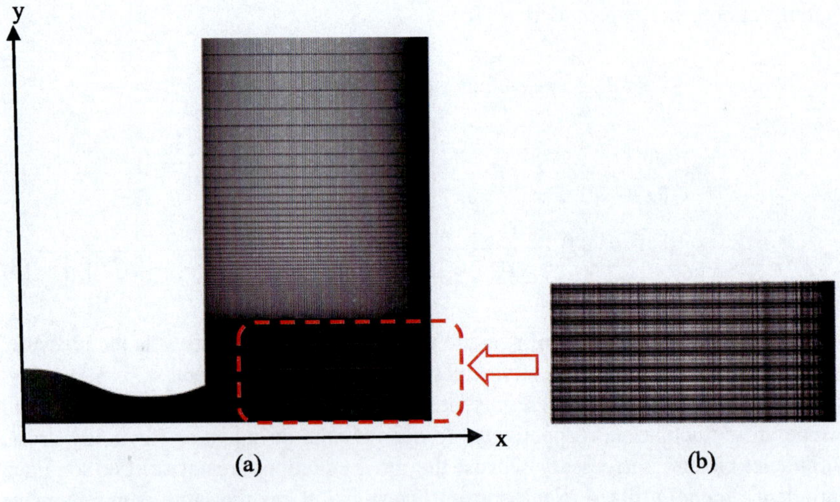

Fig. 2 **a** Structure grid used in the present study and **b** Enlarged view *Source* Author

The pressure inlet boundary condition is assigned to the inlet, and the outlet boundary is treated as the pressure outlet with a different pressure gauge for different pressure ratios, and the target plate is treated as an adiabatic wall. Table 1 gives the operating conditions in the current study.

Table 1 Operating parameters for present numerical study

Parameters	Range (Pa)	Operating pressure (Pa)	P.R
P_{in}	80,000	4000	1.2
P_{in}	81,000	3000	2

Source Author

5 Results and Discussion

5.1 *Grid Independency Test*

The representative results for the grid independence research performed for the case, $\theta = 90°$, M = 2.2, and $z/d_n z/d_n = 1$, are shown in Fig. 3a (for P.R = 1.2). The findings demonstrate that when the grid size is changed from 32,468 to 1,17,468 cells and 2,18,355 cells, respectively, there are no changes in the P/P_C profile along the deflector. Therefore, 1,17,468 cells size was taken into account for the above case. To accurately understand viscous sublayer flow characteristics, the mesh has been highly refined to ensure that $y+$ is close to unity. Since the same grid cannot be utilized in these circumstances, similar grid independence tests have been carried out for many geometries, including $\theta = 70$ and 45°, which have P.R = 1.2 M = 2.2 and $z/d_n z/d_n = 3$, respectively. Figure 3b depicts grid independence despite an increase in grid cells from 1,16,523 to 4,19,286 cells. Additionally, by increasing the grid from 1,87,350 to 3,33,067 cells for Fig. 3c (for $\theta = 45°$), grid independence is seen for $z/d_n z/d_n = 3$. In Fig. 3b and c only two grid sizes were tested viz. coarse and fine mesh and in both cases essential flow features were captured accurately. For this reason, testing of another (medium) grid size was not carried out to save computational time.

5.2 *Validation and Turbulence Model Selection*

This study takes into account five different types of turbulence models for the impingement of supersonic jets on flat plate deflector surfaces. The systematic assessment of two-equation turbulence model selection has received relatively little attention. Therefore, a detailed examination of supersonic jet impingement on deflector surfaces parallel and inclined to the nozzle exit plane is simulated using various turbulence models. To select the relevant models for parametric research, the numerical results of these models should also be evaluated with the existing experimental data [5]. Using the standard (Std.) $k - \varepsilon$, RNG $k - \varepsilon$, realizable $k - \varepsilon$, SST $k - \omega$, and BSL $k - \omega$, the performance of an impact of a supersonic jet on a deflector surface was compared in this work.

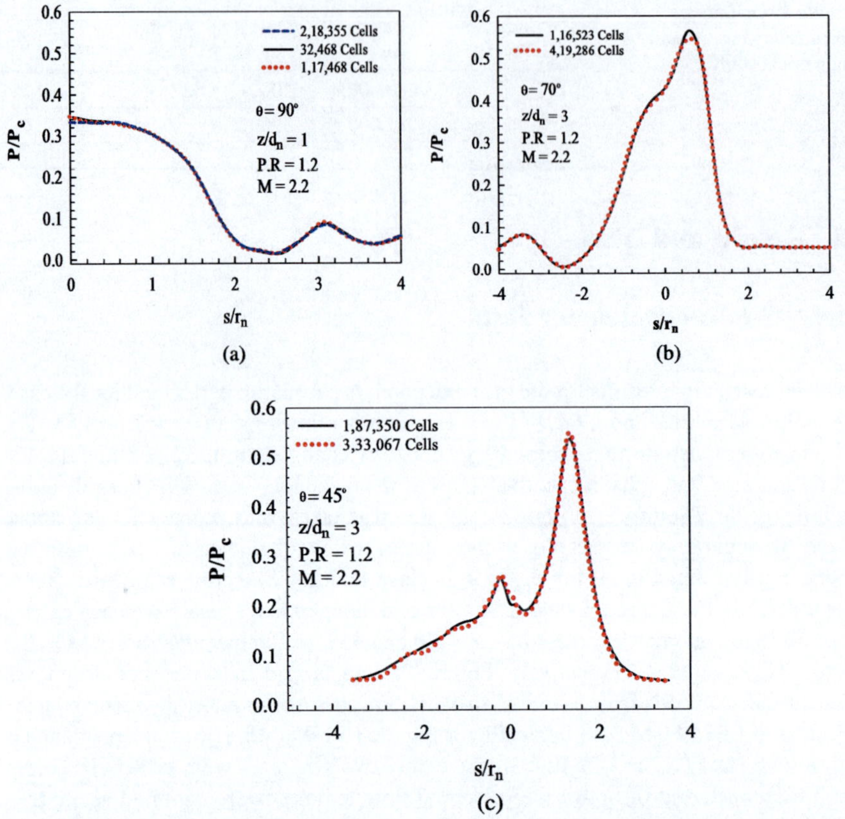

Fig. 3 Grid independence study for M = 2.2 and P.R = 1.2 at the nozzle exit for **a** $\theta = 90^{\circ}, z/d_n = 1\theta = 90^{\circ}, z/d_n = 1$ **b** $\theta = 70^{\circ}, z/d_n = 3\theta = 70^{\circ}, z/d_n = 3$ **c** $\theta = 45^{\circ}, z/d_n = 3\theta = 45^{\circ}, z/d_n = 3$ *Source* Author

The research findings that Lamont and Hunt [5] reported were used to assess the impingement flow fields. They took measurements of the static pressure ratio distribution of the jet on the deflector surfaces at various separations from the nozzle, i.e., $z/d_n z/d_n = 1$ and 3. The static pressure ratio was measured at the center line of the deflector plane. In their experimental research, the flow has a P.R of 1.2, 2, and a Mach number of 2.2 at the nozzle exit. As a result, a numerical study was done for verification while retaining the experimental setting at the nozzle's exit.

The pressure ratio distribution generated along the s/r_n at non-dimensional distances of $z/d_n z/d_n = 1$ and 3 for various deflector orientations, such as $\theta = 90$, 70, and 45°, is analyzed with various two-equation turbulence models, as illustrated in Fig. 4. The five distinct two-equation turbulence models' predictions of the pressure ratio distribution at $z/d_n z/d_n = 1$ for $\theta = 90°$, P.R = 1 and 2 in Fig. 4a, b, are in good agreement with the experimental results and so is for, $\theta = 70°$ at $z/d_n z/d_n = 1$ in Fig. 4d, which can be inferred that the numerical simulation captured all the essential flow features viz., Mach disk formation, plate shock and also recirculation region. But as the non-dimensional distance increases to $z/d_n z/d_n = 3$, as seen in Fig. 4c, d, the turbulence model is unable to accurately estimate the pressure ratio distribution. The similarity between the two cases of $z/d_n z/d_n = 1$ and 3 can be shown in Fig. 4f, g for $\theta = 45°$, where the entire two-equation turbulence model fails to adequately predict the experimental findings. A numerical simulation's findings for the case of parallel flat plate deflectors show a lesser value of pressure ratio distribution on the center line of the deflector. From Fig. 4 it can be seen that by decreasing the angle of plate orientation, the pressure ratio on the plate increases. Also, the recirculation region, which was observed in the case of $\theta = 90°$ for both the case of P.R = 1.2 and 2, disappear in the case of plate angle $\theta = 45$ and 75°. The pressure ratio distribution inside the region of jet impingement by all studied turbulence models in numerical studies is in good agreement with the measured data except for the RNG $k - \varepsilon$ findings as shown in Fig. 4e. The SST $k - \omega$ turbulence model, out of all the others, consistently matches the experimental findings in all the cases. The conclusions drawn from the current numerical study show that the SST $k - \omega$ turbulence model is best suited for the simulation of supersonic jet impingement.

6 Conclusion

The impingement of an axisymmetric under-expanded supersonic jet was numerically simulated using ANSYS FLUENT (20). The formation of the recirculation region was captured along with the Mach disk in the case of a parallel plate deflector. The numerical simulation crisply captured all the essential flow features. Along with evaluating several turbulence models, the impact of pressure ratio distribution on the deflector surface is investigated. The SST $k - \omega$ models offer consistency in all of the investigated cases, although all turbulence models predicted the pressure ratio distribution on the deflector plate.

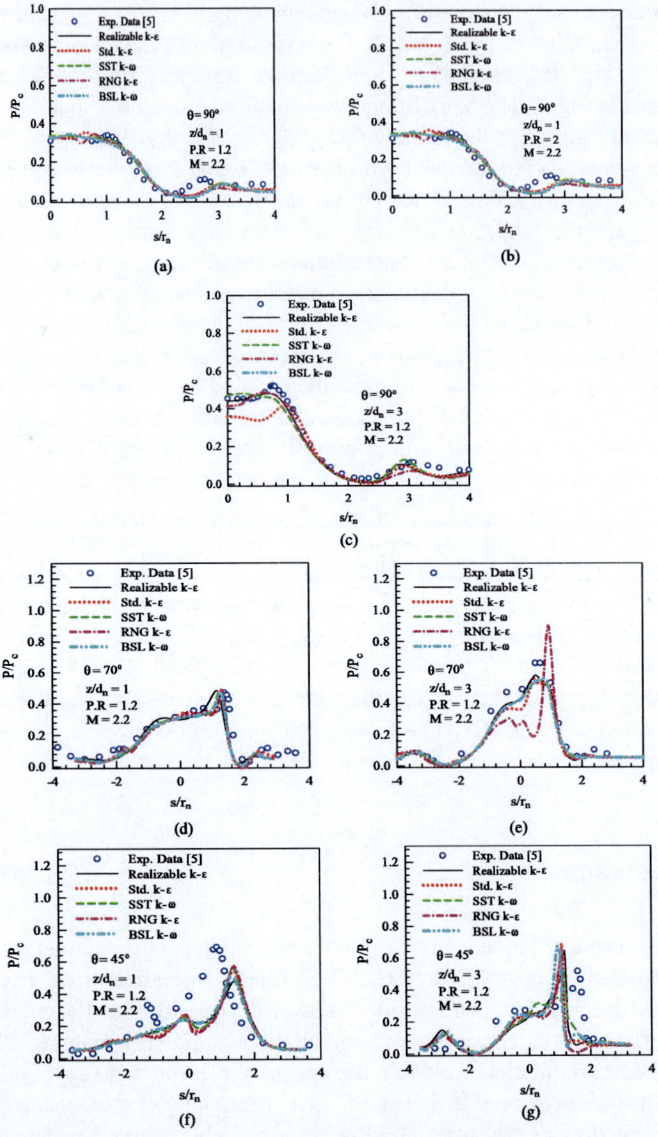

Fig. 4 Various turbulence models for the distribution of surface center-line pressure ratios on a deflector plate **a** $\theta = 90^{\circ}$, $z/d_n = 1$ P.R = $1.2\theta = 90^{\circ}$, $z/d_n = 1$ P.R = 1.2 **b** $\theta = 90^{\circ}$, $z/d_n = 1$ P.R = $2\theta = 90^{\circ}$, $z/d_n = 1$ P.R = 2 **c** $\theta = 90^{\circ}$, $z/d_n = 3$ P.R = $1.2\theta = 90^{\circ}$, $z/d_n = 3$ P.R = 1.2 **d** $\theta = 70^{\circ}$, $z/d_n = 1$ P.R = $1.2\theta = 70^{\circ}$, $z/d_n = 1$ P.R = 1.2 **d** $\theta = 70^{\circ}$, $z/d_n = 3$ P.R = $1.2\theta = 70^{\circ}$, $z/d_n = 3$ P.R = 1.2 **e** $\theta = 45^{\circ}$, $z/d_n = 1$ P.R = $1.2\theta = 45^{\circ}$, $z/d_n = 1$ P.R = 1.2 **f** $\theta = 45^{\circ}$, $z/d_n = 3$ P.R = $1.2\theta = 45^{\circ}$, $z/d_n = 3$ P.R = 1.2 *Source* Author

References

1. Jiang C, Han T, Gao Z, Lee C-H (2019) A review of impinging jets during rocket launching. Prog Aeronaut Sci 109:100547
2. Henderson LF (1966) Experiments on the impingement of a supersonic jet on a flat plate. J Appl Math Phys (ZAMP) 17:553–569
3. Carling JC, Hunt BL (1974) The near wall jet of a normally impinging, uniform, axisymmetric, supersonic jet. J Fluid Mech 66(1):159–176
4. Kalghatgi GT, Hunt BL (1976) The occurrence of stagnation bubbles in supersonic jet impingement flows. Aeronaut Q 27(3):169–185
5. Lamont PJ, Hunt BL (1980) The impingement of underexpanded, axisymmetric jets on perpendicular and inclined flat plates. J Fluid Mech 100(3):471–511
6. Nakai Y, Fujimatsu N, Fujii K (2012) Flow classification of the under-expanded super sonic jet impinging on a flat plate. In: 33rd AIAA Fluid dynamics conference and exhibit. AIAA 2003-3467. American Institute of Aeronautics and Astronautics Inc.
7. Fujii K, Oyama A, Tsuboi N, Tsukada M, Ouchi H, Ito M, Hayashi K (2005) Flow field analysis of under-expanded supersonic jets impinging on an inclined flat plate: analysis with PSP/schlieren images and CFD simulations. In: Proceedings of the ASME 2005 fluids engineering division summer meeting. Volume 1: Symposia, Parts A and B. ASME, pp 213–221
8. Kim K-H, Chang K-S (1994) Axisymmetric impingement of a hot jet on a flat plate: equilibrium flow analysis of high-temperature air. Shock Waves 4:155–162
9. Goto Y, Mcllroy K, Nonomura T, Fujii K (2009) Detailed analysis of flat plate pressure peaks created by supersonic jet impingements. In: 47th AIAA Aerospace sciences meeting including the new horizons forum and aerospace exposition. AIAA 2009–1289. American Institute of Aeronautics and Astronautics, pp 1–16
10. Wu J, Tang L, Luke EA, Tong X-L, Cinnella P (2002) Comprehensive numerical study of jet-flow impingement over flat plates. J Spacecraft Rockets 39(3):357–366
11. Dharavath M, Chakraborty D (2013) Numerical simulation of supersonic jet impingement on inclined plate. Defence Sci J 63(4):355–362
12. Singhal A, Tharakan TJ, Thomas RP () CFD analysis of water cooled flame deflector in rocket engine test facility. In: Fluid mechanics and fluid power—contemporary research. Springer, New Delhi, pp 517–528. https://doi.org/10.1007/978-81-322-2743-4_50
13. Oh H, Lee J, Um H, Huh H (2017) Numerical study for flame deflector design of a space launch vehicle. Adv Space Res 59(7):1833–1847. https://doi.org/10.1016/j.asr.2016.12.038
14. Zhou Z, Zhang L, Le G (2020) Numerical study for the flame deflector design of four-engine liquid rockets. Eng Appl Comput Fluid Mech 14(1):726–737. https://doi.org/10.1080/19942060.2020.1761453
15. ANSYS fluent theory guide (2020)
16. Al-Hemyari M, Hamdan MO, Orhan MF (2020) Optimization of a confined jet geometry to improve film cooling performance using response surface methodology (RSM). Processes 8(232):1–16
17. Singh D, Premachandran B, Kohli S (2012) Numerical simulation of the jet impingement cooling of a circular cylinder. Numer Heat Transfer A Appl 64(2):153–185. https://doi.org/10.1080/10407782.2013.772869
18. Balabel A, Hegab AM, Nasr M, El-Behery SM (2011) Assessment of turbulence modeling for gas flow in two-dimensional convergent—divergent rocket nozzle. Appl Math Model 35(7):3408–3422. https://doi.org/10.1016/j.apm.2011.01.013
19. Yu Z, Xu T, Li J, Xiu H, Li Y (2013) Numerical simulation on the effect of turbulence models on impingement cooling of double chamber model. Math Probl Eng 2013(170317):1–8
20. N. H. Transfer and P. A. Applications, Chalmers Publication Library
21. Chan L, Chin C, Soria J, Ooi A (2012) Numerical simulation of supersonic impinging jet flows using Reynolds averaged navier–stokes and large eddy simulation. In: 18th Australasian fluid mechanics conference, pp 3–7

Mathematical Model of Pressurized Solid Oxide Fuel Cell-Based Trigeneration System

Suman Pramanik and Aritra Ganguly

1 Introduction

The fossil fuel sources are depleting rapidly. At the same time, the combustion of fossil fuels is responsible for environmental degradation. So, developing eco-friendly energy systems is the need of the hour. Fuel cells are one of the viable options for energy. Fuel cells have high energy conversion efficiency and less pollutant generation [1, 2].

In recent years, Solid Oxide Fuel Cells (SOFCs) have emerged as a suitable alternative for efficient cogeneration of heat and power with reduced emission. High working temperature results in optimized kinetics and higher efficiencies than other fuel cell types. SOFC performs at high temperatures (500–1000°C). Thus, it typically works in concert with a gas turbine or some other bottoming cycle as it produces large waste heat. SOFC thus works with the ORC, Kalina cycle, Transcritical CO_2, and Brayton cycle for better energy conversion performance [2]. Several research works are available in the literature related to SOFC systems for various applications.

Yu et al. [3] investigated an integrated SOFC using an absorption chiller system. Sghaier et al. [4] studied the thermo-economic and environmental impacts of a gas turbine-integrated internal reforming SOFC. Costamagna et al. [5] studied the internal reforming fuel cells. Yari et al. [6] did a comparative study of two SOFC based cogeneration systems fed by municipal waste. They provided a new foundation for decision-making and technical advancement from an energy perspective. Exergy efficiency, energy output, and the system's exergy loss rate were taken into consideration while evaluating energetic performance. Colpan et al. [7] analyzed a Rankine cycle, SOFC, and gasification system thermodynamically for a 100 kW hybrid plant. Shirazi et al. [8] carried out thermo-economic and environmental analyses of an

S. Pramanik (✉) · A. Ganguly
Mechanical Engineering Department, Indian Institute of Engineering Science and Technology, Shibpur, Howrah, West Bengal 711103, India
e-mail: mintupramanik00@gmail.com

B. P. Swain (ed.), *Recent Advances in Materials*, Springer Proceedings in Materials 25,
https://doi.org/10.1007/978-981-99-3844-5_29

internal reforming SOFC integrated with a GT. Eveloy et al. [9] performed an analysis of a SOFC integrated with a GT and an ORC for power generation. Bang-Møller et al. [10] performed an exergy analysis and optimization of biomass gasification coupled with SOFC.

This brief review shows that several research works are available on SOFC-based systems for various individual applications like power, heating, cooling, etc. However, little research work is there related to SOFC-based trigeneration systems. Also, a mathematical model based on the first and second laws of thermodynamics for the SOFC-based trigeneration system is scarcely available in the literature. This is the motivation behind the present work.

2 Proposed Scheme

2.1 System Description

The proposed system is composed of a topping and bottoming cycle, as shown in Fig. 1. In the topping cycle, air at state point 1 is compressed in an air compressor (AC), and the same is preheated in an air preheater (PH1) by the gas turbine's exhaust. The fuel (Natural Gas) is initially pressurized in a separate compressor and is preheated in the fuel preheater (PH2). The water (at state point 8) is then pressurized using a water pump (WP) and preheated in the preheater (PH3) by the exhaust of the fuel preheater (PH2). Next, the fuel (at state point 6) and the superheated steam (at state point 9) are mixed in the mixer (M) to produce hydrogen that finally enters the anode of SOFC, while air (at state point 3) enters the cathode of SOFC. Then, there is an electrochemical reaction at the Three-Phase Boundary (TPB) in the SOFC, and electricity is generated. The exhaust gases (at state points 11 and 12) from the SOFC enter an afterburner (AB). After burning, the gases first expand to the gas turbine (GT) to have electrical power output. Subsequently, the gases move through preheaters to warm up the fuel, air, and water at the SOFC inlet. Then, the exhaust gases, respectively (at state point 17), enter Heat Recovery Vapor Generator 1 operating the Transcritical Carbon dioxide cycle. Later on, the exhaust gases enter the HRVG2 of the Organic Rankine cycle before getting released into the atmosphere. Transcritical Carbon dioxide Cycle (TRCC) and Organic Rankine Cycle (ORC) have six major components, as shown in the figure. The exhaust gas (at state point 26) evaporates carbon dioxide in HRVG1. Heat is released in the regenerator (RE1). The turbine exhaust in the TRCC exchanges heat with water. Similarly, heat exchange takes place in ORC. In condenser 1 and condenser 2, working fluid (Carbon dioxide at state point 23) and (Propane at state point 29) are condensed, respectively. LNG can recover its cold energy by consuming heat. After that, it is pressurized using a pump before entering HRVG1 and HRVG2 to recover heat. LNG is compressed using pump 3, and the same is delivered to the condenser to condense the fluid of TRCC and ORC. After heat exchange in the heat exchanger during refrigeration storage (HE2), Natural Gas in vapor form (at state point 37) is delivered to

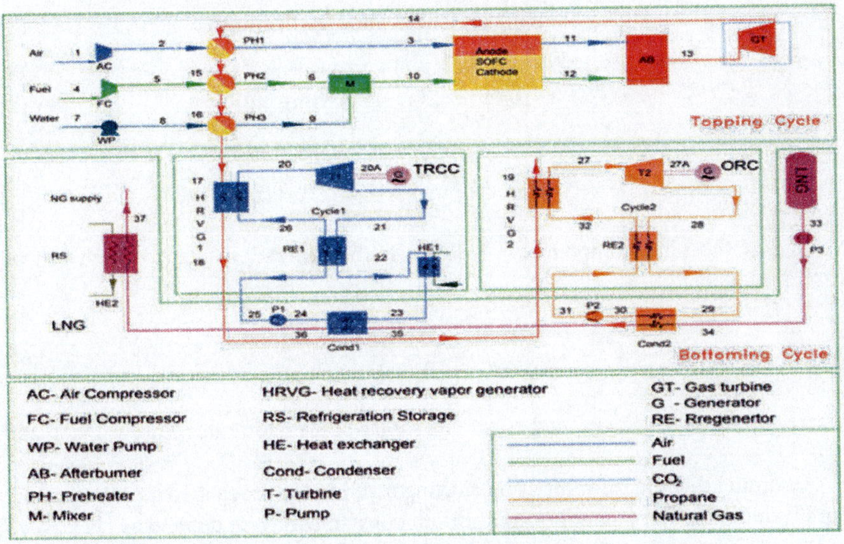

Fig. 1 Schematic diagram of the integrated trigeneration system

the terminal for Natural Gas receipt. In the TRCC and ORC cycle, the turbine is coupled to the generator unit, which further generates electricity due to the relative motion of the shaft (state points 20A and 27A).

3 Modeling

3.1 Assumption

The modeling is done with the following assumptions [11]:

1. The system is assumed to operate at steady-state conditions.
2. Pressure drop at each of the points in the system is neglected.
3. Unreacted fuels exiting SOFC burn entirely in the afterburner.

3.2 Thermodynamic Model Development

Topping Cycle

SOFC Model

The processes that take place in the SOFC's anode and cathode can be expressed as [12]:

$$CH_4 + H_2O = CO + 3H_2 \text{ (Reforming)} \tag{1}$$

$$CO + H_2O = CO_2 + H \text{ (Shifting)} \tag{2}$$

$$H_2 + \frac{1}{2}O_2 = H_2O \text{ (Electrochemical)} \tag{3}$$

One of the vital components of SOFC is the utilization factor, which can be expressed as [12]:

$$U_f = \frac{z_r}{3x_r + y_r} \tag{4}$$

where, accordingly, x_r, y_r, and z_r are the molar conversion rates Eqs. (1), (2), and (3).

Assuming that the reforming and shifting reactions have reached thermodynamic equilibrium, the corresponding equilibrium constants may be defined as [11]:

$$K_{pr} = \frac{P_{CO} \cdot P_{H_2}^3}{P_{CH_4} \cdot P_{H_2O}} \tag{5}$$

$$K_{ps} = \frac{P_{CO_2} \cdot P_{H_2}}{P_{CO} \cdot P_{H_2O}} \tag{6}$$

where K_{pr} and K_{ps} are the equilibrium constants for reforming and shifting, respectively.

The SOFC temperature directly correlates with the equilibrium constants K_{pr} and K_{ps}, which can be represented as:

$$\log k_p = AT^4 + BT^3 + CT^2 + DT + E \tag{7}$$

The SOFC voltage output can be expressed as [11]:

$$V = V_n - V_{loss} \tag{8}$$

where the voltage loss and the Nernst voltage, respectively, are V_n and V_{loss}.

The Nernst voltage can be represented by [11]:

$$V_n = -\frac{\Delta G^o}{n_e} + \frac{RT}{n_e F} \ln\left(\frac{PH_{2}\sqrt{PO_2}}{P_{H_2O}}\right) \tag{9}$$

The total of the ohmic overvoltage (V_{ohm}), the concentration overvoltage (V_{conc}), and the activation overvoltage (V_{act}) is the voltage loss (V_{loss}), which can be expressed as [12]:

$$V_{\text{loss}} = V_{\text{ohm}} + V_{\text{act}} + V_{\text{conc}} \tag{10}$$

The activation loss can be expressed as [11]:

$$V_{\text{act}} = \frac{\text{RT}}{\text{an}_e F} \sin^{-1} h\left(\frac{j}{2j_{o,a}}\right) + \frac{\text{RT}}{\text{an}_e F} \sin^{-1} h\left(\frac{j}{2j_{o,c}}\right) \tag{11}$$

Ionic and electronic resistance in the respective fuel cell components causes the ohmic overvoltage, which can be given by [12]:

$$V_{\text{ohm}} = R_c + \rho_c L_c + \rho_a L_a + \rho_e L_e + \rho_{\text{int}} L_{\text{int}} \tag{12}$$

In Eq. (12), L is the fuel cell component's thickness, R_c is the resistivity contact, and ρ is the electrical resistivity of a cell component.

The concentration overvoltage can be determined by [11]:

$$\begin{aligned} V_{\text{conc}} &= \frac{\text{RT}}{\text{an}_e F} \ln\left(1 - \frac{j}{j_{\text{as}}}\right) \\ &+ \frac{\text{RT}}{\text{an}_e F} \ln\left(1 + \frac{p_{\text{H}_2}^{\text{an}}}{p_{\text{H}_2\text{O}}^{\text{an}} j_{\text{as}}}\right) \\ &+ \frac{\text{RT}}{\text{an}_e F} \ln\left(1 - \frac{j}{j_{\text{cs}}}\right) \end{aligned} \tag{13}$$

where the anodic and cathodic limiting power densities, respectively, are j_{as} and j_{cs}.

The current and current density (I, j) can be written as [11]

$$I = j \cdot A_a \tag{14}$$

$$j = \frac{n_e \cdot F \cdot z_r}{N_{\text{cell}} \cdot A_a} \tag{15}$$

The SOFC's power output can be estimated as [11]

$$W_{\text{SOFC}} = i \cdot A_{\text{act}} \cdot V_{\text{cell}} \cdot N_{\text{cell}} \tag{16}$$

Bottoming Cycle

The energy and mass conservation concepts are used to examine the bottoming cycle's performance. The same can be described as [13]:

$$Q + \sum m_{\text{in}} h_{\text{in}} = W + \sum m_{\text{out}} h_{\text{out}} \tag{17}$$

$$\sum m_{\text{in}} = \sum m_{\text{out}} \tag{18}$$

Exergy Analysis

Exergy, which reflects the quality of energy, is the highest useful energy obtained when the system reaches equilibrium. [12]. Exergy analysis can determine which components have a high exergy loss, quantify the irreversibility of performance decline, and provide improved guidance. The proposed system's exergy balance at a steady state can be given by [11].

$$\left(\sum E^{\mathrm{ph,\,ch}}\right)_{\mathrm{in}} + \sum \mathrm{EX}_Q = \left(\sum \mathrm{EX}^{\mathrm{ph,\,ch}}\right)_{\mathrm{out}} + W_{\mathrm{u}} + I \quad (19)$$

In Eq. (20), EX_Q is the system's input heat exergy, while I is exergy destruction. Input exergy can be expressed as:

$$\mathrm{EX}_Q = Q\left(1 - \frac{T_L}{T_H}\right) \quad (20)$$

The system exergy is composed of the physical and chemical exergy, which can be represented as:

$$E^{\mathrm{ph,\,ch}} = \mathrm{EX}^{\mathrm{ph}} + \mathrm{EX}^{\mathrm{ch}} \quad (21)$$

The physical and chemical exergy can be given by [11].

$$Ep^h = m[(h_{\mathrm{in}} - h_{\mathrm{out}} - T_o(s_{\mathrm{in}} - s_{\mathrm{out}}))] \quad (22)$$

$$\mathrm{EX}^{\mathrm{ch}} = \sum_i^n n_i \mathrm{ex}_i^{\mathrm{ch},0} + \mathrm{RT}_o \sum_i^n n \ln x_i \quad (23)$$

4 Results and Discussion

4.1 Model Validation

Based on the mathematical model discussed in the previous section, a computer code has been developed using EES version 11.165. The program considers the optimum SOFC inlet temperature, operating pressure, number of cells, and various other parameters as inputs. It predicts the voltage output and the output power for a pressure of 6 bar. The input parameters are given in Table 1.

Figure 2 depicts the variation of SOFC net voltage and activation voltage with SOFC temperature for the present and the reference models [11]. It is observed that there is an increase in the voltage with the SOFC temperature. It is found that

Table 1 Input parameters for the model validation

Parameters	Value	Unit
P_{SOFC} (operating pressure)	600	kPa
T_{in} (SOFC inlet temperature)	500	°C
ΔT_{rack} (stack temperature)	200	°C
N_{cell} (number of cell)	10,000	–
A_a (active surface area)	100	cm^2
U_f (fuel utilization factor)	82	%
L_a (thickness of the anode)	$5 * 10^{-2}$	cm
L_c (thickness of the cathode)	$5 * 10^{-3}$	cm
L_{int} (thickness of the interconnect)	0.4	cm
L_e (thickness of the anode electrolyte)	$1 * 10^{-3}$	cm
J (current density)	0.5	A/cm^2

the change in activation voltage with SOFC temperature is marginal, while the net voltage varies considerably with SOFC temperature. This is because Nernst voltage (V_N) varies substantially with temperature, thus, significantly affecting the net SOFC output voltage. It may be noted that the differences in the results obtained from the two models are marginal, with a mean square error of only 2.380e-6. The present model predicts a marginally higher voltage value than the voltage indicated by the reference model [11].

Figure 3 depicts the effect of SOFC temperature on the ohmic and concentration voltage for the present model and the reference model [11]. It is seen that the ohmic voltage of the system all through decreases when SOFC inlet temperature increases

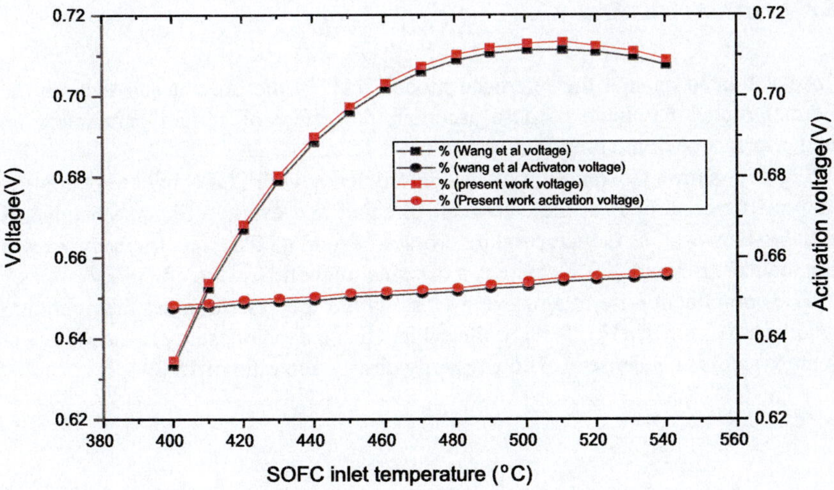

Fig. 2 Effect of the SOFC inlet temperature on the SOFC voltage output

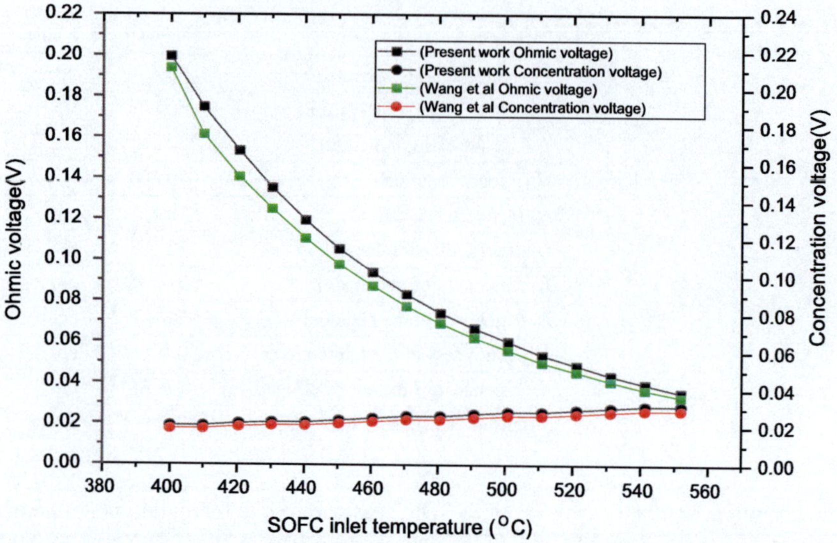

Fig. 3 Effects of the SOFC inlet temperature on ohmic voltage and concentration voltage

for both models. At the same time, SOFC concentration voltage increases with the increase in the inlet temperature. It is observed that the maximum and minimum values of the concentration voltage and ohmic voltage are 0.02871 and 0.03531 V, respectively, corresponding to a SOFC temperature of 550°C.

4.2 *Parametric Analysis*

Once validated against the reference model [11], in the present sub-section, the thermal model has been used to ascertain the effect of certain parameters on the performance of the proposed system.

Figure 4 shows the variation of system efficiency with SOFC inlet temperature. As the temperature rises, the electrical, thermal, and exergy efficiency augments initially. However, as the temperature crosses beyond 474°C, the efficiency growth rate reduces. Beyond 500°C, there is a marginal reduction in the efficiency values.

The component-wise rate of exergy loss within the system under the operating conditions is shown in Fig. 5. It is quite evident that the major exergy loss component in the system is condenser 1 with an exergy destruction ratio of 15.29.

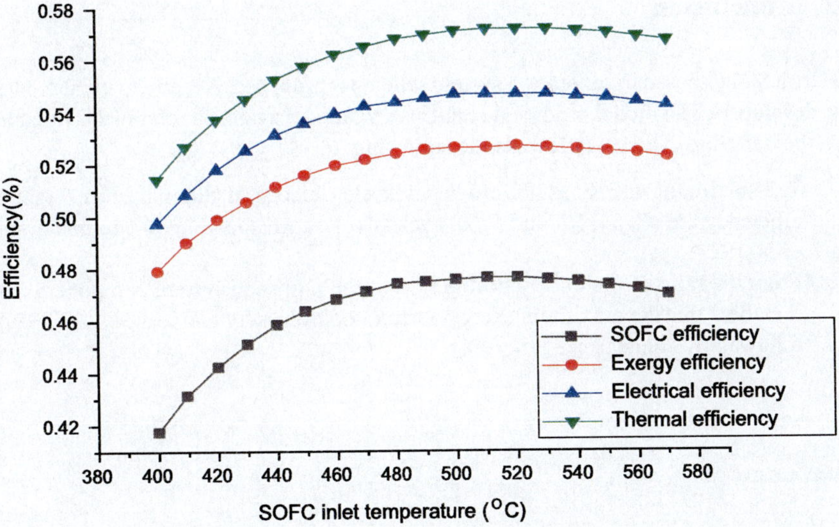

Fig. 4 Effect of the SOFC inlet temperature on the efficiency

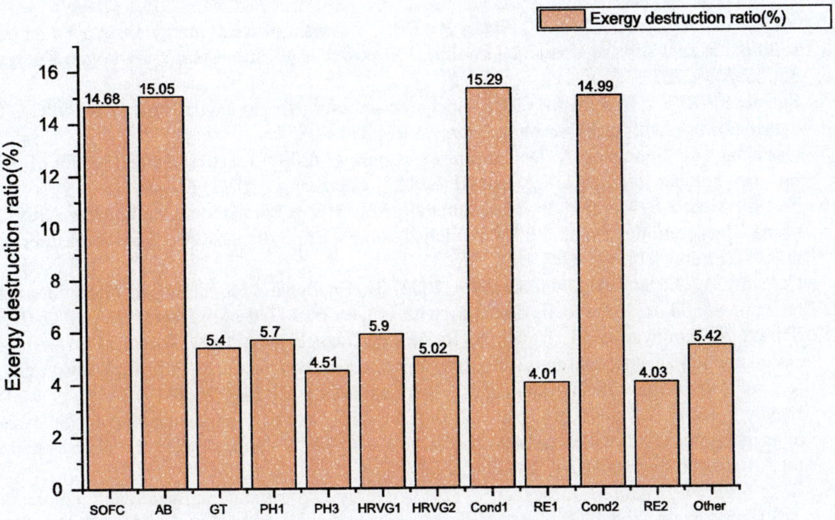

Fig. 5 Ratio of exergy destruction for each component of the pressurized system

5 Conclusion

Here a SOFC-based trigeneration system scheme is proposed. A mathematical model is developed. The model predicted results are validated against a reference available in the literature. The study reveals the following:

1. The maximum value of SOFC efficiency, system electrical, thermal, and exergetic efficiency are 47, 54, 57, and 52%, respectively, corresponding to a temperature of 500°C.
2. From the ratio of the exergy destruction of the proposed system components, it is evident that the maximum exergy destruction occurs for condenser 1 (15.29%) followed by condenser 2 (14.99%).

References

1. World Energy Statistics. Retrieved from https://www.iea.org/reports/key. Accessed on 10 Aug 2020
2. Energy in India Today. Retrieved from https://www.iea.org/reports/india-energy-outlook-2021. Accessed on 19 Feb 2018
3. Yu Z, Han J, Cao X, Chen W, Zhang B (2010) Analysis of total energy system based on solid oxide fuel cell for combined cooling and power applications. Int J Hydrogen Energy 35(7):2703–2707
4. Sghaier SF, Khir T, Brahim AB (2018) Energetic and exergetic parametric study of a SOFC-GT hybrid power plant. Int J Hydrogen Energy 43(6):3542–3554
5. Costamagna P, Selimovic A, Del Borghi M, Agnew G (2004) Electrochemical model of the integrated planar solid oxide fuel cell (IP-SOFC). Chem Eng J 102(1):61–69
6. Yari M, Mehr AS, Mahmoudi SMS, Santarelli M (2016) A comparative study of two SOFC based cogeneration systems fed by municipal solid waste by means of either the gasifier or digester. Energy 114:586–602
7. Colpan CO, Dincer I, Hamdullahpur F (2007) Thermodynamic modeling of direct internal reforming solid oxide fuel cells operating with syngas. Int J Hydrogen Energy 32(7):787–795
8. Shirazi A, Aminyavari M, Najafi B, Rinaldi F, Razaghi M (2012) Thermal- economic-environmental analysis and multi-objective optimization of an internal-reforming solid oxide fuel cell-gas turbine hybrid system. Int J Hydrogen Energy 37(24):19111–19124
9. Eveloy V, Karunkeyoon W, Rodgers P, Alili AA (2016) Energy, exergy and economic analysis of an integrated solid oxide fuel cell—gas turbine—organic Rankine power generation system. Int J Hydrogen Energy 41(31):13843–13858
10. Bang-Møller C, Rokni M, Elmegaard B (2011) Exergy analysis and optimization of a biomass gasification, solid oxide fuel cell and micro gas turbine hybrid system. Energy 36(8):4740–4752
11. Wang H, Yu Z, Wang D, Li G, Xu G (2021) Energy, exergetic and economic analysis and multi-objective optimization of atmospheric and pressurized SOFC based trigeneration systems. Energy Convers Manage 239:114183
12. Khani L, Mehr AS, Yari M, Mahmoudi SMS (2016) Multi-objective optimization of an indirectly integrated solid oxide fuel cell-gas turbine cogeneration system. Int J Hydrogen Energy 41(46):21470–21488
13. Sadat SMS, Ghaebi H, Lavasani AM (2020) 4E analyses of an innovative polygeneration system based on SOFC. Renew Energy 156:986–1007

Review of Polymers and Coagulants Used for Flocculation of Drilling Fluid

Kunal Kishor Chandan and Vikash Kumar Singh Chauhan

1 Introduction

The drilling fluids are essential for the drilling operations. Drilling fluid is introduced in the down hole tubes for the drilling of petroleum. This fluid then returns to the surface via the annular casing of the system. This returned fluid gets contaminated with pollutants (drilling cuttings and chemicals from the earth's crust) during the operation and is returned to a pit near the drill site. Such drilling fluid wastes generated by petroleum industries are highly undesirable and unsafe and need to be treated before their final discharge into the atmosphere. Common types of drilling fluids used for the drilling process are based on (1) water, (2) oil, (3) artificial chemicals, and (4) pneumatic or air-based drilling fluids. Among the above, drilling fluid based on water substrate is quite common in drilling operations in petroleum fields. As reported in the literature, about 80% of drilling of petroleum is carried out with drilling fluids based on water as they are more economical than oil- or artificial-based drilling fluids [1]. The control of the above physical properties of drilling fluids is very important for maintaining efficient drilling operations [2]. Typically, the fine particles of solids present in the fluid play a major role in modifying the physical properties such as density, viscosity, and filtration loss. These properties directly affect the drilling rate, dilatation rate, torque, and drag. Drilling fluids based on water is very sensible in both physical and chemical properties.

During the drilling of petroleum wells, drilling fluids and drill cuttings are generated as waste products [3]. Besides produced water, the oilfield drilling fluids' waste contains a large volume of waste [4] containing minerals, clay, sulfur, heavy metals and other solids' contaminants which create a significant effect on the groundwater and agricultural land area and pollute the environmental system [5]. The effect of these pollutants can cause the poisoning of drinking water which affects the physical

K. K. Chandan (✉) · V. K. S. Chauhan
Department of Mechanical Engineering, Sandip University, Madhubani, Bihar 847235, India
e-mail: kunal1042@gmail.com

B. P. Swain (ed.), *Recent Advances in Materials*, Springer Proceedings in Materials 25,
https://doi.org/10.1007/978-981-99-3844-5_30

and mental health of living things and cause tremendous damage to the ecosystem. Waste minimization, treatment, and safe disposal are the three-prong approaches to controlling and managing drilling fluid wastes [3]. The separation of solid particles from a liquid by agglomeration and flocculation processes is widely used to minimize drilling fluid wastes. The agglomeration and flocculation can be used near the rig site for enhancing the efficiency of the total solid removal. The above processes can also be used at disposal sites and waste treatment plants.

2 Agglomeration and Flocculation

This instability causes the particles to come together and form larger clusters, leading to a noticeable separation between the solid and liquid phases, which is known as flocculation [6–9]. The mechanism of flocculation includes the formation of the electrical double layer in the solution possessing various ions. This double layer is composed of the stern layer (inner) and Gouy-Chapman layer (outer) [10]. Charged particles in a solution tend to attract oppositely charged ions in the dispersion or move toward a higher concentration of oppositely charged zone. The plot of charge magnitude versus the interparticle distance is depicted in Fig. 2. It is evident from the plot that the trend follows an exponential decay, and thus, a shear layer will emerge. This further develops a potential gap between the layers known as zeta potential [6, 11, 12].

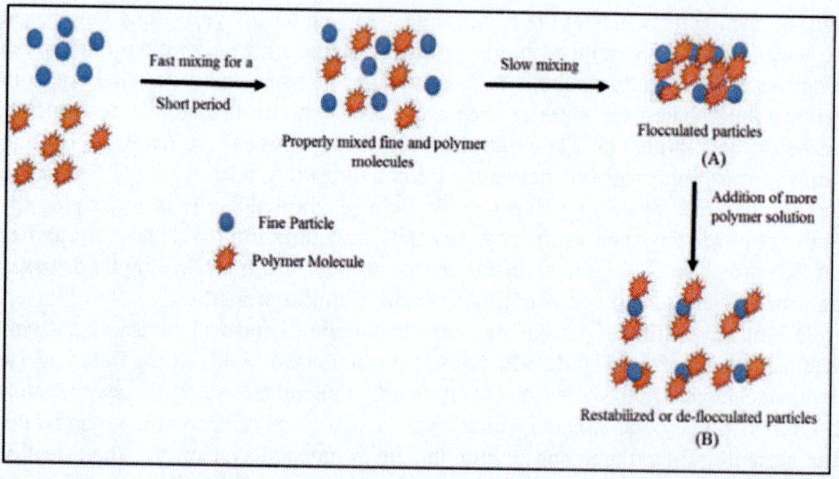

Fig. 1 Schematic diagram showing flocculation and deflocculation of fine articles by addition of polymers [9]

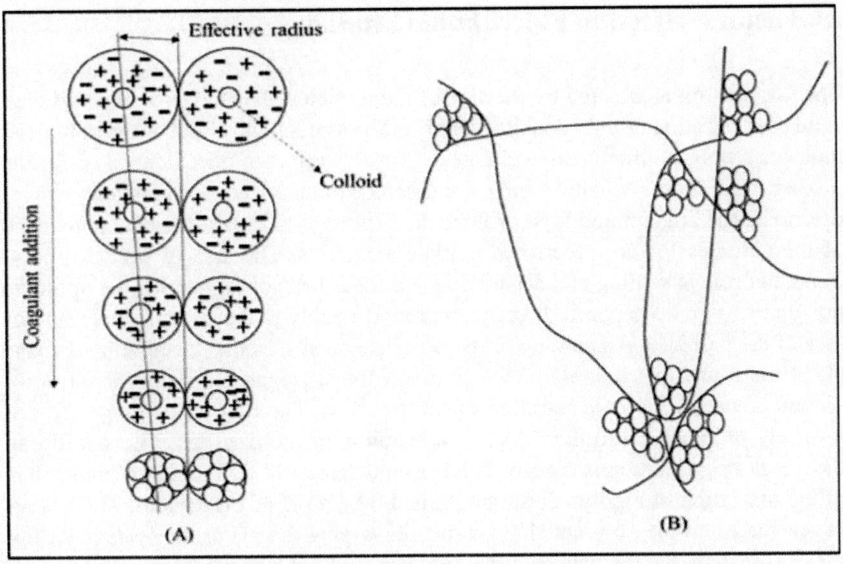

Fig. 2 Schematic representation of the double-layer structure at the solid/electrolyte interface [8]

3 Flocculation Mechanisms

When the electrostatic repulsive force overcomes van der Waal's force in a colloidal solution, natural flocculation takes place. There are chances of flocs breakdown due to weak bonding forces, and hence, such agitation results in very limited flocculation [13]. When the oppositely charged ions are added to the particles in the solution, the reduction in charge enables agglomeration and then flocculation to take place. It has been found that NaCl, $CaCl_2$, and $Al_2(SO_4)_3$ produce Na^+, Ca^{++}, and Al^{+++} ions in their aqueous solutions with Al^{+++} ions being the most effective. Agglomeration or flocculation takes place when the suspended particles carry no charge with respect to the suspending medium [9]. This mechanism is generally called the "Bridging Mechanism" of polymer in flocculation (Fig. 1). For an effective flocculation to take place, the polymer adsorbed on the surface should be able to extend from one particle to other particles. Re-stabilization of the colloidal particles takes place if the polymer solution added becomes excessive (Fig. 2) [14]. When the particles with a fairly low density of immobile surface charge are treated with a polyelectrolyte of fairly high charge density, each surface site becomes different to be neutralized by a segment of the polymer chain [15–17].

4 Factors Affecting Flocculation Studies

The flocculation is affected by the size of the particles, the charge on the particles, solid concentration or pulp density, the pH of the system, the effect of the polymer's molecular weight, structure and charge density, the effect of flocculants dosage, and mixing conditions. The nature and the magnitude of the suspended charged particles determine the amount and type of flocculants needed to satisfy the charge demand of the particles for an effective flocculation process. The size of particles plays important role in settling and dewatering [9]. It has been observed that the optimum polymer/solid ratio depends directly on proportionality to the overall active surface area of the particles. An increase in particle size decreases the flocculation demand [12]. It is found that the rate of formation of the aggregates is proportional to the second power of the concentration of the particles. The time of agglomeration is inversely proportional to the initial concentration of the particles. The stability of the floc during agitation is higher at higher pulp density [13]. The degree and nature of the ionization of a polymer are controlled by the pH of the system. The change in orientation of the polymer chain results in varying the pH of the system. At low pH (<3) system, the ionization of the polymer does not take place, but at higher pH (>7) system, the decoiling of the polymer takes place due to ionization, and thus, enhancement of flocculation efficiency is noticed [15–18]. Polymer molecular weight affects the flocculation via bridging and electrostatic path mechanism [19–22]. It has been observed that flocculation efficiency decreases due to repulsive forces between polymer molecules and particle surfaces present in the system [13]. The flocculation efficiency is affected by the dosing and mixing conditions of the flocculants [19, 23]. As the polymer dosage is increased, the destabilization of the colloidal system takes place. Further, an increase in the dosage of the polymer results in the second phase of destabilization [24, 25].

5 Flocculating Materials

Two types of flocculating agents are generally used for wastewater treatment: inorganic and organic. Apart from these two, a novel class of graft copolymers is developed for flocculants applications. The graft copolymer is prepared using both natural and artificial (synthetic) polymers [26–35]. Inorganic flocculating materials are mainly metal ion coagulants [9]. These flocculating agents are mainly based on Fe/Al hydrolysable salt. Ferric salts are more commonly in use as a coagulating agent, e.g., ferric sulfate, ferric chloride, and ferrous sulfate. These compounds perform extremely well in acidic conditions for agglomeration application. For better results, these compounds may be used with alum [6]. The organic flocculating materials used for flocculants applications are mostly linear polymers of hydrophilic nature having very high molecular weight [36]. The effectiveness of natural polymeric flocculants such as starch, amylose, amylopectin, gelatine, and alginate is well known. The latest developments in polymeric flocculants are artificial polymers. The most widely used artificial polymer flocculants are non-ionic polyacrylamides and their derivatives [9].

6 Artificially Grafted Copolymers

Polyacrylamide-based artificial polymers are well known for their application as efficient flocculating agents with the disadvantage of shear degradability. On the other hand, natural polymers are mainly based on polysaccharides' substrate. Natural polymers are relatively inexpensive, shear stable, and easily available from farm-produced waste materials. [35–48]. The graft copolymerization method is used to incorporate new properties in natural and artificial polymers. Graft copolymerization results in negligible loss to the natural properties of the substrate (Fig. 3).

A graft copolymer consists of a polymeric backbone with a lateral covalently linked side chain of a different polymer [40]. Recent advancements have been made to graft acrylamide into the natural polysaccharides like xanthan gum, guar gum/hydroxypropyl guar gum, starch, carboxymethyl cellulose (CMC), amylopectin, amylose, sodium alginate, etc. [26–35].

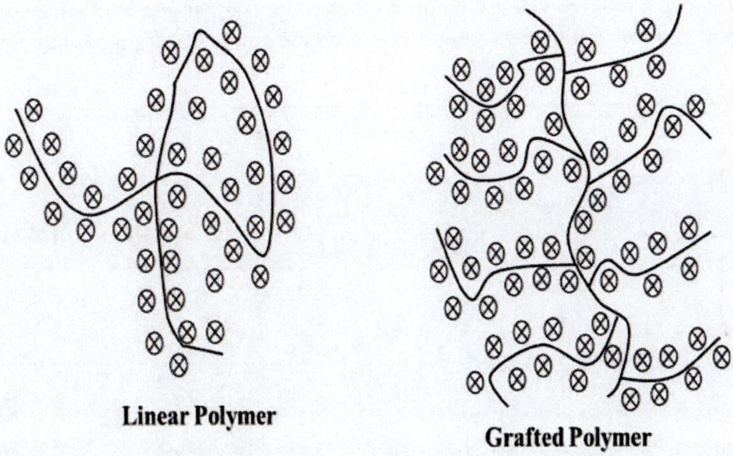

Fig. 3 Singh's easy approachability model [49]

7 Flocculation Methods

There are two major flocculation test methods practiced in the industry:

7.1 Jar Test

The jar test consists of six jars of similar volume and concentration of feed (Fig. 4). Each jar is attached with a series of stirrers synchronized with an electric drive through gears or belts. The jars can be stirred at once at a predetermined speed. Dosing of flocculants or salts is introduced directly in the jar, and the stirring pace is controlled by a speed selector [23]. The flocculants are added from the top as per requirement and mixed rapidly for two minutes at a predetermined speed. Thereafter, the mixture is stirred at a relatively lower speed for approx. 15 min. This process completes as the flocs are formed in the solution when left for settling for around 30 min. The supernatant fluid is taken out of the flocs for turbidity analysis. Based on this analysis, the optimum amount of the flocculants dose is predicted for the liquid [24].

The turbidity value is defined as particulate matter suspended in the suspension through optical measurement based on interference analysis of light going through the medium. It does not predict the number of particles in the solution; rather it just provides a relative value. In a turbidimeter, the comparison is made between a standard solution and the test sample based on the amount of light passing through

Fig. 4 Jar test apparatus [49]

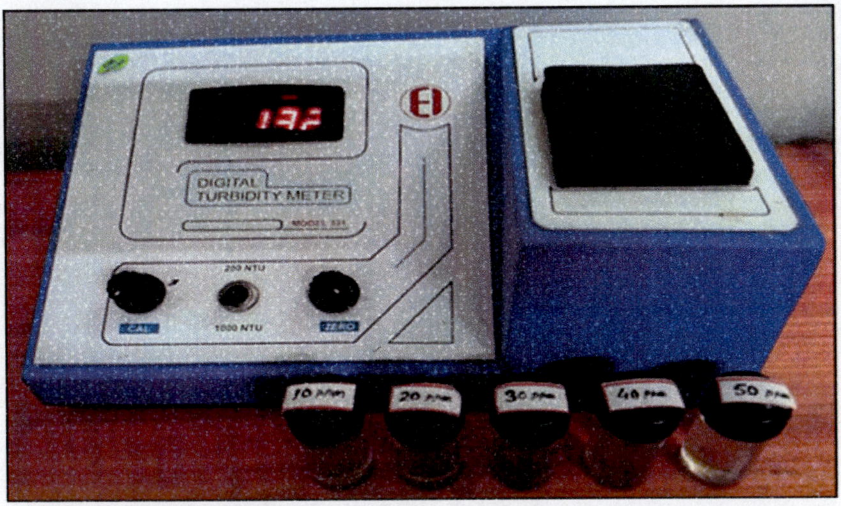

Fig. 5 Digital turbidity meter [49]

the medium (Fig. 5) [23]. Turbidimetry and nephelometry are popular methods for turbidity measurement.

7.2 *Cylinder Test (Settling Tests)*

The turbidity measurements are followed by observing the time of settling of the mud line in the suspension. This is achieved by recording the time rate of fall in the measuring cylinder. In the settling test, the particles flocculate in a standard-size jar with graduated measuring arrangements as shown in Fig. 6. Now, the motion of the mud line in liquid is recorded as dependent on time breaks was recorded using a stopwatch. 1 wt. % slurry sample is placed in the cylinder and specified dosages of the polymer were added. Further, the cylinder is inverted at least five times [50–52]. These trials were repeated with fresh samples every time to prepare a plot of height and the settling periods through which the settling rate is calculated.

8 Applications of Flocculation in Oil and Gas Fields

The oilfield waste drilling fluid consists of various fine solid particles and liquid additives which are toxic chemical substances. These waste materials negatively affect biodiversity by polluting the land surface and groundwater. It also affects agricultural land areas as well as pollutes the environmental system [53]. This waste must

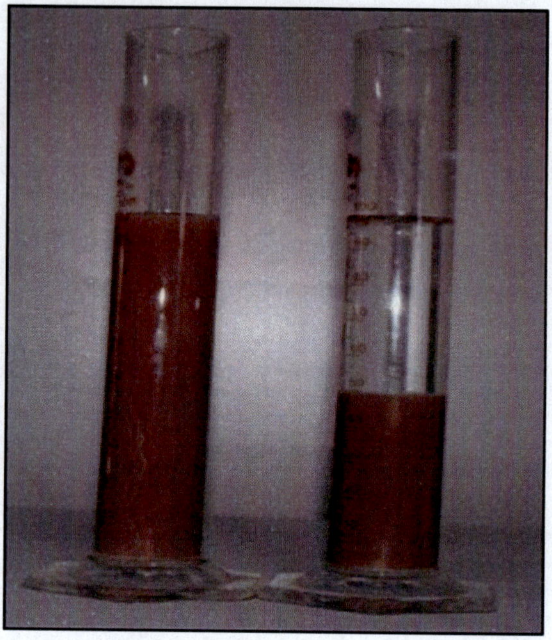

Fig. 6 Settling tests of water-based drilling fluid [49]

be disposed of after proper treatment. The treatment methods include sedimentation (gravity separation), vacuum separation, and impact and baffle methods. The treatment of drilling waste poses the most important challenge to the oil and gas drilling industry in modern times [54]. Therefore, the separation of drill cutting or solids material from the waste drilling mud is very difficult through conventional methods. Flocculation and agglomeration are the most common types of waste management techniques used by modern petroleum-producing companies. Other methods of waste management include proper disposal, solid control, waste minimization, and recycling. There are various effective drilling waste management approaches for waste minimization, solid control, and mud system monitoring [18, 55]. Ayotamuno et al. [56] had shown that powdered/activated carbon (PAC) can be used as an adsorbent in a batch adsorption process improving the flocculation of the effluent. A representative sample of waste drilling fluid was collected from oil wells in Nigeria. The crude dense liquid-type oil-based drill cuttings' sample initially had shown a Cr^{6+} concentration of 52.6 mg/liter. This sample was first flocculated with the conjugating agents of aluminum sulfate and sodium chloride. This resulted in around a 4% reduction of Cr^{6+} concentration. Further, a 45% reduction was observed when the flocculated material was treated with PAC. They showed that a large amount of chromium was removed from the waste drilling fluid after the flocculation of the effluent. Wang et al. [57] used the starch, acrylamide (AM), 2-trimethyl ammonium ethyl methacrylate chloride (DMC) and potassium persulfate as raw material to prepare a starch substrate-based cationic copolymer flocculants. The starch-based

Fig. 7 **a** Drilling waste sample before flocculation **b** Drilling waste sample after flocculation [58]

cationic copolymer has shown better flocculation performance compared to commercially available coagulants in the market for thick drilling fluid waste. A composite of organic and inorganic compounds was used as a flocculants for solid/liquid separation of discarded drilling fluid [20]. This showed an improved capacity of flocculation over other options at similar quantities. A double-step flocculation method was developed for treating oil/sand tailings with polymers chitosan and Magafloc-1011. These double-step methods dramatically reduce the particles in the wastewater by significantly improving the supernatant's turbidity. This further reduces the particulate concentration of the supernatant while maintaining a lower settling time [57]. Grafted copolymers based on polysaccharides are resulted to be efficient flocculant aids for reducing solid particulates in effluxes' drilling fluid having quantities of oceanic salt (0.1 M) under alkaline conditions [49, 58] (Fig. 7)

9 Conclusion

Agglomeration and flocculation are chemical processes, which are utilized for the treatment of waste. Various types of flocculants used during flocculation are inorganic and organic materials including natural and artificial graft copolymers. Recently, various types of synthesized graft copolymers have been developed. A flocculation test has been carried out using these graft copolymers. The graft copolymers are found to be shear stable, biodegradable, and efficient flocculants. It has seen found that the flocculation quality of graft copolymers is better than commercially available natural and artificial flocculants. It is also observed after mixing suitable coagulants in oilfield fluid wastes, polysaccharide-grafted copolymers can be utilized as an aid for flocculants of flocculation of contaminants/pollutants present in the waste drilling fluid.

Acknowledgements The Authors of the Article thank Sandip University, Madhubani, for providing the motivational and financial support to conduct this review work at their premises. I also thank Dr. Sandip Jha for his consistent support in favor of research-related activities inside the University campus and nearby areas for the betterment of the civilian society.

References

1. Sadeghalvaad M, Sabbaghi S (2015) The effect of the Tio_2/polyacrylamide nanocomposite on water-based drilling fluid properties. Powder Technol 272:113–119
2. Abdou MI, Ahmed HE-S (2011) Effect of particle size of bentonite on rheological behavior of the drilling mud. Pet Sci Technol 29(21):2220–2233
3. Onwukwe SI, Nwakaudu MS (2012) Drilling wastes generation and management approach. Int J Environ Sci Dev 3(3):252–257
4. Ismail AR, Alias AH, Sulaiman WRW, Jaafar MZ, Ismail I (2017) Drilling fluid waste management in drilling for oil and gas wells. Chem Eng Trans 56:1351–1356
5. Zou J, Zhu H, Wang F, Sui H, Fan J (2011) Preparation of a new inorganic-organic composite flocculant used in solid-liquid for waste drilling fluid. Chem Eng J 171(1):350–356
6. Tripathy T, De BR (2006) Flocculation: a new way to treat the wastewater. J Phys Sci 10:93–127
7. Montgomery JM (1985) Water treatment: principles and design. Wiley, New York, p 116
8. Bratby J (1980) Coagulation and flocculation. Uplands Press Limited, Croydon CR9 1LB, England
9. Karmakar GP (1994) Flocculation and rheological properties of grafted polysaccharides. Ph.D. thesis, IIT, Kharagpur, India
10. Kruyt HR (eds) (1952) Volume 1: irreversible systems. In: Colloid science. Elsevier Publishing Co., Amsterdam, NY
11. Shamlou PA, Hooker NT () Chapter 1: turbulent aggregation and breakup of particles in stirred vessels. In: Shamlou PA (ed) Proceeing of solid-liquid suspensions. Butterworth–Heinemann Limited
12. Riddick MT (1961) Zeta potential and its application to difficult waters. J AWWA 53(8):1007–1030
13. Moss N (1978) Theory of flocculation. Mine Quarry 7(5):57–61
14. Gregory J (1992) Flocculation of fine particles. In: Mavros P, Matis KA (eds) Innovation in flotation technology. Kluwer Academic Publishers, pp 101–124
15. Kasper DR (1971) Kasper, Dennis Robert (1971) Theoretical and experimental investigations of the flocculation of charged particles in aqueous solutions by polyelectrolytes of opposite charge. Ph. D. thesis, California Institute of Technology
16. Gregory J (1987) Chapter 8: flocculation by polymers and polyelectrolytes. In: Tadros TF (eds) Solid-liquid dispersions. Academic Press Limited, London
17. Gregory J (1985) The use of polymeric flocculants. In: Proceedings of the engineering foundation conference of flocculation sedimentation and consolidation, The Cloister Sea Islands, Georgia, USA, p 125
18. Moody G (1992) The use of polyacrylamides in mineral processing. Miner Eng 79(3–5):479–492
19. Sandell LS, Luner P (1974) Flocculation of microcrystalline cellulose with cationic ionene polymers. J Appl Polym Sci 18(7):2075–2083
20. Adachi Y, Stuart MAC, Fokkink R (1995) Initial rates of flocculation of polystyrene latex with polyelectrolyte: effect of ionic strength. J Colloid Interface Sci 171(2):520–521
21. Michaels AS (1954) Aggregation of suspensions of polyelectrolytes. Ind Eng Chem 46(7):1485–1490

22. Young YS, Chiang SH, Morsi BI (1991) Enhancement of filtration and dewatering of fine coal using flocculants. Mines Qarriers 3:39
23. Ward AS (1996) Pre-treatment techniques. In: Rushton A, Ward AS, Holdich RG (eds) Solid-liquid filtration and separation technology, vol 74, 1st edn. Wiley-VCH Publishers Inc.
24. Guibai L, Gregory J (1991) Flocculation and sedimentation of high-turbidity waters. Water Res 25(9):1137–1143
25. Hogg R (1993) Process design for flocculation, clarification and thickening of fine-particle suspensions. In: Proceeding of 18th international mineral processing congress, Sydney, Australia, pp 1315–1319
26. Rath SK, Singh RP (1997) Flocculation characteristics of grafted and ungrafted starch, amylose, and amylopectin. J Appl Polym Sci 66(9):1721–1729
27. Karmakar GP, Singh RP (1998) Flocculation studies using amylose-grafted polyacrylamide. Colloids Surf A 133(1–2):119–124
28. Tripathy T, Pandey SR, Karmakar NC, Bhagat RP, Singh RP (1999) Novel flocculating agent based on sodium alginate and acrylamide. Eur Polymer J 35(11):2057–2072
29. Tripathy T, Karmakar NC, Singh RP (2000) Grafted CMC and sodium alginate: a coparision in their flocculation performance. Int J Polym Mater Polym Biomater 46(1–2):81–93
30. Tripathy T, Bhagat RP, Singh RP (2001) The flocculation performance of grafted sodium alginate and other polymeric flocculants in relation to iron ore slime suspension. Eur Polymer J 37(1):125–130
31. Nayak BR, Singh RP (2001) Development of graft copolymer flocculating agents based on hydroxypropyl guar gum and acrylamide. J Appl Polym Sci 81(7):1776–1785
32. Pal S, Mal D, Singh RP (2005) Cationic starch: an effective flocculating agent. Carbohyd Polym 59(4):417–423
33. Pal S, Mal D, Singh RP (2006) Synthesis, characterization and flocculation characteristics of cationic glycogen: a novel polymeric flocculant. Colloids Surf A 289(1–3):193–199
34. Pal S, Mal D, Singh RP (2008) Characterization of cationic starch: an efficient flocculating agent. J Appl Polym Sci 108(4):2674–2681
35. Pal S, Ghorai S, Dash MK, Ghosh S, Udayabhanu G (2011) Flocculation properties of polyacrylamide grafted carboxymethyl guar gum (CMG-g-PAM) synthesised by conventional and microwave assisted method. J Hazard Mater 192(3):1580–1588
36. Singh RP, Karmakar GP, Rath SK, Karmakar NC, Pandey SR, Tripathy T, Panda J, Kanan K, Jain SK, Lan NT (2000) Biodegradable drag reducing agents and flocculants based on polysaccharides: materials and applications. Polym Eng Sci 40(1):46–60
37. Singh RP, Nayak BR, Biswal DR, Tripathy T, Banik K (2003) Biobased polymeric flocculants for industrial effluent treatment. Mater Res Innovations 7(5):331–340
38. Singh RP, Pal S, Krishnamoorthy S, Adhikary P, Ali SA (2009) High-technology materials based on modified polysaccharides. Pure Appl Chem 81(3):525–547
39. Singh RP, Pal S, Ali SA (2014) Novel biodegradable polymeric flocculants based on cationic polysaccharides. Adv Mater Lett 5(1):24–30
40. Stannett V (1981) Grafting. Radiat Phys Chem 18(1–2):215–222
41. Bhagat RP, Karmakar GP, Singh RP (1997) Filtration of iron ore slime using synthesized copolymer. In: Proceeding of the international conference on "recent advance in metallurgical processes". New Age International Publishers, New Delhi, pp 89–93
42. Tripathy T, Singh RP (2000) High performance flocculating agent based on partially hydrolysed sodium alginate-g-polyacrylamide. Eur Polymer J 36(7):1471–1476
43. Biswal DR, Singh RP (2006) Flocculation studies based on water-soluble polymer of grafted carboxymethyl cellulose and polyacrylamide. J Appl Sci 102(2):1000–1007
44. Sen G, Pal S (2009) Microwave initiated synthesis of polyacrylamide grafted carboxymethylstarch (CMS-g-PAM): application as a novel matrix for sustained drug release. Int J Biol Macromol 45(1):48–55
45. Ghosh S, Sen G, Jha U, Pal S (2010) Noval biodegradable polymeric flocculant based on polyacrylamide-grafted tamarind kernel polysaccharide. Bioresour Technol 101(24):9638–9644

46. Khalek MAA, Mahmoud GA (2011) Flocculation studies of fine coal using acrylamide and acrylic acid graft copolymer by gamma irradiation. Aust J Basic Appl Sci 5(11):271–278
47. Ghorai S, Sarkar A, Panda AB, Pal S (2013) Evaluation of the flocculation characteristics of polyacrylamide grafted xanthan gum/silica hybrid nanocomposite. Ind Eng Chem Res 52(29):9731–9740. ACS Publications
48. Das R, Ghorai S, Pal S (2013) Flocculation characteristics of polyacrylamide grafted hydroxypropyl methyl cellulose: an efficient biodegradable flocculant. Chem Eng J 229:144–152
49. Chandan KK (2019) Synthesis, characterization of polysaccharide based graft copolymers and their applications in oil and gas industry, Ph. D. thesis, Indian Institute of Technology (IIT), Kharagpur, West Bengal
50. Soliman EA, El-Sayer Mohamed El-Sayed Mansour, Hassan HAM, Hassan NAM (2014) Optimization of graft polymerization and performance of carboxymethyl chitosan/polyacrylamide flocculants. J Res Devel Chem 2014(351498):1–18
51. El-Latif MMA, Eldin MSM, El-Kady MF (2005) Settling of high concentration of clay suspended in water by non-ionic polyacrylamide flocculants. Alexandria Eng J (AEJ) 44(2):325–338
52. Nasser MS, James AE (2008) Degree of flocculation and viscoelastic behaviour of kaolinte-sodium chloride dispersions. Colloids Surf A 315(1–3):165–175
53. Kim M, Kim S, Kim J, Kang S, Lee S (2013) Factor affecting flocculation performance of synthetic polymer for turbidity control. J Agric Chem Environ 2(1):16–21
54. Zhang A, Li M, Lv P, Zhu X, Zhao L, Zhang X (2016) Disposal and reuse of drilling solid waste from a massive gas field. Procedia Environ Sci 31:577–581
55. Shon C-S, Estakhri CK, Lee D, Zhang D (2016) Evaluating feasibility of modified drilling waste materials in flexible base course construction. Constr Build Mater 116:79–86
56. Ayotamuno MJ, Okparanma RN, Ogaji SOT, Probert SD (2007) Chromium removal from flocculation effluent of liquid-phase oil-based drill-cuttings using powdered activated carbon. Appl Energy 84(10):1002–1011
57. Wang F, Zou J, Zhu H, Han K, Fan J (2010) Preparation of high effective flocculant for high density waste drilling mud. J Environ Prot 1:179–182
58. Chandan KK, Karmakar GP (2019) Polysaccharides-based flocculants for industrial effluents. In: Encyclopaedia of renewable and sustainable materials. Elsevier Publications, USA, pp 1–13

Condition Monitoring of Mechanical and Electrical Faults in Stationary and Rotating Equipments: A Review

Prabhat Kumar

1 Introduction

The rotor-bearing system is extensively associated with several types of mechanical faults, which include unbalance, misalignment, crack [1], internal damping, rotor-to-stator rub, mechanical looseness, etc. [2]. These faults may exist in the system during its operation or from starting of the manufacturing process or installation period. Other than mechanical faults, there are various electrical faults that appear in the static and dynamic electric equipments such as transformers, generators, induction motors, and electric pumps. The insulation fault, capacitor breakdown, winding issue, phase unbalance fault, power system fault (i.e., open-circuit and short-circuit faults, symmetrical and unsymmetrical faults) come under the category of electrical faults. The stationary and rotating machines may also fail owing to deficient working environment and heavy duty cycles. Failure of machines causes a great loss to factory owners and clients due to less and ineffective production. Moreover, a sudden break of any machine components may also result in hazardous accidents to industry's workers. Therefore, researchers have been proposing different kinds of fault detection and diagnosis techniques to overcome multiple issues by timely detecting the category, magnitude as well as location of the malfunctions. Mouzakitis [3] has classified and briefly explained the fault diagnosis techniques on the basis of model-based, hardware-based, and history-based. The flow chart for the classification is shown in Fig. 1. He has also given an overview on strength and weakness of model-based, hardware-based as well as history-based techniques. Model-based techniques can impart the most accurate results depending upon the complexity in mathematical models and consideration of noise signal and modeling errors [4–6].

P. Kumar (✉)
Department of Mechanical Engineering, NIT Manipur, Imphal West, Manipur 795004, India
e-mail: ysprabhat.pamho@gmail.com; prabhat.kumar@nitmanipur.ac.in

B. P. Swain (ed.), *Recent Advances in Materials*, Springer Proceedings in Materials 25,
https://doi.org/10.1007/978-981-99-3844-5_31

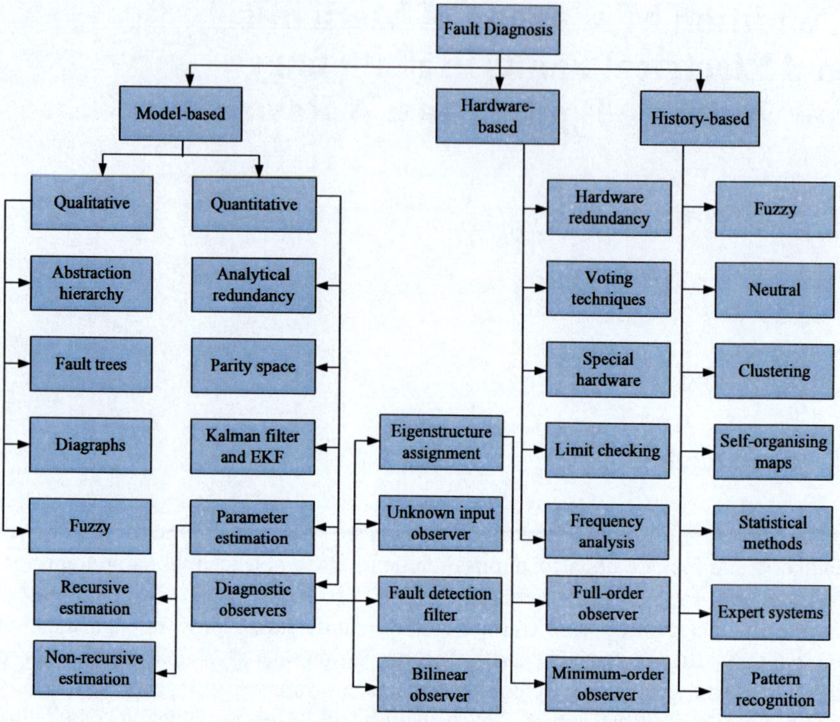

Fig. 1 Classification of fault diagnosis techniques [3]

2 Review on Mechanical Faults

The unbalance in rotor, crack, coupling and bearing misalignment, internal damping, rotor-to-stator rub as well as mechanical looseness comes under the listing of mechanical faults in rotating systems. Unbalance is one of the faults which always exist in the rotating components as it is extremely difficult to have a perfectly balanced system. There is a complete chance of dislocation of geometric center from the center of rotation. Unbalance force is mathematically calculated as product of mass of the component, eccentricity, and square of rotational speed. Therefore, the unbalance force is higher in magnitude at high speed of the rotor [7]. A very long back, Bishop and Gladwell [8] have proposed a high-speed balancing technique for an unbalanced flexible rotor. They developed equations of motion using the concept of Macaulay's method and some boundary conditions. Further, the flexible rotor was balanced without using the trial weights for the two cases, in which the first was the unconstrained shaft and the second was the shaft having two supports at both ends through hydrodynamic bearings [9]. A transfer matrix method was utilized by Shih and Lee [10] to estimate distributed unbalances in flexible shaft and discrete unbalances in rigid discs. A quite simple model consisting of a shaft with disc at

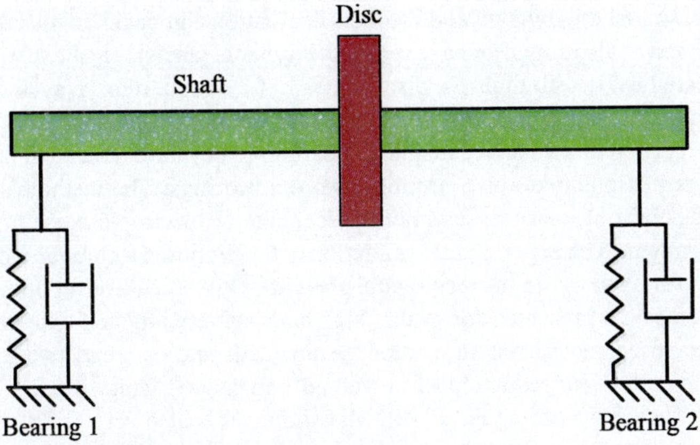

Fig. 2 Rotor system utilized by Shih and Lee [10] for numerical illustration

the middle and supported on flexible bearings (refer Fig. 2) was employed for the numerical illustration purpose. They claimed that the proposed technique can be used in future for acquiring better balancing in real systems. In the year 2000, Edwards et al. [11] developed a technique to estimate unbalance in the rotor from a single run-down. The rotor system was comprised a shaft with two flexible supports and rigid foundation.

Zhou and Shi [13] presented a detail survey on active rotor balancing and vibration control in a rotating system. The vibration control technique utilized the application of a lateral control force to the simple rotor with middle disc. However, they did not include the gyroscopic couple moment. The supported bearings were modeled as isotropic linear spring as well as damper. A low-order system was used to approximate the higher order system, so that the signal-to-noise ratio of the vibration measurement would be lower. For future work, they suggested to establish an active vibration control technique for the cost-effective maintenance and installation of the rotating components. A linear and nonlinear regression methods were also developed for estimating unbalance in a separator model. It was observed that the estimation accuracy was higher for the case of nonlinear regression [14]. Further, Liu et al. [12] have given a review on different types of artificial intelligence techniques for fault diagnosis in rotating machines. They have also provided advantages and limitations of the techniques. A Venn diagram (as shown in Fig. 3) was proposed to explain the relationship among distinct artificial intelligence disciplines. Menshikov applied an inverse problem concept to identify unbalance characteristics in a rotor [15]. The rotor was considered to be deformable and having two flexible supports. He performed experiments and used the rotor vibrational responses in two mutual perpendicular directions for the identification purpose. Three years back only, a numerical (based on finite element method) and experimental study was executed on sudden unbalance in a rotor and transient response arising due to that. The conclusion was made that the magnitude of sudden unbalance should not be too small relative to the initial unbalance, so that the dynamical deflection fluctuation can be distinguished clearly. Moreover, it was also suggested to execute the dynamic analysis and experiments on a complicated rotor with multiple discs instead of a simple rotor. Nayek et al. [16] explored a mathematical model-based estimation of unbalance parameters in a rotor system suspended on two active magnetic bearings. They have done simulation with various types of polluted noise coming from the bearings. The method was found to be even robust against the maximum percentage of electronic noise in the electronic instrument. The experimental validation of the proposed technique would be an interesting future work. In the recent published literature [17], a novel optimization-based scheme was presented for evaluating (both numerically and experimentally) the unbalance characteristics in a rotor system with one disc and two discs. The control application for reduction of unwanted unbalanced fault vibration can be a great work for the upcoming days [18]. Misalignment is also very serious malfunction, so researchers have done a lot of hard work to investigate the dynamic nature and identification of misalignment fault. Misalignment can be in the coupled shaft at the coupling position or between the rotor and the supported bearings (conventional bearings, foil bearings, Active magnetic bearings, etc.). Various papers [19–25] have been published in these misalignment areas. The experimental work needs to be performed for the analysis and identification of the magnetically levitated rotor under the effect of supported AMBs misalignment. Papers are also available in the field of analyzation and estimation of crack [26], bow [27–30], internal damping [31–34], rotor-to-stator rub [35–38], and mechanical looseness [39–42] characteristics in faulty rotor systems.

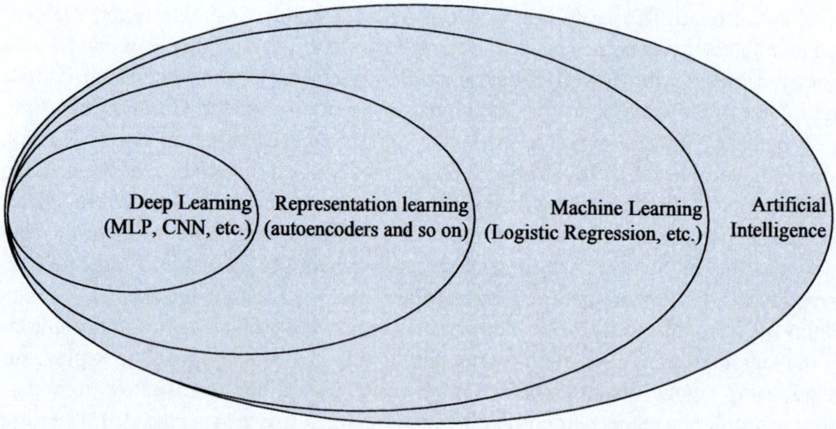

Fig. 3 Venn diagram for relations among artificial intelligence techniques [12]

3 Review on Electrical Faults

In this section, a summary of papers is discussed, which is related to condition monitoring of rotating electrical equipments. A brief survey was presented on monitoring of electrical failures such as core insulation failure; stator winding or insulation failure; rotor winding or insulation failure; brushgear failure; slip ring failure; commutator failure; electrical trip [44, 45]. The root causes for these failures are error in design, manufacturing error, machine installation issue, overspeed, overload, undesirable heating in windings, extreme steady and transient dielectric stresses, etc. At the last, it was suggested to focus on the development of more advanced techniques for health monitoring of electrical machines. Mehrjou et al. [46] discussed different methods such as acoustic emission, air gap torque, stator current-based for fault detections in squirrel cage induction machine. They have also discussed electromagnetic field monitoring, detection based on observation of voltage induced in the stator after induction motor disconnection, and signal processing techniques in time domain, frequency domain as well as time–frequency domain. In future, it is required to design a cost-effective technique for the development of accurate and efficient algorithm to detect faults in sensors and actuators. Literature is found in the field of examining fault characteristics in aerospace systems utilizing model-based techniques [47]. They have discussed mainly three kind of faults, i.e., sensors' fault, process fault, and actuators' fault in aircrafts. Spyropoulos and Mitronikas [48] described various faults in electrical propulsion system for commercial as well as cargo ships. The propulsion system includes prime mover, generator, converter, and motor. The faults existed in the machines were stator faults, broken rotor bar, static and dynamic air gap irregularities, failures in gearbox, short circuits in the rotor field windings, defective permanent magnet in motors. They have also provided characteristic frequencies of the current spectrum for every fault to illustrate their presence and severity. Later, Sharma et al. [43] described the different condition monitoring techniques in induction motor. They have given classification of probable faults in induction machines, which include internal and external faults as depicted in Fig. 4.

Goel et al. [50] presented a methodical survey on several health monitoring schemes for faults in electrical machines through variant temperature, vibro-acoustic signal, ultrasound technique, infrared thermography, and lubrication oil analysis. They have discussed different electric components and types of failure in them. For example, a transformer can fail due to loose connection, insulation defect, lack of oil, and dampness. Similarly, the electric pumps and hydro-turbines can have failure as a result of cavitation and erosion. However, there is a substantial research opportunity in the field of progression in sensing, signal processing as well as artificial intelligence methods. Afterward, a review on existing health monitoring schemes and industrial automation for plant-wide condition monitoring of rotating electrical machines was also presented by Kande et al. [49]. The technology trends, plant life cycle as well as monitoring system evolution was also described by them which included different sections, i.e., planning, commissioning, engineering, operations,

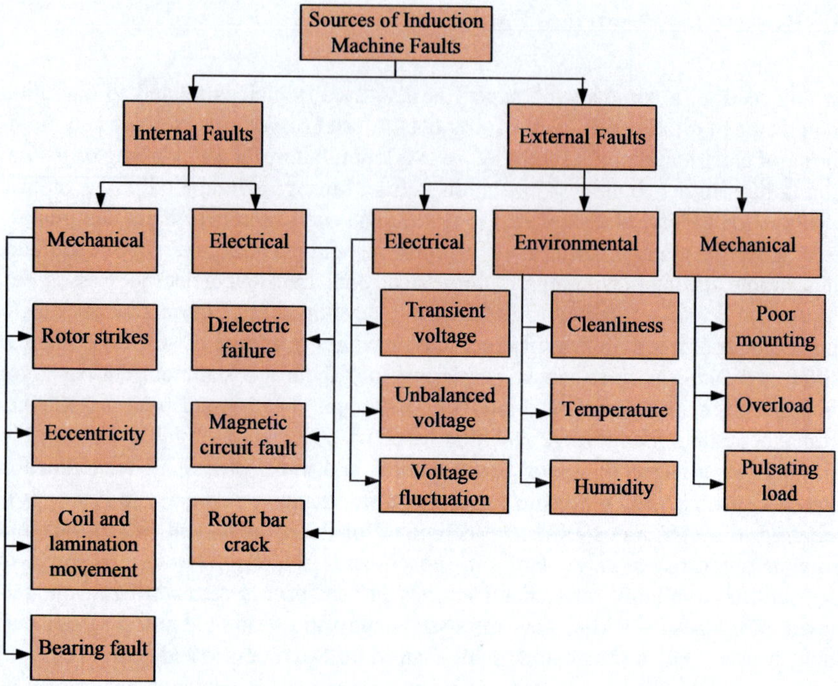

Fig. 4 Different types of faults in induction machine [43]

and maintenance. A flow diagram (refer in Fig. 5) was given, which described the condition-based plant maintenance process.

Dineva et al. [51] utilized multi-label classification method for multiple faults' identification and health monitoring of systems. They have performed experiments on a test rig consisting of three kinds of motors, viz. brushless motor, permanent magnet synchronous motor, double-fed asynchronous motor, coupling and inverter-driven cage induction machine, etc. The time-domain and frequency-domain electrical signals were analyzed to explore the issues of simultaneous presence of noisy unbalance as well as misalignment faults. In the recent publication [52], a detailed review was made which explored various research done in the field of multiplicative faults, compound faults and their identification techniques. For future work, they have given suggestion to perform research on feature extraction techniques for motor fault based on acoustic signals and electrical signature analysis.

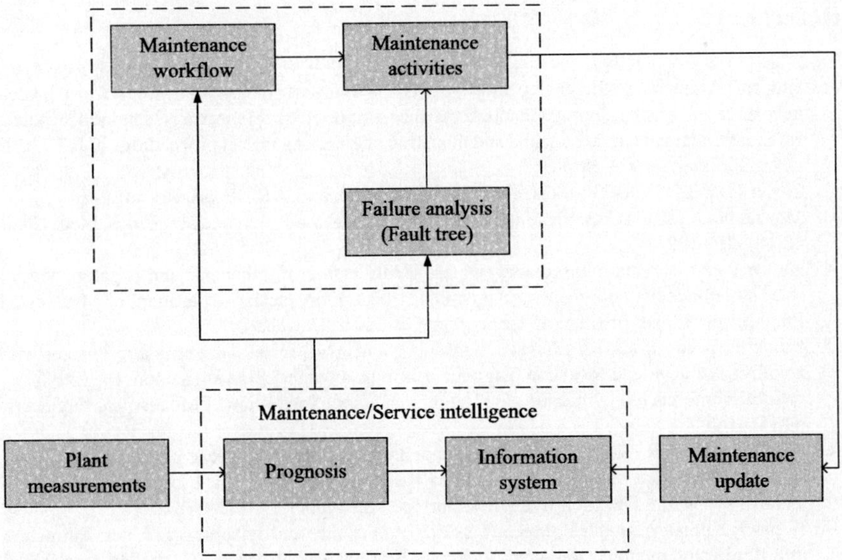

Fig. 5 Flow diagram describing the condition-based plant maintenance [49]

4 Concluding Remarks and Future Directions of Research

In this review paper, a survey on different types of condition monitoring and mechanical as well as electrical faults in rotating and stationary system has been presented by going through various research and review articles. There are some of the following works which have not been done yet and there is great chance of doing them in future:

1. The rotor balancing techniques can be performed in real and complex systems in applications based industries.
2. An active vibration control technique can be established for the cost-effective maintenance and installation of the rotating components.
3. Experimental work can be performed in the field of dynamic analysis and identification of crack and AMBs' misalignment in a practical rotor system.
4. More numerical and experimental investigations can also be done in the area of rotor-to-stator rub and mechanical looseness faults.
5. There is a substantial research opportunity in the field of progression in sensing, signal processing as well as artificial intelligence methods in detection and diagnosis of electrical faults.
6. Research can be executed further on feature extraction techniques for motor fault based on acoustic signals and electrical signature analysis.

References

1. Kumar P, Arambam C, Singh LD, Singh ND (2023) Analysing the dynamic interaction between unbalance and crack responses in a Jeffcott rotor supported by foil bearings: a numerical study. In: Emerging trends in mechanical and industrial engineering: select proceedings of ICETMIE 2022. Springer, pp 971–980
2. Tiwari R (2017) Rotor Systems: analysis and identification. CRC Press, Boca Raton
3. Mouzakitis A (2013) Classification of fault diagnosis methods for control systems. Meas Contr 46(10):303–308
4. Kumar P (2021) Model based analysis and identification of unbalance and misalignment in rotor systems levitated by active magnetic bearings. PhD, thesis, Department of Mechanical Engineering, Indian Institute of Technology Guwahati, Guwahati
5. Kumar P, Tiwari R (2020) Effects of unbalance and AMB misalignment in a rigid rotor with an offset disc levitated by active magnetic bearings: a numerical investigation. In: 12th International conference on vibrations in rotating machinery. Institution of Mechanical Engineers, pp 151–168
6. Kumar P, Tiwari R (2021) Dynamic response analysis of an unbalanced and misaligned rotor supported on active magnetic bearings and touchdown bearings. In: Proceedings of the 6th national symposium on rotor dynamics. Springer, Singapore, pp 407–418
7. Kumar P, Kumar V, Kumar K, Meena LS (2020) Unbalance and dynamic parameters estimation in a rigid rotor mounted on active magnetic bearings. In: Advances in applied mechanical engineering. Springer, Singapore, pp 363–371
8. Bishop RED, Gladwell GML (1959) The vibration and balancing of an unbalanced flexible rotor. J Mech Eng Sci 1(1):66–77
9. Morton PG (1985) Modal balancing of flexible shafts without trial weights. Proc Inst Mech Eng C J Mech Eng Sci 199(1):71–78
10. Shih Y-P, Lee A-C (1997) Identification of the unbalance distribution in flexible rotors. Int J Mech Sci 39(7):841–857
11. Edwards S, Lees AW, Friswell MI (2000) Estimating rotor unbalance from a single run-down. In: IMechE conference transactions, vol 6. Professional Engineering Publishing, pp 323–334
12. Liu R, Yang B, Zio E, Chen X (2018) Artificial intelligence for fault diagnosis of rotating machinery: a review. Mech Syst Signal Process 108:33–47
13. Zhou S, Shi J (2001) Active balancing and vibration control of rotating machinery: a survey. Shock Vib Digest 33(4):361–371
14. Nauclér P, Söderström T (2010) Unbalance estimation using linear and nonlinear regression. Automatica 46(11):1752–1761
15. Menshikov Y (2013) Identification of rotor unbalance as inverse problem of measurement. Adv Pure Math 3:20–25
16. Nayek B, Das AS, Dutt JK (2021) Model based estimation of inertial parameters of a rigid rotor having dynamic unbalance on active magnetic bearings in presence of noise. Appl Math Model 97:701–720
17. Abbasi A, Firouzi B, Sendur P, Ranjan G, Tiwari R (2022) Identification of unbalance characteristics of rotating machinery using a novel optimization-based methodology Soft Comput 26:4831–4862
18. Rachananjali K, Pavani N, Suman S, Chaitanya DVSB (2014) Damping of subsynchronous resonance using SSSC with hysteresis current control. In: 2014 International conference on green computing communication and electrical engineering (ICGCCEE). IEEE, pp 1–5
19. Gibbons CB (1976) Coupling misalignment forces. In: Proceedings of the 5th Turbomachinery Symposium, pp 111–116
20. Hori Y, Uematsu R (1980) Influence of misalignment of support journal bearings on stability of a multi-rotor system. Tribol Int 13(5):249–252
21. Lal M, Tiwari R (2018) Experimental identification of shaft misalignment in a turbo-generator system. Sādhanā 43(80):1–20

22. Kumar P, Tiwari R (2020) Development of a novel approach for quantitative estimation of rotor unbalance and misalignment in a rotor system levitated by active magnetic bearings. Iran J Sci Technol-Trans Mech Eng 45:769–786
23. Kumar P, Tiwari R (2021) Dynamic analysis and identification of unbalance and misalignment in a rigid rotor with two offset discs levitated by active magnetic bearings: a novel trial misalignment approach. Propul Power Res 10(1):58–82
24. Kumar P, Tiwari R (2021) Finite element modelling, analysis and identification using novel trial misalignment approach in an unbalanced and misaligned flexible rotor system levitated by active magnetic bearings. Mech Syst Signal Process 152:107454
25. Tiwari R, Kumar P (2022) An innovative virtual trial misalignment approach for identification of unbalance, sensor and active magnetic bearing misalignment along with its stiffness parameters in a magnetically levitated flexible rotor system. Mech Syst Signal Process 167(A):108540
26. Sekhar AS (2008) Multiple cracks effects and identification. Mech Syst Signal Process 22(4):845–878
27. Rao JS, Sharma M (2000) Dynamic analysis of bowed rotors. In: Proceedings of VETOMAC-1 conference, Bangalore, India. Adv Vib Eng 2(2):1–14
28. Pennacchi P, Vania A (2004) Accuracy in the identification of a generator thermal bow. J Sound Vib 274(1–2):273–295
29. Song GF, Yang ZJ, Ji C, Wang FP (2013) Theoretical–experimental study on a rotor with a residual shaft bow. Mech Mach Theory 63:50–58
30. Shin D, Suh J, Palazzolo AB (2022) Parametric study of flexure pivot bearing induced thermal bow-rotor instability (Morton effect). J Tribol 144(7):071801-1–071801-15
31. Ehrich FF (1964) Shaft whirl induced by rotor internal damping. J Appl Mech 31(2):279–282
32. Ku D-M (1998) Finite element analysis of whirl speeds for rotor-bearing systems with internal damping. Mech Syst Signal Process 12(5):599–610
33. Vatta F, Vigliani A (2008) Internal damping in rotating shafts. Mech Mach Theory 43(11):1376–1384
34. Wang L, Wang A, Jin M, Yin Y, Heng X, Ma P (2021) Nonlinear dynamic response and stability of a rod fastening rotor with internal damping effect. Arch Appl Mech 91:3851–3867
35. Smalley AJ (1989) The dynamic response of rotors to rubs during startup. J Vib Acoust Stress Reliab 111(3):226–233
36. Abuzaid MA, Eleshaky ME, Zedan MG (2009) Effect of partial rotor-to-stator rub on shaft vibration. J Mech Sci Technol 23:170–182
37. Liang H, Lu H, Feng K, Liu Y, Li J, Meng L (2021) Application of the improved NOFRFs weighted contribution rate based on KL divergence to rotor rub-impact. Nonlinear Dyn 104:3937–3954
38. Zhang X, Yang Y, Shi M, Zhang Y, Wang P (2022) An energy track method for early-stage rub-impact fault investigation of rotor system. J Sound Vib 516:116545
39. Chu F, Tang Y (2001) Stability and non-linear responses of a rotor-bearing system with pedestal looseness. J Sound Vib 241(5):879–893
40. Wu TY, Chung YL, Liu CH (2010) Looseness diagnosis of rotating machinery via vibration analysis through Hilbert–Huang transform approach. J Vib Acoust 132(3):031005
41. Wang H, Guan X, Chen G, Gong J, Yu L, Yuan S, Zhu Z (2019) Characteristics analysis of rotor-rolling bearing coupled system with fit looseness fault and its verification. J Mech Sci Technol 33:29–40
42. Zhang H, Lu K, Zhang W, Fu C (2022) Investigation on dynamic behaviors of rotor system with looseness and nonlinear supporting. Mech Syst Signal Process 166:108400
43. Sharma A, Chatterji S, Mathew L, Khan MJ (2015) A review of fault diagnostic and monitoring schemes of induction motors. Int J Res Appl Sci Eng Technol (IJRASET) 3(IV):1145–1152
44. Tavner P, Ran L, Penman J, Sedding H () Condition monitoring of rotating electrical machines, vol 56. IET, Power and Energy Series
45. Tavner PJ (2008) Review of condition monitoring of rotating electrical machines. IET Electr Power Appl 2(4):215–247

46. Mehrjou MR, Mariun N, Marhaban MH, Misron N (2011) Rotor fault condition monitoring techniques for squirrel-cage induction machine—a review. Mech Syst Signal Process 25(8):2827–2848
47. Marzat J, Piet-Lahanier H, Damongeot F, Walter E (2012) Model-based fault diagnosis for aerospace systems: a survey. Proc Inst Mech Eng G J Aerosp Eng 226(10):1329–1360
48. Spyropoulos DV, Mitronikas ED (2013) A review on the faults of electric machines used in electric ships. Adv Power Electron 2013(216870):1–8
49. Kande M, Isaksson AJ, Thottappillil R, Taylor N (2017) Rotating electrical machine condition monitoring automation—a review. Machines 5(24):1–15
50. Goel S, Ghosh R, Kumar S, Akula A (2014) A methodical review of condition monitoring techniques for electrical equipment. In: National seminar and exhibition on non-destructive evaluation (NDE 2014). NDE-India
51. Dineva A, Mosavi A, Gyimesi M, Vajda I, Nabipour N, Rabczuk T (2019) Fault diagnosis of rotating electrical machines using multi-label classification. Appl Sci 9(5086):1–18
52. Kumar P, Tiwari R (2023) A review: multiplicative faults and model-based condition monitoring strategies for fault diagnosis in rotary machines. J Braz Soc Mech Sci Eng 45(5):282. https://doi.org/10.1007/s40430-023-04203-z

Autonomous Robotic Underwater Welding—A Review

Milan Kumar Maity, Saurav Suman, and Pankaj Biswas

1 Introduction

Welding is a century-old technique for joining metals; evidence of welding was also found in the Bronze Age. And the advancement of welding throughout the century has made it an efficient and affordable method of combining metals. Practically, in every manufacturing industry and for structural applications, either on land or in water, welding methods have grown importance [9]. Underwater welding (UWW) is a type of welding carried out in an aqueous environment. Even though there are various methods available for welding in the atmosphere, many of them cannot be used in marine and offshore applications because the presence of water poses a serious threat. Underwater welding is classified into two main types—dry underwater welding and wet underwater welding. The cost and time required for repairing in dry welding conditions are twice as compared to wet welding conditions [7]. Therefore, it is crucial to develop on-site repair methods to bring the damaged offshore equipment back to full functionality.

The underwater welding technique is widely considered to be the most important technique for repairing submerged heavy structures, which are either required to be welded in an aquatic environment or damaged equipment that cannot be moved into a dry dockyard. It is also used to maintain and repair underwater oil transmission pipelines, offshore oil platforms, ship components, port construction, and nuclear power plant structures. As marine equipment's is highly prone to develop surface damage, including water corrosion, abrasion, and surface fatigue cracks due to unpredicted failures.

M. K. Maity · S. Suman (✉)
Department of Mechanical Engineering, NIT Mizoram, Aizawl, India
e-mail: saurav.suman@nitmz.ac.in

P. Biswas
Department of Mechanical Engineering, IIT Guwahati, Assam, India

B. P. Swain (ed.), *Recent Advances in Materials*, Springer Proceedings in Materials 25,
https://doi.org/10.1007/978-981-99-3844-5_32

Different types of underwater welding have been developed, and their applications, advantages, and limitations have been discussed by many authors from countries like India, USA, Germany, Japan, and China for the past forty years [7, 13, 23]. The main challenges that are found in underwater welding are the higher pressure created by the water head under which the welding occurs, the cooling effects of water on weld metal which could alter its metallurgical properties, the potential for hydrogen and oxygen arc mixtures to form in pockets which might cause an explosion, and common risk of divers having electric shock and diffusion of nitrogen in their blood in the dangerous amount [13]. Thus, robotic help to human cognition is necessary to function effectively in such a demanding environment with all human limitations. The effective use of robots in manufacturing and space industries is a crucial illustration of why underwater welding can only be done with robotic help.

In the modern era, the teleportation of unmanned production and drilling platforms, Autonomous Underwater Vehicle (AUV), Remotely Operated Vehicles (ROVs), Robotic underwater inspection and welding and underwater manipulators have all made it easier for offshore rigs to move from shallow depth to ultra-depth water. The majority of robotics technologies in use today in marine and offshore applications primarily focus on inspecting, maintaining, and repairing (IMR) plant facilities more frequently and accurately. Using fully autonomous robots is still a far-off option when taking into account the risks associated with these operations, especially before obtaining very high reliability. As a result, semi-autonomous robots are a great alternative for this industry as a solution for the foreseeable future, where activities are carried out by robots, but cognitive judgements are still made by skilled operators [20]. Recent developments in robotic underwater welding are discussed in this paper. Also, a thorough survey of current, commercial, and prototype manipulators used for the underwater environment, including essential information such as design elements, sensors, their merits and capabilities and making detailed assessments, are discussed in the following sections.

2 Evolution of Robots in Underwater Applications

The ocean, which makes up roughly two-thirds of the planet, significantly impacts how long humans will continue to exist. In the world, 37% of people reside within 100 km of the ocean [6]. The oceans play a significant part in our daily lives since they have long served as a means of transporting products and providing food resources. We typically pay more attention to the land and the atmosphere than the water. Thus, the vast ocean's living and non-living resources have yet to be fully explored by us. For example, after the discovery of petroleum oil in 1859, first offshore drilling was done in 1897 to extract crude oil from shallow seawater after 38 years [28]. Underwater robots can effectively use ocean resources for human well-being, safeguard them from contamination, and aid us in understanding marine and other environmental challenges. However, due to health, safety, and environment (HSE) concern, human travel into the deep ocean is still challenging.

Perhaps Aristotle first came up with the concept of an underwater machine; he made the *skape Andros* (boat-man) that allowed Alexander the Great of Macedon (356–323 b.c.) to stay underwater for at least twelve hours during the Trio war in 325 b.c. It may sound unrealistic, as Archimedes's law had yet to be formulated (around 250 b.c.) [2]. The majority of early scientific work done for the underwater machine was manned underwater submersibles, but these manned robots made by companies like Rockwell, General Dynamics and Westinghouse eventually saturated.

Dimitri Rebikoff developed the first tethered Remotely Operated Vehicle (ROV) in 1953, known as PODDLE. An ROV is an underwater vehicle that is tethered via an umbilical cable to an operator who may be on a ship, on land or in a submarine. Later in the 1980s, many commercial ROVs came into the offshore industry to take over manned submersible jobs. However, because of high operational cost, its effectiveness decreases with depth, difficulties in handling long umbilical cables and operators fatigue, usage of ROVs is now restricted to some specific applications like site survey, drilling assistance, inspection, maintenance, and repair (IMR) [19, 26]. A list of ROVs used for scientific exploration is given in Table 1 and Fig. 1 [1].

As a result of the restrictions placed on ROVs, Untethered Underwater Vehicles (UUV) or Autonomous Underwater Vehicles (AUV) began to appear in industries in the 1970s with the help of advanced computer processors because of the semiconductor revolution. AUVs are entirely autonomous and rely on their onboard intelligence and power source because underwater communication bandwidth is minimal. Currently, AUVs are used for under-ice surveying, underwater archaeology, oceanographic sampling, and scientific surveying. Additionally, AUVs are currently used in military applications like mine detection and landing site surveys, while more

Table 1 ROVs used for scientific exploration [5]

Manufacturer	Vehicle	Depth (m)
Japan Marine Science and Technology Centre (JMSTEC)	Dolphin-3 K	3300
	Hyperdolphin	4500
	Kaiko Mk-IV	4500
	Kaiko 7000II	7000
	Kaiko	11,000
Woods Hole Oceanographic Institution (WHOI)	Jason	6500

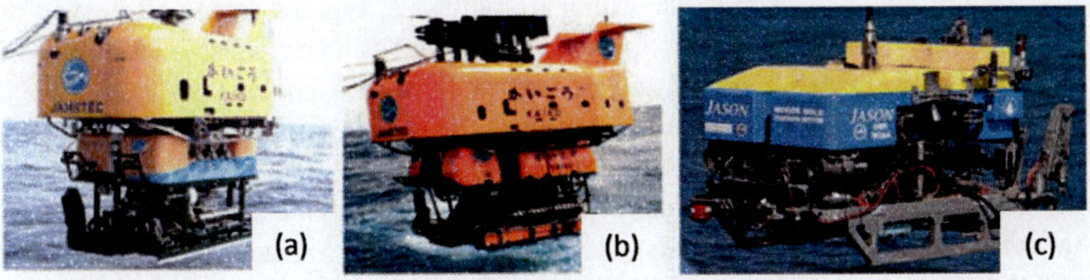

Fig. 1 Some commercial ROVs—**a** Kaiko 7000II, **b** Kaiko, **c** Jason [22, 33]

challenging applications like long-term deep-sea surveillance are being developed by engineers [1]. Currently, about 100 experimental AUVs are available made by research institutions and military organizations. Recently, commercial companies from several countries are offering commercial AUVs for well-defined, specific tasks [26]. ROVs have some rotating parts, such as propellers, thrusters, and turbines, which help to generate the necessary thrust and maneuverability to navigate and operate in underwater environment. Active Magnetic Bearings (AMBs) are used in these rotating parts now a days rather than conventional mechanical bearings [10, 24].

3 Underwater Mechanical Manipulators

A robotic arm, also known as a manipulator, is an instrument used to carry out undersea intervention (interaction with the environment) operations like inspection, maintenance, and repair (IMR). Unmanned underwater vehicles (UUV) like ROVs are coupled with one or two arms; this combined setup is also known as Underwater Vehicle Manipulator System (UVMS) [2]. Whereas the majority of AUVs don't have manipulators, so they are restricted to applications like surveys [26]. As the manipulators are attached to constantly moving vehicles, and thus it becomes very difficult to accurately operate the underwater manipulators. Most of the manipulators coupled with UUVs available in the market are anthropomorphic (i.e. made to look like a human arm). These robotic arms consist of a series of rigid links connected by revolute joints (binary joint or more) with stepper motors to get desired angular displacement, and at the end-effector, some interchangeable tools and grippers are attached. Also, to observe the surroundings of UUVs, one or more numbers of spotlights and cameras are installed either on the vehicle or on the manipulator [21].

Subsea activities like marine civil construction industries, marine renewable energy (MRE) sector and marine military and civil applications are the core application areas of underwater manipulators. Underwater manipulators are designed for a variety of applications like inspection of submerged pipes, underwater welding, mine disposal, subsea bed cleaning, valve controlling, cable laying, rope cutting, biological and geological sample collection, lifting heavy objects, and so on [2, 21]. Manipulators are equipped with actuators, grippers, vacuum cups, cameras, and different sensors to execute these kinds of jobs. Fifty different kinds of manipulators made by different manufacturers are listed in [21].

Manipulators are designed according to their size, type of actuator (electric or hydraulic), a number of kinematic links, wrist torque, lifting capacity, gripping force, maximum working depth, power source, material, sensors, and control system [21]. Most of the manipulators are made of titanium alloys (Ti 6–4), stainless steel alloys, and anodized aluminium alloys as they are corrosion-resistant materials and have high strength to weight ratio. The weight of the manipulator is more significant as it helps the base vehicle to stay underwater. Earlier in the 1990s, seawater-driven actuators are used for underwater manipulators as they have low viscosity, no environmental impact, and high-power density. But, this type of actuator is extenuated because of corrosion, sealing problem, and a very low range of working temperature. Nowadays, most of the manipulators are driven by either hydraulic power or electric power, as they have certain advantages and limitations, one over the other [5]. Apart from these, manipulators with control systems like rate control (used for the hydraulic manipulators which don't have position sensors), position control (by sensors like digital optical encoder, potentiometers), force feedback sensors (to simplify subsea operations), gripper control, manipulator motion control (like decentralized control with feedback mechanisms, centralized control, and neural and fuzzy control) are briefly discussed by [21].

Extreme climatic conditions always leave submerged welded structures susceptible to a variety of defects, including fatigue and corrosion. Multiple non-destructive testing (NDT) sensors are installed on ROVs to monitor those underwater and floating structures, pipes, and tanks continuously. Underwater inspection mechanisms fall primarily into three basic categories: close visual inspection by CCTV, general visual inspection, and NDT-based inspections [8]. But inspection operation faces various difficulties because of the environment, so researchers have made multiple NDT sensors for inspection, like AC current field measurement (ACFM) system, to mark out underwater weld cracks [19]. And it is sometimes necessary to weld in order to repair the damaged underwater structures but doing so underwater becomes the most difficult operations for human divers. As a solution, researchers have made ROVs coupled with a robotic arm for holding welding tools to the required weld zone, along with various sensors for constant monitoring. UUVMs used in marine industries generally have two manipulators, one to anchor the UUVM onto the underwater structure for stability, the other for testing and maintenance. Some commercially available underwater manipulators are listed in Table 2 and Fig. 2.

Table 2 Commercially available underwater manipulators [21]

Manufacturer	Manipulators	Specifications
TechnipFMC [30]	Titan 4	4–7 functions, Position/ Rate control, Hydraulic powered, 4000–7000 m sea water (msw) Depth rating, can lift 68–250 kg at full reach
	Atlas 7P and 7R	
	Conan 7P	
	RigMaster	
	Orion 7P, 7R or 4R	
ROV Innovations [32]	Arm 5E, micro and mini	5–7 functions, up to 6000 m depth resistance, maximum reach up to 1790 mm,
	Arm 7E, mini	Lift up to 40 kg at full extension
Hydro-Lek [29]	HLK-4000, 4200, 43,000	4 to 6 functions, dry weight from 8 to 28 kg, up to 6000 m depth resistance, maximum reach up to 1428 mm
	HLK-5600, 5680, 5300,	

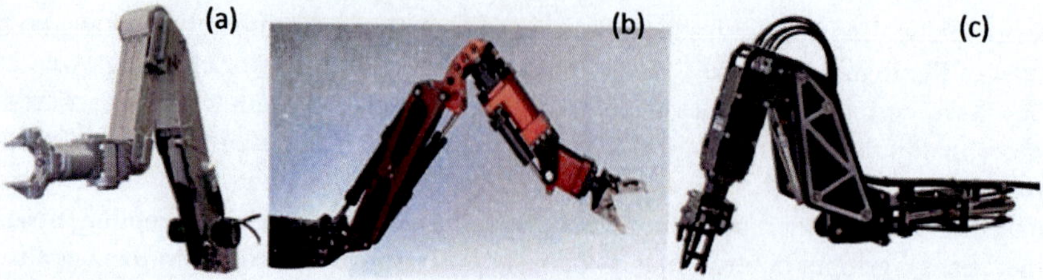

Fig. 2 Commercial subsea manipulators **a** Titan 4, **b** Arm 7E mini, **c** HLK 5600 [29, 30, 32]

4 Underwater Robotic Welding

When a Remotely Operated Underwater Vehicle (ROV) is outfitted with a welding manipulator to perform welding activities at depths underwater, the technique is known as Underwater Robotic Welding. Different types of conventional underwater welding like SMAW, FCAW, TIG welding, and some advanced underwater welding like friction welding (FRW), laser welding are discussed in [13]. The main risk that is associated with manual underwater welding is to the diver of electric shock. Therefore, precautions like electrical insulation of equipment's, limiting the open circuit voltage must be taken. As a result, researchers are now focusing more on robotic underwater welding rather than manual. Another advantage of using underwater robotic welding is that it allows for the repair and maintenance of underwater structures without the need for divers, which can be dangerous and time-consuming. Additionally, it can also be used to construct new underwater structures, such as offshore wind turbines. Some robotic underwater welding done by some researchers and industries are discussed in Table 3.

Table 3 Chronological synopsis of underwater robotic welding

Year	Welding method	Findings
1989 [3]	Hyperbaric MAG welding	Subsea hyperbaric condition emulated in a pressurized chamber. Two 5-axis and 6-axis robots used to conduct MAG welding in He, O_2, H_2 atmosphere between a range of operating pressures of 1 to 110 bar. The tests showed good welds were produced that adhere to the standards
1992 [18]	Hyperbaric orbital Gas Tungsten Arc Welding (GTAW) and Gas Metal Arc Welding (GMAW)	Orbital GTAW module used for hot passes and GMAW module used for high deposition of filler layer. The welding system tested under 500 msw hyperbaric condition. A driverless repair station (underwater habitat) conceptualized. Various subsystems, operational procedures discussed to actualize the welding module

(continued)

Table 3 (continued)

Year	Welding method	Findings
2001 [15]	Friction Hydro Pillar Processing (FHPP)	European Commission funded project—Affordable Underwater Robotic Welding Repair System (ROBHAZ) (1997–2000). Specially designed multipurpose mobile friction welding head (HMS 3000) mounted on a 6-axis robot (TRICEPT) tested in dry and wet condition. Tool changing system for different machining operations used to successfully achieve the objectives
2001 [16]	Hyperbaric Plasma Transferred Arc (PTA) welding and local dry PTA welding using Water Exclusion Device (WED)	Hyperbaric PTA welding done using PTA torch mounted on a 3-axis manipulator to simulate the hyperbaric welding condition at 30 m depth. Multi-axis UW manipulator attached with newly developed WED and torch to conduct local dry welding developed, that can be used in underwater Boiling Water Reactor (BWR) repairing sector
2004 [14]	YAG-laser welding	Remote UW YAG-laser repair-welding robot developed. L-type laser welding torch used to supply laser beam and shielding gas coaxially. UW fillet weld of 2 mm throat thickness done. Liquid penetrant, macroscopic, tensile, and bend test conducted on weld bead to check the properties
2014 [12]	NA	3D model of and UW welding robot having 8 thrusters designed that would be used in Spent Fuel Pool (SFP). Robot dynamic model and fuzzy PID controller simulated in MATLAB. Later Physical model made by Harbin Institute of Technology, China in 2016
2015 [4]	Hyperbaric MIG welding	Statoil developed the welding machine for repairing large diameter subsea pipelines (>30 inch) up to 1300 m depth. The welding system has three modules—Remote welding habitat, welding Power and Control (POCO) and remote welding tool. Sleeve joining of pipes can only be done
2016 [17]	Laser Welding	A remotely operated underwater laser welding vehicle with two-link robotic arm is designed to repair pipelines and ships outer surface
2016 [27]	Local dry arc welding	ROV based local dry arc welding robot patent filed by Harbin Engineering University, China. Welding arm is installed on ROV. Sonar imaging sensor, light, camera system, pipeline fixing claw are attached with the ROV
2017 [25]	Local dry arc welding	Specially designed welding torch mounted on LBBBD robot for automatic seam tracking and welding. Pulsed current power supply and wire feeding mechanism are also developed for the torch to get stable arc
2018 [11]	Ultrasonic assisted underwater wet welding process (U-FCAW)	Tethered underwater welding robot developed to weld walls of SFP with high accuracy. The robot consists of mobile system, UW welding system and control system. 3-DOF mobile platform used to weld the crack with an accuracy up to 0.1 mm
2021 -2024 [31]	Robotic Friction Stir Welding (FSW)	RESUGRAM project is now in progress to develop an FSW robot that can be used in underwater as well as under-oil to join steel. This AI-enabled robot will be able to conduct inspection and perform FSW in restricted spaces

5 Conclusion

Using the advanced underwater engineering technologies, it is possible to robotically weld underwater up to depth of 2000 m in coming 30 years. As due to diver safety, manual underwater welding is only permitted up to the depth of 180 m. MIG welding and laser welding are getting popularized now a days which can be done with the help of a robotic manipulator controlled from a remote location. These two robotic welding processes have various advantages over the other depending upon the type of robot, depth of water, and field of application. Like robotic laser welding mainly used in nuclear power plants and local dry arc welding is used in subsea pipe repairing, ship, vessel, and underwater structure surface crack welding. However, there are still challenges that need to be addressed to further improve and expand its use. This review paper provides an overview of the current state of the art in underwater robotic welding and its applications, and highlights the potential for future developments in this field.

6 Future Scope for Research

The future of underwater robotic welding is promising, with increasing demand for advanced and efficient welding solutions in various industries. As technology continues to advance, it is likely that underwater robotic welding will become more widely used and its applications will expand. Some areas where the future scope of underwater robotic welding may be promising include:

- Development of new and advanced welding techniques.
- Improving safety and efficiency.
- Improved cost-effectiveness.
- Inspection of weld bead.
- Development of new nozzle for local dry welding.

References

1. Antonelli G et al (2008) Underwater R 43. Springer Handb. Robot. 987–1008
2. Antonelli G (2018) Underwater robots. Springer Tracts in Advanced Robotics. Springer International Publishing AG
3. Aust ER et al (1989) Subsea work with robots in hyperbaric environments. In: Proceedings of the annual offshore technology conference, pp 437–442
4. Berge JO et al (2015) Welding robot repairing subsea pipelines. Proc Annu Offshore Technol Conf 5:3599–3610. https://doi.org/10.4043/25969-MS
5. Capocci R et al (2017) Inspection-class remotely operated vehicles—a review. J Mar Sci Eng 5(1):13. https://doi.org/10.3390/jmse5010013

6. Cohen JE et al (1997) Estimates of coastal populations. Science (80-.). 278(5341):1209–1213. https://doi.org/10.1126/science.278.5341.1209c
7. Grubbs CE, Reynolds TJ (1998) Underwater welding: seeking high quality at greater depths. Weld J (Miami, Fla). 77(9):35–39
8. Hedayati MR et al (2010) Intelligent ship hull inspection and NDT using ROV based flux leakage expert system. In: Proceedings on 2nd international conference on computational intelligence, modelling and simulation, CIMSim 2010, pp 412–415. https://doi.org/10.1109/CIMSIM.2010.68
9. Khanna OP (2013) A textbook of welding technology
10. Kumar P, Tiwari R (2021) Finite element modelling, analysis and identification using novel trial misalignment approach in an unbalanced and misaligned flexible rotor system levitated by active magnetic bearings. Mech Syst Signal Process 152:107454. https://doi.org/10.1016/J.YMSSP.2020.107454
11. Luo Y et al (2018) A new underwater robot for crack welding in nuclear power plants. In: 2018 IEEE international conference on robotics and biomimetics (ROBIO) 2018, pp 77–82 (2018). https://doi.org/10.1109/ROBIO.2018.8665279
12. Lv X et al (2014) Design of underwater welding robot used in nuclear plant. Key Eng Mater 620:484–489. https://doi.org/10.4028/www.scientific.net/KEM.620.484
13. Majumdar JD (1970) Underwater welding—present status and future scope. J Nav Archit Mar Eng 3(1):38–47. https://doi.org/10.3329/jname.v3i1.927
14. Makihara Y et al (2004) The application of the welding technique at fillet groove by the YAG-laser repair-welding robot for underwater environment. In: Proceedings of the international conference on nuclear engineering (ICONE12). American Society of Mechanical Engineers Digital Collection, pp 149–156. https://doi.org/10.1115/icone12-49394
15. Meyer A et al (2001) Subsea robotic friction-welding-repair system. Presented at the April 30. https://doi.org/10.4043/13250-ms
16. Offer HP et al (2001) Recent developments in underwater repair welding
17. S K (2016) Remote operated underwater welding vehicle/SNAMETOS/proceedings-abstract/TOS16/1-TOS16/3870
18. Dos Santos JF et al (1992) Automatic and diverless underwater welding: new systems and concepts, /ISOPEIOPEC/proceedings-abstract/ISOPE92/All-ISOPE92/21815
19. Shukla A, Karki H (2016) Application of robotics in offshore oil and gas INDUSTRY-A REVIEW Part II. Rob Auton Syst 75:508–524. https://doi.org/10.1016/j.robot.2015.09.013
20. Shukla A, Karki H (2016) Application of robotics in onshore oil and gas industry-a review part i. Rob Auton Syst 75:490–507. https://doi.org/10.1016/j.robot.2015.09.012
21. Sivčev S et al (2018) Underwater manipulators: a review. https://doi.org/10.1016/j.oceaneng.2018.06.018
22. Sugiyama Y (2015) Japan agency for Marine-Earth science and technology (JAMESTEC), https://www.jamstec.go.jp/e/
23. Sun G et al (2022) Underwater laser welding/cladding for high-performance repair of marine metal materials: a review. https://doi.org/10.1186/s10033-021-00674-0
24. Tiwari R, Kumar P (2022) An innovative virtual trial misalignment approach for identification of unbalance, sensor and active magnetic bearing misalignment along with its stiffness parameters in a magnetically levitated flexible rotor system. Mech Syst Signal Process 167:108540. https://doi.org/10.1016/J.YMSSP.2021.108540
25. Wang L et al (2017) Innovative methodology and database for underwater robot repair welding: a technical note. ISIJ Int 57(1):203–205. https://doi.org/10.2355/isijinternational.ISIJINT-2016-407
26. Yuh J, West M (2001) Underwater robotics. Adv Robot 15(5):609–639. https://doi.org/10.1163/156855301317033595
27. CN103785923A—Local dry-method underwater welding robot based on ROV—google patents, https://patents.google.com/patent/CN103785923A/en. Last accessed 01 Feb 2023
28. History of offshore drilling units—PetroWiki, https://petrowiki.spe.org/History_of_offshore_drilling_units. Last accessed 01 Feb 2023

29. Hydro-Lek Remote Handling—Hydraulic Manipulators, http://www.hydro-lek.com/manipulators.php. Last accessed 01 Feb 2023
30. Manipulator systems—TechnipFMC plc, https://www.technipfmc.com/en/what-we-do/subsea/robotics/manipulator-systems/. Last accessed 01 Feb 2023
31. Resurgam|Overview, https://www.resurgamproject.eu/overview.html. Last accessed 01 Feb 2023
32. ROV Innovations|ROV manipulator arms in Australia, New Zealand, and the Pacific—ROV innovations—underwater ROV inspections and surveys in Australia, New Zealand, and the Pacific, http://www.rovinnovations.com/manipulator-arms.html. Last accessed 01 Feb 2023
33. ROV Jason/Medea—woods hole oceanographic institution. https://www.whoi.edu/. Last accessed 01 Feb 2023

Estimation of Life and Wear Rate of Tibial Insert for Total Knee Arthroplasty (TKA) Using Finite Element Method Approach

Ajeesh M. Kurup and Rajesh Kumar Bhushan

1 Introduction

Total knee arthroplasty (TKA) is a surgical method of replacing load-bearing surfaces present in the knee joint to heal pain bearing section and inability areas of the joint. Knee replacement is a surgical technique that has been carried out mostly during cases like osteoarthritis. The condition of infected knee joint components in patients is described in various medical literature [1]. Generally, important reasons for knee prosthesis damage are fatigue fractures of metal components and wear of ultra-high-molecular-weight polyethylene (UHMWPE) components [2, 3]. In-vitro simulations can be a very effective tool for estimating the behavior of newly designed prostheses and materials during the time of wear. Pre-clinical analysis of components is necessary to know the probability of failure and wear rate with time [4]. In real life, it is difficult to get the same flexibility and dynamic movement of knee joints after implantation.

Although the success of TKR is very well, in some patients, there is still some pain that remains during motion. This might be due to some surgical errors, which can further lead to a different biomechanical behavior than what the prosthesis was designed for. [5, 6]. Because of wear and fracture of the tibial insert material, it is obvious that the stability of the knee joint is of limited time. If total knee replacement is done on an elderly patient whose life expectancy is shorter than the life span of that prosthesis, then total knee replacement is a better solution for them but if TKR is done for a youngster then after some time when the prosthesis material starts degrading due to wear then further surgery is required. Therefore it is necessary to estimate the life of knee joint prosthesis, and reasons for the wear rate and the suitable material for a prosthesis with less wear and tear properties.

A. M. Kurup (✉) · R. K. Bhushan
Department of Mechanical Engineering, NIT Manipur, Imphal, India
e-mail: ajeeshmkurup@gmail.com

B. P. Swain (ed.), *Recent Advances in Materials*, Springer Proceedings in Materials 25,
https://doi.org/10.1007/978-981-99-3844-5_33

Table 1 Material properties used for analysis [7, 8]

Material	Density (g/cm^3)	Young's modulus(Pa)	Poisson's ratio	Yield strength(Pa)	Ultimate strength(Pa)
UHMWPE	0.94	6.73E + 08	0.31	2.10E + 07	4.90E + 07
UHMWPE(cross-linked, moderated and stabilized)	0.93	4.63E + 08	0.46	2.20E + 07	5.20E + 07
Ti6Al4V	4.43	1.10E + 11	0.342	8.85E + 07	9.60E + 08
CoCrMo	8.30	2.30E + 11	0.3	6.15E + 07	9.80E + 07

2 Materials and Method

2.1 Materials

Here we have chosen bio-materials; ultra-high-molecular-weight polyethylene (UHMWPE) and highly cross-linked, stabilized UHMWPE for tibial insert and Ti6Al4V and CoCrMo alloy for femoral component for our analysis (Table 1).

2.2 Method

The prosthesis geometry has a significant influence on its performance, so we modeled the prosthesis according to standard procedure. A knee prosthesis comprises a femoral component, tibial component and insert. Insert lies between the femoral component and the tibial component. 3D solid models of knee joint prosthetics were modeled by using average human data (weight 70 kg). 3D models are modeled in CATIA V5R20 after referring to the design standards prescribed by Mallesh and Sanjay, 2012. [9]. Here, we have taken design variables such as femoral distal sagittal radii of curvature of the condyle (FDSR), femoral posterior sagittal radii of curvature of the condyle (FPSR) and distance from the most distal point on condyles to midline in the frontal plane (W) as 40 mm, 15 mm, 15 mm, respectively (Figs. 1 and 2).

2.3 Design Modification and Convergence Test

In finite element meshing, for femur part 10-node quadratic tetrahedron element, and for tibia part 8-node linear brick, reduced integration, hourglass control element type is chosen. Femur and tibia part consist of 8179 elements, 14,746 nodes and 8449 elements, 10,144 nodes, respectively. Fine meshing is necessary for accurate

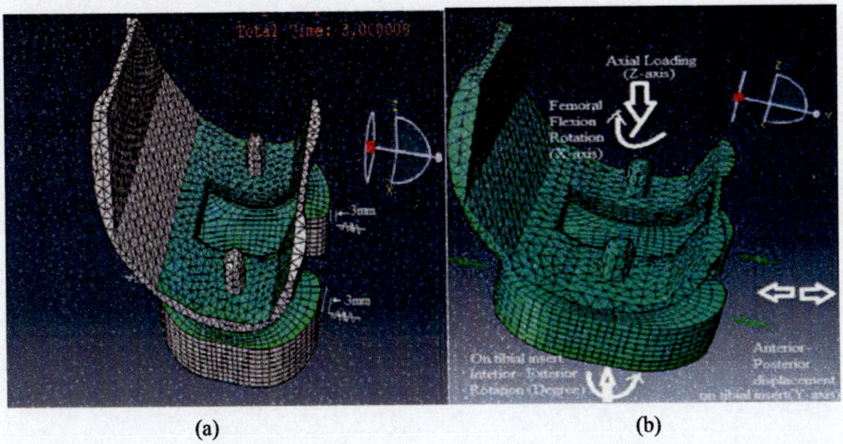

Fig. 1 **a** Assembly of the femur and tibial instances after meshing, **b** Loading and boundary conditions in the simulation model

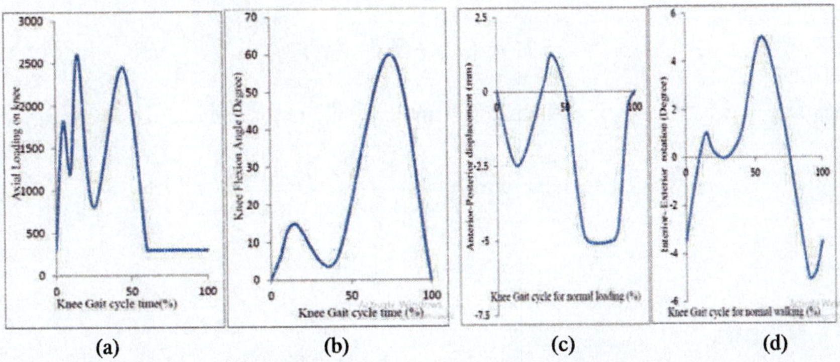

Fig. 2 Inputs variables applied on the femoral part **a** Axial loading (*N*) applied on the knee and **b** Flexion angle applied on the femoral part and Input variables on tibial insert part **c** AP loading (in the y-direction) (in mm) **d** IE rotation on the tibia (Degree) [10]

results of contact pressure so element size was reduced from 3 to 1 mm in steps of 0.1 mm. These steps are very important to know; whether results are correct or not. If variation in results are more than 5% then the obtained results are considered as incorrect. So mesh size of 0.5 mm was taken from this convergence test.

Table 2 Contact pressure at ISO loadings

ISO Gait loading (N)	800	1200	2600	800	2450
Contact pressure (MPa)	17.38	24.09	40.91	17.38	39.74

2.4 *Wear Calculation*

The wear rate calculations are based on a modified form of classic Archard's law. This model was proposed from the classic Archard model, which model was not considering the frictional effect on the wear rate. It is a simple abrasive wear model, and the equation is [11];

$$H = K \times C_p \times S \times \left(1 + 3 \times \mu^2\right)^{0.5} \quad (2.1)$$

where, K = wear coefficient (mm^3/Nmm), S = sliding distance or slip (mm), H = linear wear (mm), C_p = contact pressure (MPa), μ = coefficient of friction.

Here, wear coefficient K is given by

$$K = (a + b \times \mathrm{CSR})^{\frac{1}{c}} \quad (2.2)$$

where a = 8.5173e-65, b = 9.3562e-60 and c = -6.7454, CSR = cross-shear ratio.

$$\text{cross shear ratio} = \frac{\text{work in} \times \text{direction}}{\text{work in both direction}}$$

3 Results

3.1 *Contact Pressure (C_p)*

To calculate the wear rate in prosthesis, analysis of contact pressure is required, so through finite element analysis we have to find out the contact pressure (c_p). The contact pressure induced on tibia insert due to various load is given in Table 2.

3.2 *Wear Depth (H)*

We have calculated the wear rate in prosthesis by using modified wear Archard's equation given in 2.1 and contact pressure given in Table 2. The wear rate is given in Tables 3 and 4.

Table 3 Wear rate by modified Archard law for UHMWPE

Duration (million cycle)	1	1.5	2	2.5	3	3.5	4	4.5	5
Wear rate (mm)	0.0289	0.0450	0.0637	0.0848	0.1083	0.1336	0.1597	0.1875	0.2173

Table 4 Wear rate by modified Archard law for UHMWPE (moderate and stabilized)

Duration (million cycle)	1	1.5	2	2.5	3	3.5	4	4.5	5
Wear rate (mm)	0.0200	0.0306	0.0494	0.0634	0.0789	0.0950	0.1121	0.1304	0.1501

4 Discussion

The simulations under the ISO loading conditions is very useful to predict the behavior of new prosthesis designs and materials during the time of wear and the selection of material for the insert plays a significant role in reducing the wear rate. Here, by numerical analysis the contact pressure is obtained on the tibial part which is shown in Fig. 3 that shows the pressure distribution on meniscus, and by Fig. 3 we can conclude that the contact pressure is more toward the lateral meniscus than the medial meniscus. Modified Archard's law shows the friction effect on wear rate of tibial insert because during motion or movement between two bodies, there is always a frictional term that opposes the movement in between the bodies. As per the Figs. 4 and 5, the wear rate on both materials shows by the modified Archard law theory. Here Figs. 6 and 7 show the comparison of wear rate between the two bio-materials. The wear rate on UHMWPE (moderate and stabilized) material shows the different patterns here, during the first one million cycle wear rate follows the curve shape path, from one million cycles to two million cycles it follows another curve path. Further, from two million cycles to five million cycles, the path of the wear rate is increased. Similarly, UHMWPE material also follows the curve path up to the first one million cycles. Further, from one million cycles to five million cycles, it follows the almost linear curve that has a steeper curve compared with the UHMWPE (moderate and stabilized). After two million cycles, both materials show divergence slope of wear rate. In these studies, the figure shows the wear rate on UHMWPE has more compared with UHMWPE (moderate and stabilized) material.

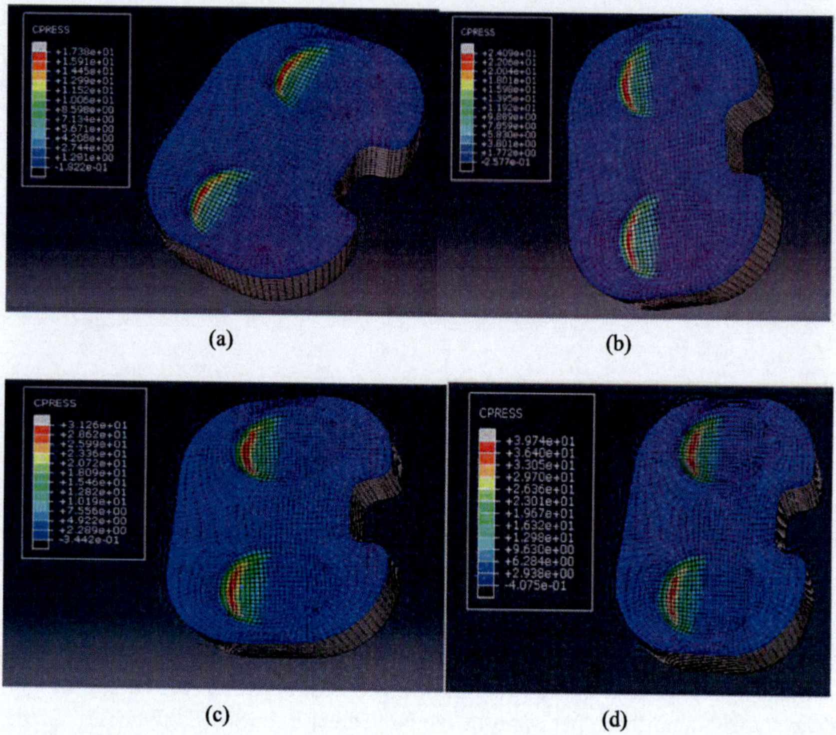

Fig. 3 Contact pressure obtained with different gait loading, **a** 800 loading, **b** 1200 loading, **c** 1800 loading **d** 2450 loading

Fig. 4 Wear rate for UHMWPE

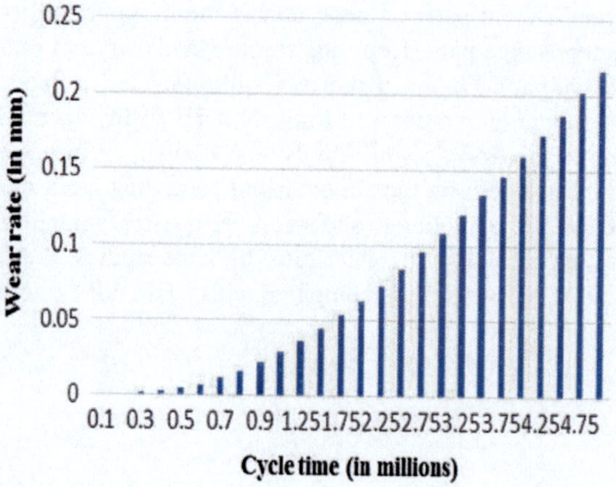

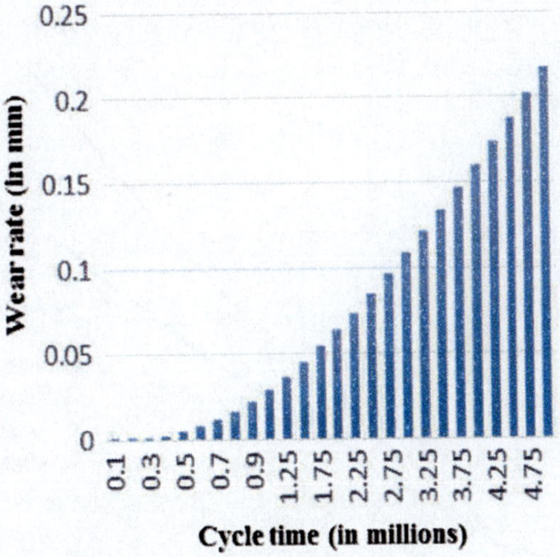

Fig. 5 Wear rate for UHMWPE (modified and stabilized)

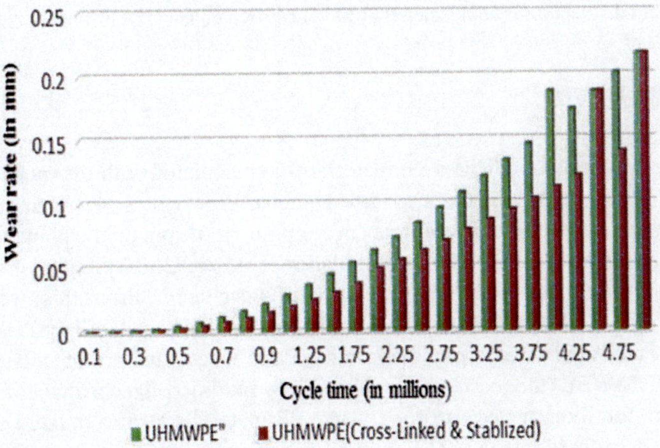

Fig. 6 Wear rate with time

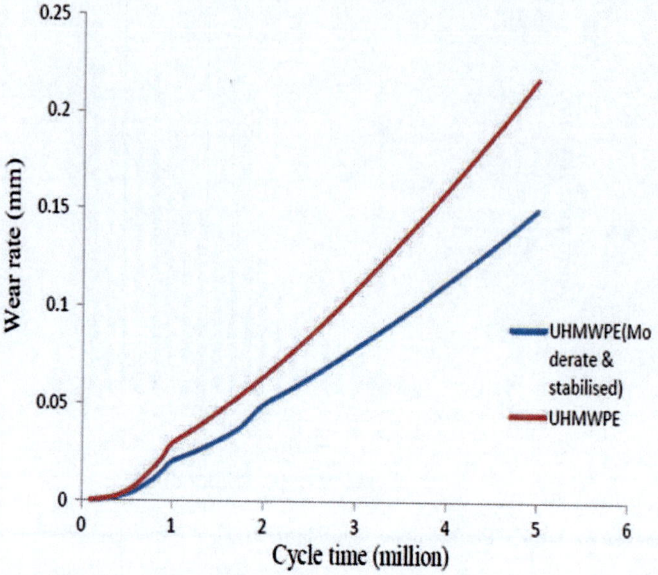

Fig. 7 Wear rate (depth) for both material by modified Archard law

5 Conclusion

The linear wear rate on UHMWPE materials has calculated with the modified wear methods. Wear analysis through modified Archard law needs wear factor or wear coefficient for calculation of wear rate and volume; as per pin on the plane test, basic way to analyze the wear factor is by simulation in ABAQUS/Explicit v6.14 and find out the cross-shear ratio that is the function of wear factor. From this, wear factor for conventional UHMWPE and cross-linked (moderated and stabilized) UHMWPE have been analyzed on our prosthetic model and wear rates are calculated. Wear rates of UHMWPE (moderate and stabilized) by modified Archard wear method is 0.1501 mm/5million cycles and for UHMWPE materials also calculated wear rate by modified Archard law is 0.2173 mm/5 million cycles are observed. After all the analysis, we concluded that UHMWPE materials experience total damage rate that occurs on tibial insert is more compared with UHMWPE material that moderated by sterilization process and by cross-linking and stabilization; material properties are improved, hence it decreases the rate of total damage rate on tibial insert.

References

1. Wright TM, Goodman SB (2001) Implant wear in total joint replacement. American Academy of Orthopaedic Surgeons, Rosemont, USA
2. Abernethy PJ, Robinson CM, Foeler RM (1996) Fracture of the metal tibial tray after kinematic total knee replacement. J Bone Joint Surg 78B:220–225
3. Costa L, Brach Del Prever EM (2000) UHMWPE for arthroplasty. Minerva Medica Torino
4. Ahir SP, Blunn GW, Haider H (1999) Evaluation of a testing method for the fatigue performance of total knee tibial trays. J Biomech 32:1049–1057
5. Wasielewski RC, Galante JO, Leighty RM, Natarajan RN, Rosenberg AG (1994) Wear patterns on retrieved polyethylene tibial inserts and their relationship to technical considerations during total knee arthroplasty. Clin Orthop Relat Res 299:31–43
6. Lewis P, Rorabeck CH, Bourne RB, Devane P (1994) Posteromedial tibial polyethylene failure in total knee replacements. Clin Orthop Relat Res 299:11–17
7. AZoM™.com Pty. Ltd Copyright © 2000–2014 retrieved on 28/02/2014 'Accelerated Ageing and Characterisation of UHMWPE used in Orthopaedic Implants' from http://www.azom.com/properties.aspx?ArticleID=909 (2014)
8. Hosseini S (2012) Biomedical engineering-technical applications in fatigue of Ti-6Al-4V. INTECH Open Access Publisher Ch 3:75–91
9. Mallesh G, Sanjay SJ (2012) Finite element modeling and analysis of rosthetic knee joint. Int J Emerg Technol Adv Eng 2(8):264–269
10. Lundberg HJ, Ngai V, Wimmer MA (2012) Comparison of ISO standard and TKR patient axial force profiles during the stance phase of gait. Proc Inst Mech Eng H 226(3)
11. Hamilton MA, Sucec MC, Fregly BJ, Banks SA, Sawyer WG (2005) Quantifying multidirectional sliding motions in total knee replacements. J Tribol 127(2):280–286

Numerical Analysis on the Performance of Nitrogen Pulsating Heat Pipe

K. Satyanarayana, N. V. S. M. Reddy, P. Rosang, and S. Venugopal

1 Introduction

Conduction cooling cryocoolers are used in many industrial applications to keep equipment at a constant temperature such as MRI scanning systems and semiconductor cooling in transportation. To ensure the continuous cooling effect of equipment (accessories electronic modules), electronic devices are kept in close contact with the cryocooler. The vibrations produced by cryocoolers harm performance and shorten the life of electronic devices. A passive heat exchanging device known as a pulsating heat pipe (PHP) is required to operate as a heat exchanger between the device hotspot and cryocoolers. A PHP is a continuous and long capillary tube bent into turns to form a meandering tube, which is invented by Akachi [1] in 1990. PHP is mainly consisting of evaporator, adiabatic, and condenser sections. Choosing of working fluid is crucial to use the PHP in cryogenic applications which work at a definite range of temperatures [2]. Experimental investigations are carried out by researchers to understand the performance of PHP in different working fluids, such as nitrogen [3] and neon [4].

Computational Fluid Dynamics (CFD) simulations have gained a significant focus in examining PHP performance. Lin et al. [5] used a 2D simulation using water as the working fluid to investigate the flow and thermal performance of PHP and discovered that the oscillation phenomena are mostly caused by pressure differences. Pouryussefi et al. [6] carried out a two-turn water-based PHP numerical simulation. They investigated the chaotic flow of fluid within the PHP using a nonlinear analysis of adiabatic wall temperature. Wang J et al. [7] conducted a CFD analysis at a low heat flux boundary condition. They discovered that hydrophilic surfaces had higher heat resistance than hydrophobic ones. Xie et al. [8] performed a 2D numerical study

K. Satyanarayana (✉) · N. V. S. M. Reddy · P. Rosang · S. Venugopal
Mechanical Engineering Department, National Institute of Technology Nagaland, Dimapur, Nagaland 797103, India
e-mail: myselfsatyakommuri@gmail.com

B. P. Swain (ed.), *Recent Advances in Materials*, Springer Proceedings in Materials 25,
https://doi.org/10.1007/978-981-99-3844-5_34

to analyse the multisource heating impact in PHP with the various flow geometries. They discovered that right-angled elbow PHP performed better thermally than traditional PHP. Satyanarayana et al. [9, 10] described a similar type of work in which they did a 2D numerical analysis using ANSYS FLUENT software to enhance the unidirectional flow by introducing an extra branch (AB) in the PHP. Their findings show that the AB in the evaporator section improves starting time, flow velocity. Furthermore, the results showed that PHP with AB improves heat transport in low-tilt angles. To cool the superconductor magnet at various temperatures, Sagar et al. [11] used a 2D numerical simulation on a six-turn PHP with nitrogen as the working fluid (115, 105, 95, and 85 K). They used an adiabatic time series analysis to identify the dominating frequency and discovered a dominant between 0.25 and 1.85 Hz.

According to the preceding literature, most CFD analysis is available for water-based PHP operating at room temperature. There has been very little work documented on cryogenic PHP functioning at low temperatures. The working fluid used in cryogenic PHPs has different properties from those used in room temperature PHPs, particularly the contact angle, latent heat of evaporation, viscosity, surface tension, and so on. As a result, the available computational and experimental data at ambient temperature cannot predict the performance and fluid flow within the cryogenic PHP. Therefore, a numerical analysis was carried to better understand the thermal behaviour and fluid flow of PHP at cryogenic temperatures using nitrogen as the working fluid.

2 Physical Model

The fluid domain and geometric specifications of the PHP are shown in Fig. 1a. Nitrogen is employed as the working fluid, and a filling ratio of 60% is used in the bottom heating mode. The diameter of the PHP is calculated using the Bond number. At the current operating temperature and fluid properties, the diameter (D) of the channel is set at 1 mm to ensure adequate distribution of liquid slugs and vapour plugs. The PHP has a height of 400 mm, with the condenser, evaporator, and adiabatic lengths of $L_c = 150$, $L_e = 100$, and $L_a = 150$ mm, respectively. The simulations were carried out with a wall temperature of 76 K in the condenser and 115 K in the evaporator and with a constant wall heat flux in the adiabatic region. The quadrilateral mesh of size 0.16 mm was employed in the simulation as shown in Fig. 1b. The entire computational domain is split into two sections, the bottom one representing the liquid nitrogen phase and the top one representing the vapour phase. As shown in Fig. 1c, the bottom area is patched with 1 and the upper section with 0 to distribute the liquid and vapour phases (filling ratio of 60%). To capture the small movement of two-phase fluid, 0.0001 is considered as a time step by satisfying the Courant number 0.25. For all parameters, the residual is set to 0.00001. The simulations are carried out in the ANSYS FLUENT 17.0 software package using the CFD approach, and the VOF model is employed to represent the two-phase fluid flow inside the PHP. The initial flow velocity is set to zero. Transient time conditions are

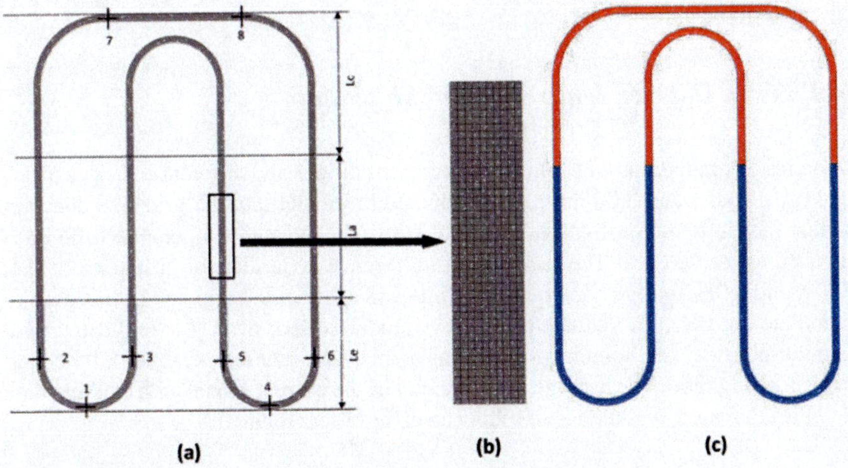

Fig. 1 **a** Physical model, **b** 0.16 mm mesh size, and **c** 60% filling ratio

used to capture fluid flow with respect to the flow time. For pressure velocity coupling, the SIMPLE algorithm is used. Under discretisation, the geo-reconstruction scheme is used for volume fraction and first-order discretization for energy and momentum equations.

3 Governing Equations

The volume of fluid model is used to capture the interface between the two phases of working fluid [12, 13]. The conservation equations of mass, momentum, and energy used to simulate the fluid flow are given as Eqs. 1, 2, and 3 respectively.

$$\frac{\partial \alpha}{\partial t} + \nabla.(\vec{v}\alpha) = \frac{S_m}{\rho} \tag{1}$$

$$\frac{\mathrm{d}(\rho\vec{v})}{\mathrm{d}t} + \nabla.(\rho\vec{v}\vec{v}) = \rho\vec{g} + \nabla.\left[\mu(\nabla\vec{v})^T\right] - \nabla p + F_\sigma \tag{2}$$

$$\frac{\mathrm{d}(\rho E)}{\mathrm{d}t} + \nabla.\left[\vec{v}(\rho E + p)\right] = \nabla.(k_{\mathrm{eff}}\nabla T) + S_h \tag{3}$$

where $\vec{v}$, ρ, and S_m are the velocity, density, and mass source terms in the conservation equation, whereas α, g, and p are the volume fraction, gravity, and pressure in the momentum equation. F_σ term induced by surface tension forces generated by cohesive interactions between fluid molecules and curved interface. Where k and S_h are the thermal conductivity and heat source terms in the energy equation.

4 Results

4.1 Grid Independence Test and Validation

A series of simulations with similar working conditions and different cell sizes (0.12, 0.11, 0.1, 0.09, and 0.08 mm) are conducted to predict that the grid size does not affect the simulation results. In all simulations, the average temperature difference is used as a reference. The temperature difference is similar for grid sizes of 0.1, 0.09, and 0.08. To save computational time and space, a 0.1 mm cell size is used to simulate the PHP. To validate the current model, a PHP of [11] is replicated with similar physical and boundary conditions at an evaporator temperature of 95 K and a 42% filling ratio. The average flow velocity of the current model with existing data [11] is compared, and the results show an error of less than 10%.

4.2 Volume Fraction and Start-Up Process

Vapour contours are examined to understand the development of two-phase fluid flow inside PHP. Blue colour represents the liquid nitrogen and red represents the vapour phase. Initially, i.e. when ($t = 0$), 60% of the PHP is filled with liquid and the remaining 40% is filled with vapour, which represents the 60% filling ratio of working fluid inside the PHP as depicted in Fig. 2a. As the wall temperature of the evaporator increases, nucleation is observed at the walls of the evaporator section. As time passes, the nucleation process turns to the formation of small bubbles inside the evaporator section. The initial nucleation process is observed at the time ($t =$ 0.045 S) as depicted in Fig. 2b. Further, the small bubbles start to coalesce with the neighbour bubbles and form a small vapour plug at time t = 0.123 S as shown in Fig. 2c. As time rises, the whole PHP is filled with the train of slugs and plugs Fig. 2d.

The increase in vapour plug size in the evaporator represents the transfer of latent heat from working fluid to increases. On contrary, in the condenser section, the vapour plugs lose their latent heat and liquid slugs lose the sensible heat. As the heat transfer increases, the pressure between the adjacent vapour plug and slug's increases, resulting in the pressure difference of the evaporator, condenser section, and adjacent limbs of the PHP. As a consequence, the working fluid starts to oscillate towards the condenser section and starts circulating in the clockwise direction; such a process is called the start-up process. In the current simulation, it took 4.2 s to initiate the start-up process.

Evaporator and condenser heat fluxes (WHF) are studied quantitatively to understand the start-up process, as shown in Fig. 3. The WHF is calculated using ANSYS CFD-Post (post-processing). The sum of T_0 and T_1 represents PHP's start-up time [8]. From Fig. 3, it is noticed that the time taken for the start-up process is 4.2 s.

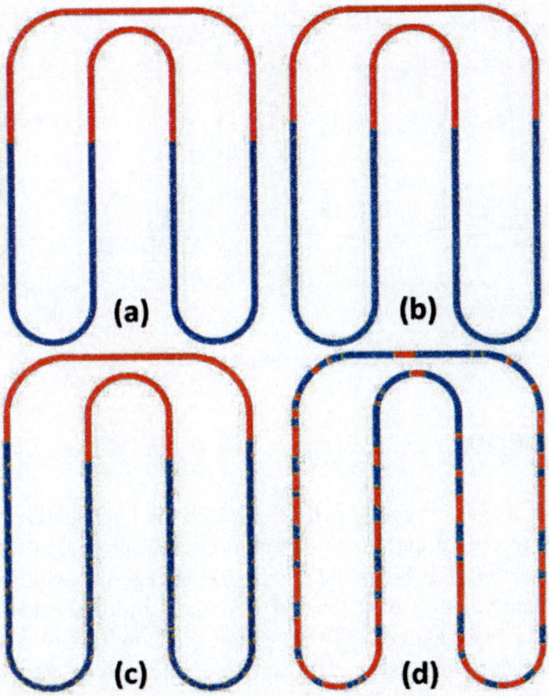

Fig. 2 **a** Initial distribution of working fluid, **b** nucleation process, **c** coalesces of vapour plugs, **d** vapour plugs and slugs in PHP

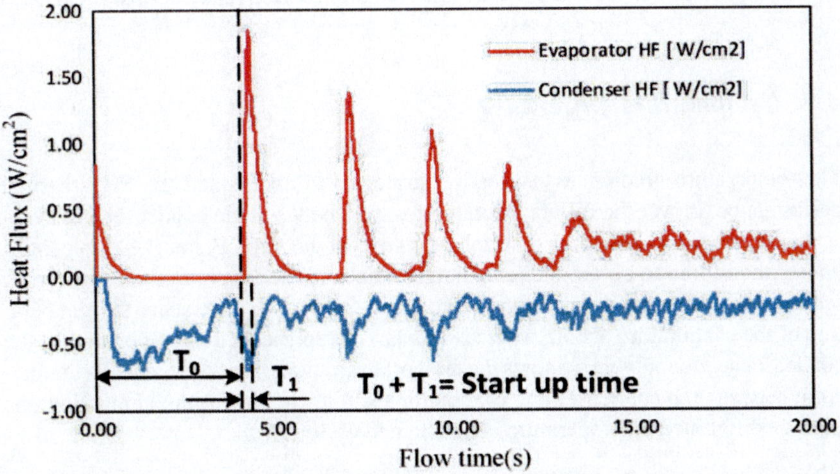

Fig. 3 Heat flux of evaporator and condenser

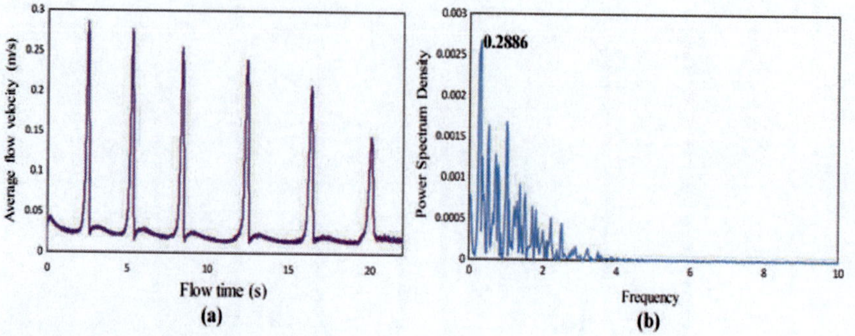

Fig.4 **a** Average flow velocity, **b** PSD of flow velocity

4.3 Flow Velocity

A quantitative analysis is carried out to understand the flow velocity in PHP. Points 1–8 are tracked to obtain the temporal variation of displacement of vapour plugs and liquid slugs in the PHP. Figure 4a depicts the average flow velocity in PHP. The peak working fluid velocity of 0.286 m/s is noticed inside the PHP. The average velocity of 0.08 m/s is maintained throughout the flow time. To understand the oscillation frequency of velocity in PHP, a nonlinear analysis of average flow velocity is conducted. The time series data of velocity at Points 1 and 4 are continuously monitored and analysed by Power Spectrum Density (PSD). Figure 4b shows the PSD analysis of average fluid flow velocity concerning the frequency. Spikes in Fig. 4b represent that flow is periodic with the highest frequency at 0.2886 Hz.

4.4 Temperature Difference

The temperature gradient is one of the strategies to understand the PHP thermal performance. Lower the difference in temperature between the condenser and evaporator sections, the lower the thermal resistance of the PHP. Points 1–8 are continuously monitored to obtain the temperature values in the evaporator and condenser sections. The average temperature of Points 1–2-3–4-5–6 represents the temperature of the evaporator section, whereas Points 7–8 represents the temperature of the condenser section. Figure 5 shows the temporal variation of the difference in temperature between the condenser and evaporator sections. In the current simulation, the average difference in temperature is maintained as 30 K.

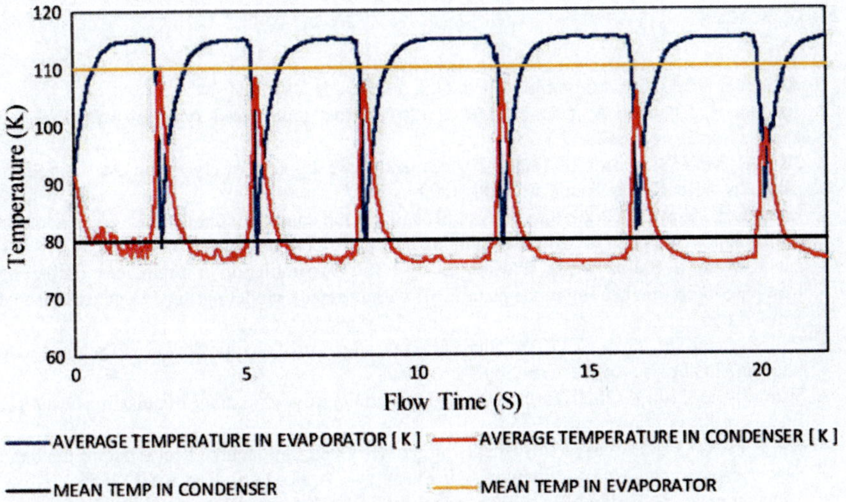

Fig. 5 Difference in temperature between condenser and evaporator sections

5 Conclusion

To understand the performance of PHP in cryogenic cooling, a 2D numerical simulation is performed using Computational Fluid Dynamics (CFD) techniques. Nitrogen is considered the two-phase fluid of PHP with a 60% filling ratio. The analysis is carried out at a cryogenic condition of 75 K and 115 K in condenser and evaporator sections, respectively. The VOF model is used to track the interface between the liquid and vapour phases of the working fluid. The primary findings of the study are as follows:

- The initial nucleation process is noticed at ($t = 0.04$ S), and the coalescence of vapour bubbles with neighbouring bubbles is observed at ($t = 0.123$ S). At time $t = 4.2$ S, the start-up process is observed.
- Following the plugs and slug's distribution, the working fluid begins to circulate clockwise. The greatest flow velocity measured is 0.28 m/s. The periodic nature of flow velocity is seen at 0.2886 Hz based on the nonlinear analysis of flow velocity.
- The temperature gradient is utilised to comprehend PHP's thermal behaviour. The PHP maintains an average temperature differential of 30 K during the flow time.

References

1. Akachi H (1990) Structure of heat pipe. U.S. patent, P.N. 4921041, U
2. Jouhara H, Chauhan A, Nannou T et al (2017) Heat pipe based systems—advances and applications. Energy 128:729–754
3. Jiao AJ, Ma HB, Critser JK (2009) Experimental investigation of cryogenic oscillating heat pipes. Int J Heat Mass Transf 52:3504–3509
4. Liang Q, Li Y, Wang Q (2018) Effects of filling ratio and condenser temperature on the thermal performance of a neon cryogenic oscillating heat pipe. Cryogenics 89:102–106
5. Lin Z, Wang S, Shirakashi R, Winston Zhang L (2013) Simulation of a miniature oscillating heat pipe in bottom heating mode using CFD with unsteady modeling. Int J Heat Mass Transf 57:642–656
6. Pouryoussefi SM, Zhang Y (2016) Numerical investigation of chaotic flow in a 2D closed-loop pulsating heat pipe. Appl Therm Eng 98:617–627
7. Wang J, Xie J, Liu X (2020) Investigation of wettability on performance of pulsating heat pipe. Int J Heat Mass Transf 150:119354
8. Xie F, Li X, Qian P et al (2020) Effects of geometry and multisource heat input on flow and heat transfer in single closed-loop pulsating heat pipe. Appl Therm Eng 168:114856
9. Satyanarayana K, Reddy NVSM, Venugopal S (2021) Numerical investigation of single turn pulsating heat pipe with additional branch for the enhancement of heat transfer coefficient and flow velocity. Heat Transf Res 52:45–62
10. Satyanarayana K, Reddy NVSM, Venugopal S (2023) Numerical study to recover low-grade waste heat using pulsating heat pipes and a comparative study on performance of conventional pulsating heat pipe and additional branch pulsating heat pipe. Numer Heat Transf Part A: Appl 83(3):248–264
11. Sagar KR, Naik HB, Mehta HB (2021) Numerical study of liquid nitrogen based pulsating heat pipe for cooling superconductors. Int J Refrig 122:33–46
12. Reddy NVSM, Satyanarayana K, Venugopal S (2022) Comparative numerical study of R134a and low global warming potential refrigerants during condensation inside a smooth and dimpled tube. Heat Mass Transf 1–16
13. Reddy NVSM, Satyanarayana K, Venugopal S (2022) Influence of saturation temperature on pressure drop during condensation of R-134a inside a dimpled tube: a numerical study. Theor Found Chem Eng 56:395–406

Vibrational Nature of an Unbalanced Rigid Rotor System with Three Discs Secured by Two Active Magnetic Bearings

Prabhat Kumar, Maruvada Sanket, Suyash Srivastav, and Tanmay Dinesh Madav

1 Introduction

High-speed rotating systems are widely used in a variety of applications such as gas turbines, centrifugal pumps, aircraft engines [1, 2]. Conventional bearings are utilized in these rotating machines to stabilize the rotor by physical contact [3–6], but more recently, Active Magnetic Bearings (AMBs) have been developed that utilize electromagnetism to levitate the rotor system [7]. The advantage of this type of system is that it does not require any lubrication, due to which the load can be suspended with no friction, which can eliminate wear and avoid failure, due to which there is a cost down in maintenance [8–11]. The spinning machinery may go through a phase where the rotor starts to vibrate with a lot of amplitude and even fails while obtaining this high speed. This issue arises as a result of an unbalance. Nordmann and Aenis [12] have used AMBs with developed built-in software for the analysis of centrifugal pumps. De Queiroz [13] has used a method for identifying parameters responsible for unbalance in a Jeffcott rotor using an effective and study control mechanism. Markert et al. [14] developed a least-squares fitting approach-based and model-based method for the online detection of errors in rotor systems. For the purpose of identifying an unbalance defect in a two ball bearings supported rotor system, Sudhakar and Sekhar [10] used three distinct methodologies based on the least-squares fitting approach. Clark et al. [7] demonstrated the superiority of AMB technology over the widely utilized conventional bearings' technology in the aerospace industry. Lal and Tiwari [6] created an estimation technique to evaluate the bearing and coupling dynamic characteristics, finding unbalances at rotor planes, and misalignment forces for a test setup containing a misaligned turbogenerator. Kumar and Tiwari [15–17] put out

P. Kumar (✉) · M. Sanket · S. Srivastav · T. D. Madav
Department of Mechanical Engineering, NIT Manipur, Imphal West, Manipur 795004, India
e-mail: ysprabhat.pamho@gmail.com; prabhat.kumar@nitmanipur.ac.in

B. P. Swain (ed.), *Recent Advances in Materials*, Springer Proceedings in Materials 25,
https://doi.org/10.1007/978-981-99-3844-5_35

an advanced method to calculate the misalignment and unbalance in a stiff rotor-AMB systems. Alves and Cavalca [18] have employed a method that utilized linear hydrodynamic force or nonlinear hydrodynamic force models to find rotor-bearing system unbalance at single rotational speed. Nayek et al. [19] computed the inertial characteristics of a rotor with an unbalance in AMB in a noisy environment. Yun et al. [20] have employed a method for determining the multiplane rotor's imbalance parameters. Majumder and Tiwari [21] have used AMB and conventional bearings to analyze a geared rotor system. For the estimate of sensors misalignment, rotor unbalance, misalignment of AMBs, and their stiffness characteristics in a floating rotor-bearing system, a novel mathematical technique has recently been developed [22]. Ma et al. [23] have designed a six-pole AMB to overcome the asymmetry problems faced in three-pole magnetic bearings. Fuzzy active controller technique was utilized by them for stable levitation of the rotor in air.

Based on studying various literatures, it has been observed that researchers are greatly focusing on AMB technology as a vibration controller and fault identification tool. Papers have been also found in the development of more effective and efficient six-pole actuators. However, in both transient and steady-state settings, this work presents the vibrational properties of an inelastic rotor system that is magnetically levitated and connected to three discs.

2 Model of the Rotor-AMB System

Figure 1 depicts the rotor system used for this study. The rotor is having supports of two AMBs (i.e., AMB1 and AMB2) at their end positions. Three discs are present in the rotor system, in which two discs, i.e., OFD1 and OFD2 are at the offset positions and one disc, i.e., MD is at the middle position. It should be noted that the motion in horizontal (z-axis) and vertical (x-axis) directions is considered to be the translational degrees of freedom. The gyroscopic effect resulting from offset discs has been considered. This was due to their tilting positions during operation. However, the middle disc does not induce gyroscopic moment as it does not tilt while in rotational motion. Discs can be visualized as flywheels, pump impellers, turbine blades, etc. The symbol G represents the rotor center of gravity. Here, l_1, l_2, and l_3 are considered to be the distances of OFD1, OFD2, and MD from the point of center of gravity, respectively. The respective distance between point G and AMB1 as well as point G and AMB2 is symbolized, respectively, by a_1 and a_2. The rigid rotor translational and angular movements in the (x–z) and (y–z) planes are shown in Fig. 2. This figure also depicts the displacements at different locations of the shaft, where AMBs and discs are located.

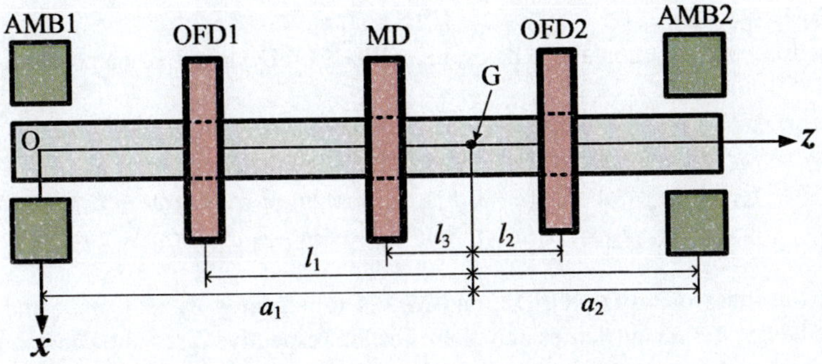

Fig. 1 Two active magnetic bearings supporting a three-disc rotor system

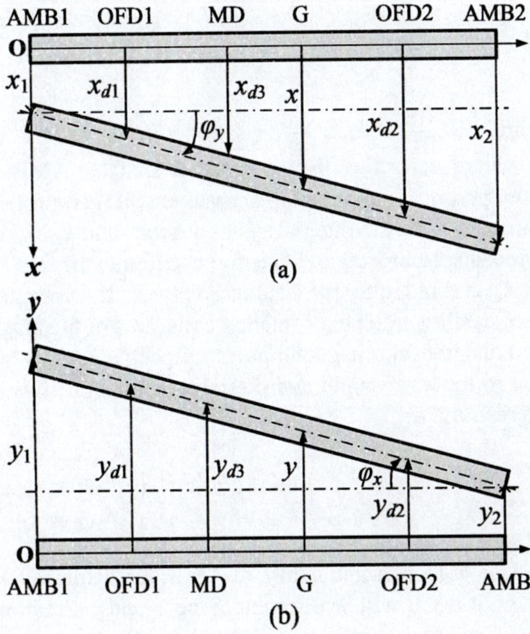

Fig. 2 Rigid rotor movements in **a** x–z plane as well as **b** y–z plane

3 Development of Equations of Motion

Following Fig. 2, AMB1 and AMB2 can be used to express the translational displacements (i.e., x, y) and rotational displacements (i.e., φ_y, φ_x) of the rigid shaft at the center of gravity,

$$\begin{aligned} x &= \bar{a}_2 x_1 + \bar{a}_1 x_2; \quad y = \bar{a}_2 y_1 + \bar{a}_1 y_2 \\ \varphi_y &= (-x_1 + x_2)/l; \; \varphi_x = (y_1 - y_2)/l \end{aligned} \tag{1}$$

with $\bar{a}_1 = \frac{a_1}{l}$; $\bar{a}_2 = \frac{a_2}{l}$.

The force due to unbalanced discs, i.e., OFD1, OFD2, and MD can be expressed as,

$$\begin{aligned} f_{ux1} &= m_{d1}e_1\omega^2 \cos(\omega t + \beta_1);\ f_{uy1} = m_{d1}e_1\omega^2 \sin(\omega t + \beta_1) \\ f_{ux2} &= m_{d2}e_2\omega^2 \cos(\omega t + \beta_2);\ f_{uy2} = m_{d2}e_2\omega^2 \sin(\omega t + \beta_2), \\ f_{ux3} &= m_{d3}e_3\omega^2 \cos(\omega t + \beta_3);\ f_{uy3} = m_{d3}e_3\omega^2 \sin(\omega t + \beta_3) \end{aligned} \quad (2)$$

where the mass of OFD1, OFD2, and MD is represented by m_{d1}, m_{d2}, and m_{d3}, respectively. The eccentricities and phases of the respective discs unbalance are (e_1, e_2 and e_3) and (β_1, β_2 and β_3). Here, the symbol ω is the rotor rotational speed. Further, the linearized force due to AMB, f_A [24], can be written as a summation of force component due to displacement stiffness and current stiffness parameters, as

$$f_A = k_s u_x + k_i i_x;\quad i_x = -\left(k_P u_x + k_I \int u_x\,\mathrm{d}t + k_D \dot{u}_x\right) \quad (3)$$

with $k_s = \frac{4ki_0^2}{s_0^3}$; $k_i = \frac{4ki_0}{s_0^2}$; $k = \frac{1}{4}\mu_0 N^2 A_a \cos\frac{\alpha}{2}$,

where u_x is the displacement of shaft at AMB, i_x is the controlling current from proportional–integral–derivative (PID) controller (k_P is the proportional gain constant, k_I is the integral gain constant, and k_D is derivative gain constant), k_s is force–displacement, and k_i is force–current stiffness factor. The variables s_0, i_0, μ_0, N, A_a, α stand in for the clearance between the rotor and stator, bias current, vacuum permeability, number of rotating coils, area of the magnetic pole, and angle between two neighboring magnetic poles, respectively. The force in the x- and y-directions due to the anisotropic and dissimilar characteristics of AMB1 and AMB2 can be represented as,

$$\begin{aligned} f_{Ax1} &= k_{sx1}x_1 + k_{ix1}i_{x1};\ f_{Ay1} = k_{sy1}y_1 + k_{iy1}i_{y1} \\ f_{Ax2} &= k_{sx2}x_2 + k_{ix2}i_{x2};\ f_{Ay2} = k_{sy2}y_2 + k_{iy2}i_{y2} \end{aligned}. \quad (4)$$

Due to anisotropic nature of AMB, the displacement and current stiffness properties of AMB will be different in the x- and y-directions. Therefore, the AMB force in the x- and y- directions will be also different (this can be seen from Eq. (4)). For developing equations of motion, the moment equilibrium method has been utilized with consideration of moment equilibrium about AMB2 and AMB1 centers in the (x–z) plane and (y–z) plane. Moments are also being considered due to the unbalance force of OFD1, OFD2, and MD, inertia moment of the rotor, gyroscopic moment due to OFD1 and OFD2, moment due to AMB force. Please note that the moment due to AMB2 force about AMB2 center will be zero and the moment due to AMB1 force about AMB1 center will be zero, as the line of action of the force passes through the same point about which moment is taken. With this concept, the equations of motion in matrix form are given as

$$\mathbf{M}\Delta\,\ddot{\mathbf{u}}(t) - \omega\mathbf{G}\Delta\,\dot{\mathbf{u}}(t) = \mathbf{f}_u + \mathbf{f}_A \tag{5}$$

with $\mathbf{M} = \begin{bmatrix} (m\bar{a}_2^2 + i_d) & 0 & (m\bar{a}_1\bar{a}_2 - i_d) & 0 \\ 0 & (m\bar{a}_2^2 + i_d) & 0 & (m\bar{a}_1\bar{a}_2 - i_d) \\ (m\bar{a}_1\bar{a}_2 - i_d) & 0 & (m\bar{a}_1^2 + i_d) & 0 \\ 0 & (m\bar{a}_1\bar{a}_2 - i_d) & 0 & (m\bar{a}_1^2 + i_d) \end{bmatrix}$; $\mathbf{G} = \begin{bmatrix} 0 & i_p & 0 & -i_p \\ i_p & 0 & -i_p & 0 \\ 0 & -i_p & 0 & i_p \\ -i_p & 0 & i_p & 0 \end{bmatrix}$; $i_d = \frac{I_d}{l^2}$; $i_P = \left(\frac{I_{P1}+I_{P2}+I_{P3}}{l^2}\right)$; $\bar{l}_1 = \frac{l_1}{l}$; $\bar{l}_2 = \frac{l_2}{l}$; $\bar{l}_3 = \frac{l_3}{l}$.

The rotor displacement vector at AMB locations is written as

$$\Delta\mathbf{u}(t) = \left\{\, x_1\; y_1\; x_2\; y_2 \,\right\}^T. \tag{6}$$

The unbalance force vector is expressed as

$$\mathbf{f}_u = \left\{ \begin{array}{l} m_{d1}e_1\omega^2\cos(\omega t + \beta_1)(\bar{a}_2 + \bar{l}_1) + m_{d2}e_2\omega^2\cos(\omega t + \beta_2)(\bar{a}_2 - \bar{l}_2) \\ \quad + m_{d3}e_3\omega^2\cos(\omega t + \beta_3)(\bar{a}_2 + \bar{l}_3) \\ m_{d1}e_1\omega^2\sin(\omega t + \beta_1)(\bar{a}_2 + \bar{l}_1) + m_{d2}e_2\omega^2\sin(\omega t + \beta_2)(\bar{a}_2 - \bar{l}_2) \\ \quad + m_{d3}e_3\omega^2\sin(\omega t + \beta_3)(\bar{a}_2 + \bar{l}_3) \\ m_{d1}e_1\omega^2\cos(\omega t + \beta_1)(\bar{a}_1 - \bar{l}_1) + m_{d2}e_2\omega^2\cos(\omega t + \beta_2)(\bar{a}_1 + \bar{l}_2) \\ \quad + m_{d3}e_3\omega^2\cos(\omega t + \beta_3)(\bar{a}_1 - \bar{l}_3) \\ m_{d1}e_1\omega^2\sin(\omega t + \beta_1)(\bar{a}_1 - \bar{l}_1) + m_{d2}e_2\omega^2\sin(\omega t + \beta_2)(\bar{a}_1 + \bar{l}_2) \\ \quad + m_{d3}e_3\omega^2\sin(\omega t + \beta_3)(\bar{a}_1 - \bar{l}_3) \end{array} \right\}. \tag{7}$$

The AMB force vector is given as

$$\mathbf{f}_A = \left\{\, f_{Ax1}\; f_{Ay1}\; f_{Ax2}\; f_{Ay2} \,\right\}^T. \tag{8}$$

On substituting Eq. (4) into Eq. (8), the AMB force vector and controlling current vector can be given as

$$\mathbf{f}_A = \mathbf{K}_s\Delta\mathbf{u}(t) + \mathbf{K}_i\mathbf{i}(t); \quad \mathbf{i}(t) = -\left\{ k_P\Delta\mathbf{u}(t) + k_I \int \Delta\mathbf{u}(t)\,\mathrm{d}t + k_D\Delta\,\dot{\mathbf{u}}(t) \right\}, \tag{9}$$

$$\mathbf{i}(t) = \left\{\, i_{x1}\; i_{y1}\; i_{x2}\; i_{y2} \,\right\}^T \tag{10}$$

$$\text{with } \mathbf{K}_s = \begin{bmatrix} k_{sx1} & 0 & 0 & 0 \\ 0 & k_{sy1} & 0 & 0 \\ 0 & 0 & k_{sx2} & 0 \\ 0 & 0 & 0 & k_{sy2} \end{bmatrix}; \quad \mathbf{K}_i = \begin{bmatrix} k_{ix1} & 0 & 0 & 0 \\ 0 & k_{iy1} & 0 & 0 \\ 0 & 0 & k_{ix2} & 0 \\ 0 & 0 & 0 & k_{iy2} \end{bmatrix}.$$

4 Results and Discussion

In this section, the equations of motion (i.e., Eq. (5)) have been solved by developing a SIMULINK™ model (refer Fig. 3) in MATLAB (Version R2020b) and employing Runge–Kutta method with step size of 0.0001 s. The values of various rotor-AMB parameters taken for numerical simulation are given in Table 1. The solution of the matrix gave the linear rotor displacement (i.e., the elements of Eq. (6)) and controlling current (i.e., the elements of Eq. (10)) at AMB1 and AMB2 locations. Linearity is existing in the system for least vibrations of the rotor [25]. Figure 4 shows the system responses (in the time frame) at 35 Hz frequency for the initial transient state at AMB1 location.

Further, Fig. 5 shows the orbital plot of the responses for the transient state at the AMB1 position (i.e., displacement in x versus displacement in y-direction as well as current in x-direction versus current in y). The rotor is given an operational speed of 35 Hz (i.e., 220 rad/s).

Further, the responses have been also studied for steady-state behavior for the rotor-AMB arrangement. Figure 6 shows the time-domain map of the displacement at AMB1 and AMB2. It was noticed that the greatest values of x- and y-directional displacements at AMB1 are 1.25×10^{-4} m and 1.02×10^{-4} m, respectively. At AMB2, the x- and y-displacements are, respectively, 6.94×10^{-5} m and 7.19×10^{-5} m. Likewise, Fig. 7 displays the time-domain plot for regulating current at AMB1 and AMB2. It is discovered that the highest x- and y-current values at AMB1 are 0.717 A and 0.569 A, respectively. The displacements at AMB2 are 0.396 A and 0.400 A in the x- and y-axes. The displacement and controlling current orbital plots at AMB1 and AMB2, separately, are depicted in Figs. 8 and 9. The orbits are observed to be elliptical in nature with some asymmetricity. This was due to discs' unbalance force with anisotropic as well as different behaviors of both AMBs.

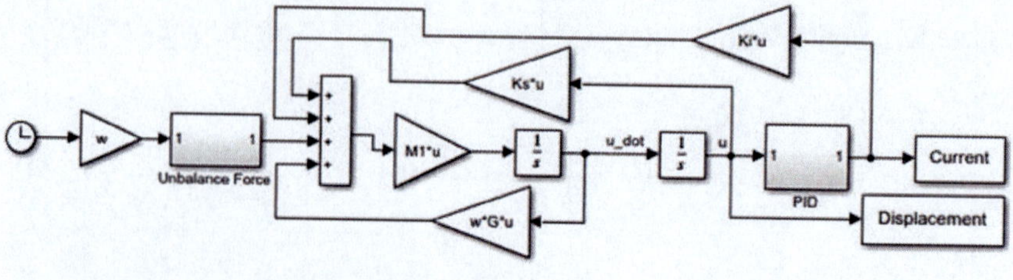

Fig. 3 Simulink model developed in MATLAB

Table 1 Assumed values of rotor-AMB system for numerical simulation

Rotor-AMB parameters	Assumed values
Mass of the complete rotor (m)	5.11 kg
Eccentricity of disc 1, disc 2, disc 3 (e_1, e_2, e_3)	80, 100, 110 μm
Unbalance phase ($\beta_1, \beta_2, \beta_3$,)	20, 30, 35 deg
Force–displacement stiffness of AMB 1 (k_{sx1}, k_{sy1})	1.7415 × 10^5, 1.95 × 10^5 N/m
Force–displacement stiffness of AMB 2 (k_{sx2}, k_{sy2})	3.6571 × 10^5, 3.83 × 10^5 N/m
Force–current stiffness of AMB 1 (k_{ix1}, k_{iy1})	34.82, 36.20 N/A
Force–displacement stiffness of AMB 2 (k_{ix2}, k_{iy2})	73.14, 75.40 N/A
PID controller parameters (k_P, k_I, k_D)	5500 A/m, 8000 A/m-s, 3 A-s/m

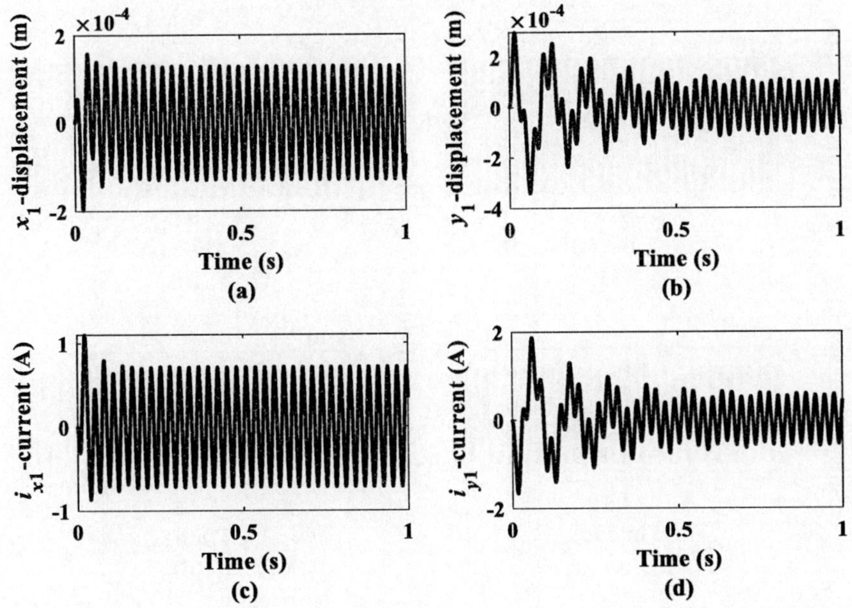

Fig. 4 Rotor system responses for initial transient state **a**, **b** x- and y-directional displacement at AMB1 location **c**, **d** x- and y-directional currents at AMB1

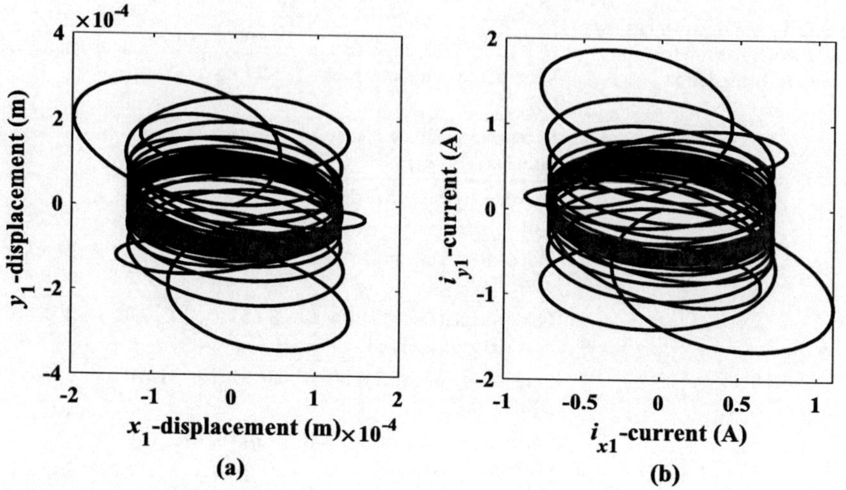

Fig. 5 For the initial transient condition at AMB1, the orbital response **a** displacement orbit **b** current orbit

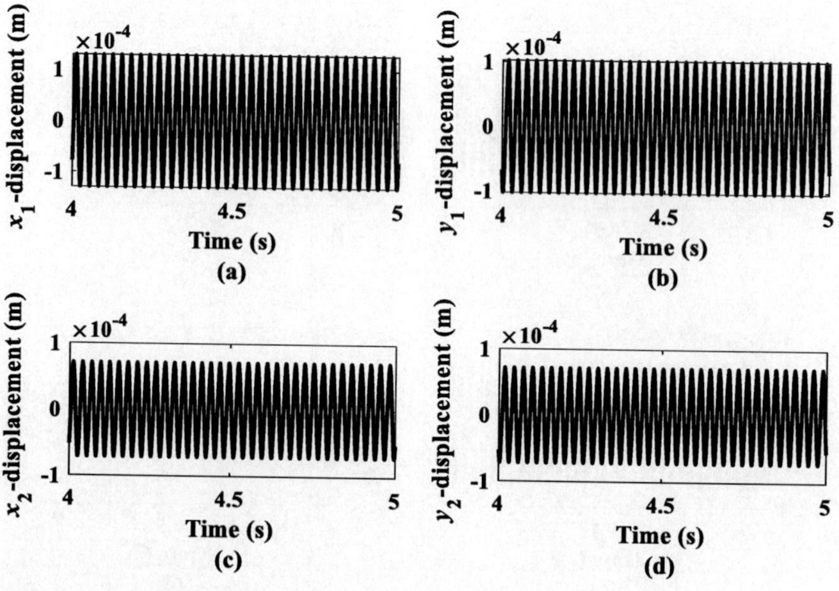

Fig. 6 **a, b** *x*- and *y*-displacement at AMB1 **c, d** *x*- and *y*-displacement at AMB2

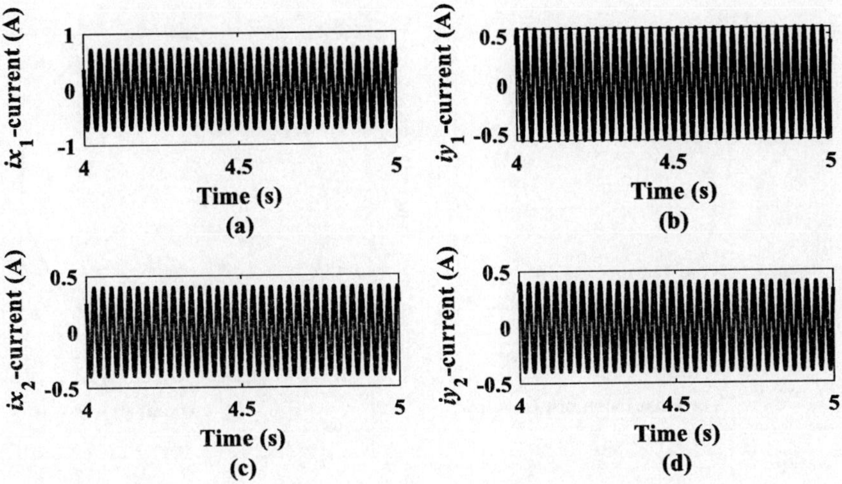

Fig. 7 **a, b** Current in *x*- and *y*-direction at AMB1 **c, d** current in *x*- and *y*-direction at AMB2

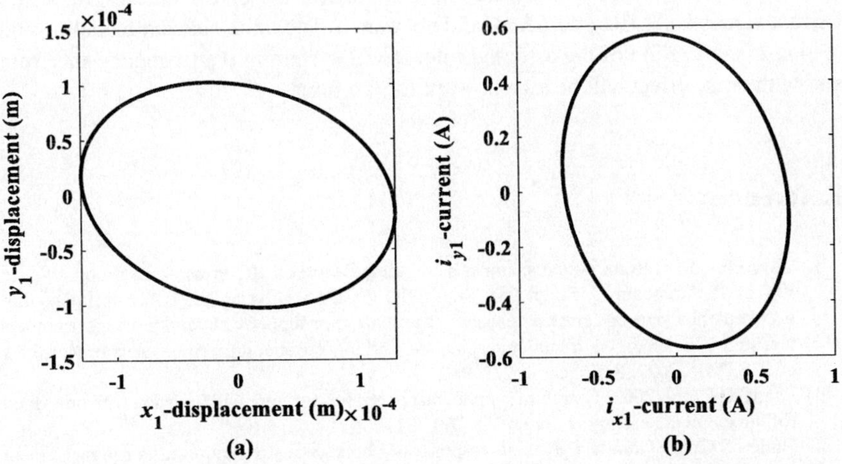

Fig. 8 Steady-state orbital plots at AMB1 for **a** displacement **b** current, at 35 Hz angular frequency

5 Conclusions

In this paper, an unbalanced rigid rotor with three discs having two AMB supports has been successfully modeled and the equations of motion were derived using moment equilibrium method. For the initial transient and steady-state circumstances, displacement and controlling current have been established at AMB1 and AMB2. The time-domain responses were of harmonic in nature due to sinusoidal nature of unbalance force. Moreover, the orbital plots were found to be elliptical in nature.

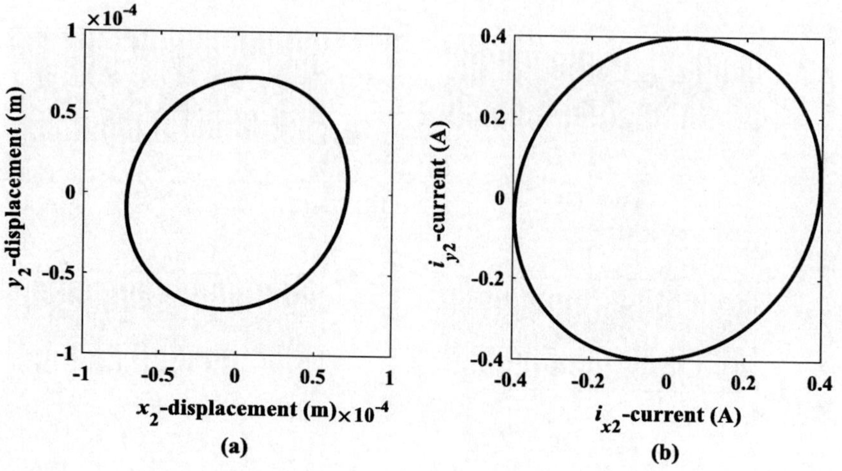

Fig. 9 Steady-state orbital plots at AMB2 for **a** displacement **b** current, at 35 Hz angular frequency

Because of their anisotropic nature and differences, the displacement values and current values at AMB1 and AMB2 did not match. The finite element modeling and response analyzation of the assessed rotor-AMB system in the frequency spectrum and with noise effect will be a great work for the future.

References

1. Tiwari R (2017) Rotor systems: analysis and identification. CRC Press, Boca Raton
2. Kumar P, Arambam C, Singh LD, Singh ND (2023) Analysing the dynamic interaction between unbalance and crack responses in a jeffcott rotor supported by foil bearings: a numerical study. In: Emerging trends in mechanical and industrial engineering: select proceedings of ICETMIE 2022. Springer, pp 971–980
3. Panda K, Dutt J (2003) Optimum support characteristics for rotor–shaft system with preloaded rolling element bearings. J Sound Vib 260:731–755
4. Harsha S (2006) Nonlinear dynamic response of a balanced rotor supported by rolling element bearings due to radial internal clearance effect. Mech Mach Theory 41:688–706
5. Tomovic R, Miltenovic V, Banic M, Miltenovic A (2010) Vibration response of rigid rotor in unloaded rolling element bearing. Int J Mech Sci 52:1176–1185
6. Sharma A, Upadhyay N, Kankar PK, Amarnath M (2018) Nonlinear dynamic investigations on rolling element bearings: a review. Adv Mech Eng 10:1687814018764148
7. Kumar P (2021) Model based analysis and identification of unbalance and misalignment in rotor systems levitated by active magnetic bearings. Ph.D. Mechanical Engineering, Indian Institute of Technology Guwahati, Guwahati
8. Kumar P, Tiwari R (2019) A numerical study on the effect of unbalance and misalignment fault parameters in a rigid rotor levitated by active magnetic bearings. In: ASME 2019 Gas Turbine India Conference
9. Kumar P, Kumar V, Kumar K, Meena LS (2020) Unbalance and dynamic parameters estimation in a rigid rotor mounted on active magnetic bearings. In: Advances in applied mechanical engineering. Springer, pp 363–371

10. Kumar P, Tiwari R (2020) Dynamic response analysis of an unbalanced and misaligned rotor supported on active magnetic bearings and touchdown bearings. In: Proceedings of the 6th national symposium on rotor dynamics, pp 407–418
11. Kumar P, Tiwari R (2020) Effects of unbalance and AMB misalignment in a rigid rotor with an offset disc levitated by active magnetic bearings: a numerical investigation. In: 12th international conference on vibrations in rotating machinery, 151–168
12. Nordmann R, Aenis M (2004) Fault diagnosis in a centrifugal pump using active magnetic bearings. Int J Rotating Mach 10:183–191
13. De Queiroz M (2009) An active identification method of rotor unbalance parameters. J Vib Control 15:1365–1374
14. Markert R, Platz R, Seidler M (2001) Model based fault identification in rotor systems by least squares fitting. Int J Rotating Mach 7:311–321
15. Kumar P, Tiwari R (2020) Development of a novel approach for quantitative estimation of rotor unbalance and misalignment in a rotor system levitated by active magnetic bearings. Iranian J Sci Technol Trans Mech Eng 45:769–786
16. Kumar P, Tiwari R (2020) Dynamic analysis and identification of unbalance and misalignment in a rigid rotor with two offset discs levitated by active magnetic bearings: a novel trial misalignment approach. Propulsion Power Res 10:58–82
17. Kumar P, Tiwari R (2021) Finite element modelling, analysis and identification using novel trial misalignment approach in an unbalanced and misaligned flexible rotor system levitated by active magnetic bearings. Mech Syst Signal Process 152:107454
18. Alves DS, Cavalca KL (2021) Investigation into the influence of bearings nonlinear forces in unbalance identification. J Sound Vib 492:115807
19. Nayek B, Das A, Dutt J (2021) Model based estimation of inertial parameters of a rigid rotor having dynamic unbalance on active magnetic bearings in presence of noise. Appl Math Model 97:701–720
20. Yun X, Pang Z, Jiang G, Mei X (2021) Research on identification of unbalance parameters of rotor with multi-plane using improved particle swarm optimization. J Braz Soc Mech Sci Eng 43:1–12
21. Majumder G, Tiwari R (2022) Application of active magnetic bearings in control and estimation of geared-rotor faults in high speed offset spur gear transmission system. Mech Syst Signal Process 176:109113
22. Tiwari R, Kumar P (2022) An innovative virtual trial misalignment approach for identification of unbalance, sensor and active magnetic bearing misalignment along with its stiffness parameters in a magnetically levitated flexible rotor system. Mech Syst Signal Process 167:108540
23. Ma Z, Liu G, Liu Y, Yang Z, Zhu H (2022) Research of a six-pole active magnetic bearing system based on a fuzzy active controller. Electronics 11:1723
24. Schweitzer G, Maslen EH (2009) Magnetic bearings: theory, design, and application to rotating machinery, vol 1. Springer, Berlin
25. Kumar P, Tiwari R (2023) A review: multiplicative faults and model-based condition monitoring strategies for fault diagnosis in rotary machines. J Braz Soc Mech Sci Eng 45(5):282. https://doi.org/10.1007/s40430-023-04203-z

Numerical Simulation of Sandwiched Composite Armor Subjected to Velocity Impact

Jitendra Rajput, Rajesh Kumar Bhushan, and Azhar Jamil

1 Introduction

Over the past decades, body armors have become a hotspot for research because of increasing the global terrorism and civil conflicts, which protect the person against impact caused by projectile fired by arms. Traditional body armors are made of ceramics, transparent glasses, and metals which are heavy and rigid [1]. Several researchers are aimed to design high-stiffness and lightweight structures to improve the impact performance and minimize the damage size under velocity impact [2]. A sandwich panel is a composite that is fabricated by attaching a thick core between two thin but stiff skin. The thin skin is usually made of metal or high-strength composite face sheets [3]. The traditional armor uses are limited due to their shortage of flexibility, and these armors can hardly protect legs or arms. So, the researchers focused more and more on reducing the weight of body armor and improving flexibility while keeping the same impact resistance [4–6]. Kevlar is an aramid fiber with having lightweight high-stiffness ratio which has been widely used as the base material in soft body armor because of its modulus, toughness, high strength, and stability [7, 8]. To further improve the impact resistance, numerous methods have developed to modify aramid fabrics. Nowadays, multilayer aramid fibers were used to develop body armor due to their lightweight high-strength ratio [9].

J. Rajput (✉) · R. K. Bhushan
Department of Mechanical Engineering, NIT Manipur, Imphal, Manipur, India
e-mail: jitendrarajput16@gmail.com

A. Jamil
Department of Mechanical Engineering, Aligarh Muslim University, Uttar Pradesh, Aligarh, India

B. P. Swain (ed.), *Recent Advances in Materials*, Springer Proceedings in Materials 25,
https://doi.org/10.1007/978-981-99-3844-5_36

2 Problem Formulation

Impacts are usually two types, low-velocity impact and high-velocity impact. The projectile's kinetic energy is directly proportional to the mass of the projectile and also depends on the barrel of the gun. For low-velocity impact short barrel gun and for high-velocity impact long barrel gun are preferred [10]. The kinetic energy of a bullet is transferred to the targeted sandwiched composite plate when impact takes place which decreases the kinetic energy of the bullet and increases the internal energy of sandwiched composite plate. The generated longitudinal and transverse waves travel to the end of the plate [11]. For different materials, different mechanism dominates. So, the energy transferred to the sandwiched plate is calculated by Eq. (1) and the ballistic limit can be determined by Eq. (2).

$$E_{\text{tr}} = \frac{1}{2} m_p \left(V_i^2 - V_r^2 \right), \tag{1}$$

$$V_{50} = \sqrt{\left(V_i^2 - V_r^2 \right)}. \tag{2}$$

3 Finite Element Modeling

A finite element model of the composite plate is developed to model the velocity impact behavior of the Kevlar-29 sandwich panel with Al6061 T6 skin. For finite element analysis, ANSYS AUTODYN is used which relates the stress–strain relation and gives optimum results. The meshed model and ACP Model are shown in Fig. 1. The composite plate is fixed at all side faces, and the projectile is movable toward the center of the plate perpendicularly.

3.1 Geometric Model

The sandwich panel is modeled with a dimension of 100 × 100 × t mm^3, where 100 mm is the length and breadth t mm is the thickness of the sandwich composite plate. This thickness is varying from 1.9 mm to 3.1 mm as the number of Kevlar layers varies. The projectile is made in cylindrical form with a hemispherical nose. The diameter of cylindrical shank and nose radius is taken as 12.8 mm and 6.4 mm simultaneously. The stand-off distance between the striking face of the plate and the projectile's front face is 13.6 mm. For simulation, all side faces of the composite plate kept a fixed velocity assigned to the projectile.

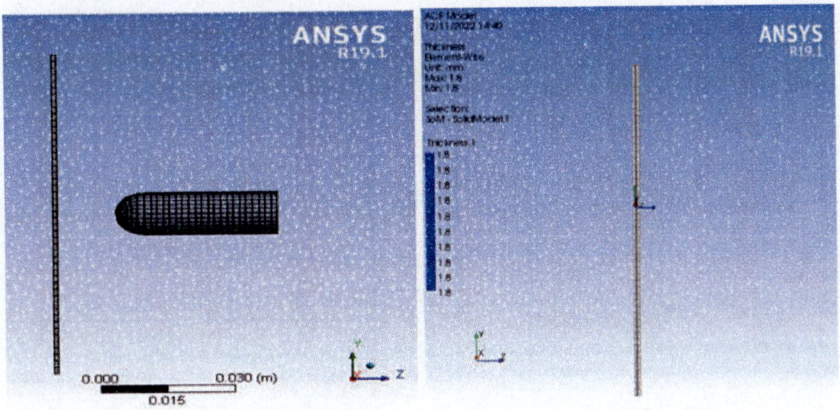

(a) (b)

Fig. 1 **a** Meshed model of plate and projectile. **b** ACP model of sandwich composite plate

3.2 Meshing

The meshing of the composite plate is done in such a way that accurate simulation results can be obtained. Three sections were created on the target composite plate, the first impact zone is where the projectile comes in immediate contact with the targeted composite plate, the second one is lesser affected zone, and the third one is the outer portion of the composite plate which is the least affected zone. Keeping the mesh refined in the impact zone so that an element aspect ratio will be close to unity as possible.

The element size is given as:

$$\text{Element Size} = \frac{\text{Thickness of plate}}{\text{Number of elements along the thickness}}.$$

3.3 Materials Used and Properties

See Tables 1, 2, 3, and 4.

Table 1 Material parameters of Kevlar-29 in ANSYS [13]

Density (kg/m^3)	E_{11}, E_{22} (MPa)	E_{33} (MPa)	G_{12} (MPa)	$G_{23}G_{13}$ (MPa)	ν_{12}	ν_{23}, ν_{13}	Compressive yield strength (MPa)	Tensile yield strength (MPa)
1440	18,500	6000	770	5430	0.25	0.33	185	1850

Table 2 Material properties of Johnson Cook (JC) material model used in ANSYS simulations for Al 6061-T6 [12]

Density (kg/m^3)	Yield stress, A (MPa)	Hardening constant, B (MPa)	Hardening exponent, n	Temperature softening exponent, m	Melting temperature, T_m (K)	Reference temperature, T_r(K)
2700	324.1	113.8	0.42	1.34	925	293.2

Table 3 Johnson Cook (JC) damage parameters of Al 6061-T6 [12]

Strain rate constant, C	Reference strain rate, $\dot{\varepsilon}_0$ (s^{-1})	Damage evolution	Failure parameters				
			D_1	D_2	D_3	D_4	D_5
0.002	1	16.5% of L_e	−0.77	1.45	-0.45	0	1.6

Table 4 Properties of epoxy resin [13]

Modulus of elasticity (GPa)	Tensile yield strength (Mpa)	Density (g/cm^3)
42	4.84	1.12

4 Result and Discussion

The result shown in Figs. 2 and 3 has been obtained from AUTODYN simulation. The model of the sandwiched composite plate has been simulated using ANSYS Explicit Dynamics with AUTODYN, and the modeling results were generated. The material of sandwiched composite plate and projectile is modeled into ANSYS Engineering data. Then, the meshing of the 3D model and contact condition between parts of the composite plate have been set up. Finally, the solver like boundary conditions and desired solution have been defined.

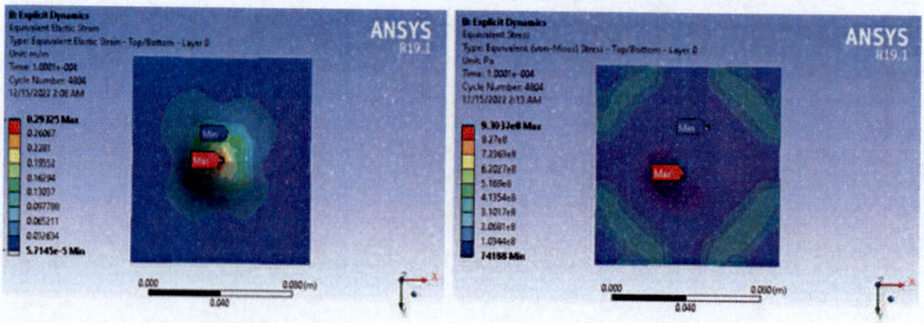

(a) (b)

Fig. 2 **a** Equivalent strain at velocity 100 m/s, at 0.1 ms **b** Equivalent stresses at velocity 100 m/s, at 0.1 ms

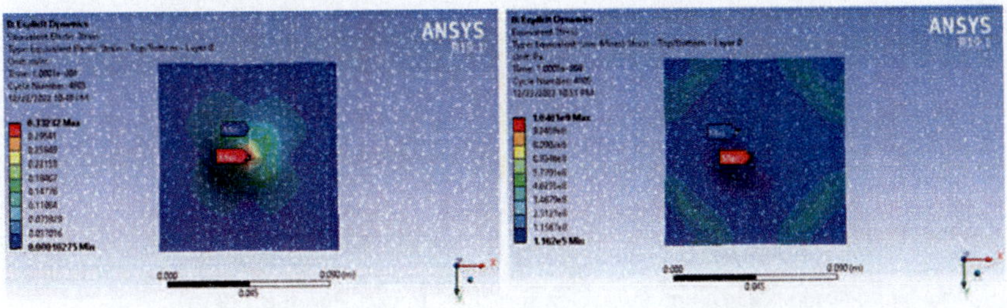

(a) (b)

Fig. 3 **a** Equivalent strain at velocity 120 m/s, at 0.1 ms **b** Equivalent stresses at velocity 120 m/s, at 0.1 ms

At some distance beyond the crack, the fibers are unbroken. They are broken in the high-stress region where the tip of the projectile is impacted. Kevlar fibers pull out of the matrix immediately behind the crack tip. It is observed that the kinetic energy drop of the projectile increases as the target plate thickness increases. The difference in kinetic energy drop in the three thicknesses of the composite plate decreases at higher-impact velocities. For the hemispherical nose projectile, the difference in energy absorption is higher at low impact velocities and reduces at higher-impact velocities. Hence studying the obtained result, it is clear that increasing the number of Kevlar layers in the sandwiched panel results in a higher reduction of kinetic energy. And also observed that the generated waves take more time to propagate in the soft sandwiched panel. As the radial part of the projectile comes in contact with the targeted composite plate, the nose of the projectile generates a minute fracture in it, and this fracture gets enlarged in the form of a star shaped crack, partly because the tensile strength is exceeded on the rear side of the target.

5 Conclusion

1. Failure mode of the composite plate by hemispherical nose projectile is shearing of the plate. A clear-cut plug is obtained from the front Al6061 T6 plate and middle Kevlar-29 plate. The plug of the Kevlar-29 plate was split into three layers. Failure of the rear Al6061 T6 plate is somewhat a combination of petalling and shearing.
2. Ballistic limit for higher thickness is higher for hemispherical nose projectile. We can say that the composite plate exhibits more resistance to the hemispherical nose projectile as number of Kevlar layers increases.
3. The energy absorbed by the thick plate with a thickness of 3.1 is higher. Hence, energy absorbed by the targeted plate increases as thickness increases.

References

1. Xu C, Wang Y, Wu J, Song S, Cao S, Xuan S, Jiang W, Gong X (2017) Anti-impact response of Kevlar sandwich structure with silly putty core. Compos Sci Technol 153:168–177
2. Zangana S, Epaarachchi J, Ferdous W, Leng J (2020) A novel hybridized composite sandwich core with Glass, Kevlar and Zylon fibres—investigation under low-velocity impact. Int J Impact Eng 137:103430
3. Liu C, Zhang YX, Ye L (2017) High velocity impact responses of sandwich panels with metal fibre laminate skins and aluminium foam core. Int J Impact Eng 100:139–153
4. Wang PF, Yang JL, Liu WS, Tang XZ, Zhao K, Lu XH, Xu SL (2017) Tunable crack propagation behavior in carbon fiber reinforced plastic laminates with polydopamine and graphene oxide treated fibers. Mater Des 113:68–75
5. Hwang HS, Malakooti MH, Patterson BA, Sodano HA (2015) Increased interyarn friction through ZnO nanowire arrays grown on aramid fabric. Compos Sci Technol 107:75–81
6. Hudspeth M, Agarwal A, Andrews B, Claus B, Hai F, Funnell C, Zheng J, Chen WN (2014) Degradation of yarns recovered from soft-armor targets subjected to multiple ballistic impacts. Compos A-Appl Sci Manuf 58:98–106
7. Manero A, Gibson J, Freihofer G, Gou JH, Raghavan S (2015) Evaluating the effect of nano-particle additives in Kevlar® 29 impact resistant composites. Compos Sci Technol 116:41–49
8. Bandaru AK, Chavan VV, Ahmad S, Alagirusamy R, Bhatnagar N (2016) Ballistic impact response of Kevlar® reinforced thermoplastic composite armors. Int J Impact Eng 89:1–13
9. Zhao C, Wang Y, Cao S, Xuan S, Jiang W, Gong X (2019) Conductive shear thickening gel/Kevlar wearable fabrics: a flexible body armor with mechano-electric coupling ballistic performance. Compos Sci Technol 182:107782
10. Pundhir N, Goyal D, Singh P, Pathak H, Zafar S (2019) Numerical simulation of composite armor subjected to ballistic impact. Mater Today: Proc 18:696–703
11. Kulkarni SG, Gao XL, Horner SE, Zheng JQ, David NV (2013) Ballistic helmets—their design, materials, and performance against traumatic brain injury. Compos Struct 101:313–331
12. Sharma A, Mishra R, Jain S, Padhee SS, Agnihotri PK (2017) Deformation behavior of single and multi-layered materials under impact loading. Thin Walled Struct 126:193–204
13. Ramadhan AA, Talib ARA, Rafie ASM, Zahari R (2013) High velocity impact response of Kevlar-29/epoxy and 6061–T6 aluminum laminated panels. Materi Des 43:307–332